Lecture Notes in Physics

The Editorial Policy for Proceedings

The series Lecture Notes in Physics reports new developments in physical research and teaching – quickly, informally, and at a high level. The proceedings to be considered for publication in this series should be limited to only a few areas of research, and these should be closely related to each other. The contributions should be of a high standard and should avoid lengthy redraftings of papers already published or about to be published elsewhere. As a whole, the proceedings should aim for a balanced presentation of the theme of the conference including a description of the techniques used and enough motivation for a broad readership. It should not be assumed that the published proceedings must reflect the conference in its entirety. (A listing or abstracts of papers presented at the meeting but not included in the proceedings could be added as an appendix.)
When applying for publication in the series Lecture Notes in Physics the volume's editor(s) should submit sufficient material to enable the series editors and their referees to make a fairly accurate evaluation (e.g. a complete list of speakers and titles of papers to be presented and abstracts). If, based on this information, the proceedings are (tentatively) accepted, the volume's editor(s), whose name(s) will appear on the title pages, should select the papers suitable for publication and have them refereed (as for a journal) when appropriate. As a rule discussions will not be accepted. The series editors and Springer-Verlag will normally not interfere with the detailed editing except in fairly obvious cases or on technical matters.
Final acceptance is expressed by the series editor in charge, in consultation with Springer-Verlag only after receiving the complete manuscript. It might help to send a copy of the authors' manuscripts in advance to the editor in charge to discuss possible revisions with him. As a general rule, the series editor will confirm his tentative acceptance if the final manuscript corresponds to the original concept discussed, if the quality of the contribution meets the requirements of the series, and if the final size of the manuscript does not greatly exceed the number of pages originally agreed upon. The manuscript should be forwarded to Springer-Verlag shortly after the meeting. In cases of extreme delay (more than six months after the conference) the series editors will check once more the timeliness of the papers. Therefore, the volume's editor(s) should establish strict deadlines, or collect the articles during the conference and have them revised on the spot. If a delay is unavoidable, one should encourage the authors to update their contributions if appropriate. The editors of proceedings are strongly advised to inform contributors about these points at an early stage.
The final manuscript should contain a table of contents and an informative introduction accessible also to readers not particularly familiar with the topic of the conference. The contributions should be in English. The volume's editor(s) should check the contributions for the correct use of language. At Springer-Verlag only the prefaces will be checked by a copy-editor for language and style. Grave linguistic or technical shortcomings may lead to the rejection of contributions by the series editors. A conference report should not exceed a total of 500 pages. Keeping the size within this bound should be achieved by a stricter selection of articles and not by imposing an upper limit to the length of the individual papers. Editors receive jointly 30 complimentary copies of their book. They are entitled to purchase further copies of their book at a reduced rate. As a rule no reprints of individual contributions can be supplied. No royalty is paid on Lecture Notes in Physics volumes. Commitment to publish is made by letter of interest rather than by signing a formal contract. Springer-Verlag secures the copyright for each volume.

The Production Process

The books are hardbound, and the publisher will select quality paper appropriate to the needs of the author(s). Publication time is about ten weeks. More than twenty years of experience guarantee authors the best possible service. To reach the goal of rapid publication at a low price the technique of photographic reproduction from a camera-ready manuscript was chosen. This process shifts the main responsibility for the technical quality considerably from the publisher to the authors. We therefore urge all authors and editors of proceedings to observe very carefully the essentials for the preparation of camera-ready manuscripts, which we will supply on request. This applies especially to the quality of figures and halftones submitted for publication. In addition, it might be useful to look at some of the volumes already published. As a special service, we offer free of charge LATEX and TEX macro packages to format the text according to Springer-Verlag's quality requirements. We strongly recommend that you make use of this offer, since the result will be a book of considerably improved technical quality. To avoid mistakes and time-consuming correspondence during the production period the conference editors should request special instructions from the publisher well before the beginning of the conference. Manuscripts not meeting the technical standard of the series will have to be returned for improvement.

For further information please contact Springer-Verlag, Physics Editorial Department II, Tiergartenstrasse 17, D-69121 Heidelberg, FRG

T. L. Wilson K. J. Johnston (Eds.)

The Structure and Content of Molecular Clouds

25 Years of Molecular Radioastronomy

Proceedings of a Conference
Held at Schloss Ringberg, Tegernsee, Germany
14-16 April 1993

Springer-Verlag
Berlin Heidelberg GmbH

Editors

Thomas L. Wilson
Max-Planck-Institut für Radioastronomie
Auf dem Hügel 69, D-53121 Bonn, Germany

Kenneth J. Johnston
US Naval Research Laboratory
Remote Sensing Division, Code 7200
Washington D.C. 20375-5000, USA

ISBN 978-3-662-13962-2 ISBN 978-3-540-49035-7 (eBook)
DOI 10.1007/978-3-540-49035-7

CIP data applied for

Originally published by Springer-Verlag Berlin Heidelberg New York in 1994

Typesetting: Camera ready by author/editor
SPIN: 10127082 55/3140-543210 - Printed on acid-free paper

Preface

This book is a collection of 24 talks given at a meeting to celebrate the twenty-fifth anniversary of the discovery of complex molecules in space. At Schloss Ringberg, near Munich, a distinguished group of 37 astronomers met for 3 days to discuss the past history, present results and future trends in molecular cloud astronomy. Most of those present submitted written contributions, which have been edited and are presented in this volume. We hope that the relevant facts about and research trends in galactic, and some aspects of extragalactic, molecular astronomy are presented in such a way that non-specialists can obtain an overview of this field.

Concerning the history of this field: Before 1968, there were vague notions that molecular clouds might exist, but there was no definite proof until the detection of interstellar NH_3 inversion line radiation by Cheung, Rank, Townes, Thornton and Welch. In the 1940s, diatomic molecules had been discovered in the optical range, and in 1963, OH had been discovered by Weinreb, Barrett and their coworkers. However, most astronomers still favored two ideas: First, that molecular hydrogen could not exist in outer space because of dissociation by the abundant 11 eV photons, and second, that complex molecules could not exist in quantity since gas densities were too small. Both of these objections were based on existing concepts about chemistry and cloud structure. However, the discovery of ammonia, and the subsequent discovery of water vapor, formaldehyde, carbon monoxide, and then a large number of molecules, showed that these concepts needed modification.

In regard to present day research, the discoveries of 1968-70 have led to a field in which there is an enormous range of topics: The objects studied range from the envelopes of stars to high-red-shift ultra-luminous galaxies. In the last 25 years, molecular astrophysics has become a central topic of modern astrophysics, especially in connection with star formation, with the content, structure and evolution of molecular clouds, the structure of galaxies and the study of molecules in extreme temperature and density conditions.

Molecular clouds consist almost entirely of molecular hydrogen, H_2. However under normal interstellar conditions H_2 produces no spectral lines. Thus, H_2 can be traced only indirectly, using molecules with permanent dipole moments. Relating the measurements of these molecules to the parameters of H_2 clouds is complicated by deviations of the excitation from thermodynamic equilibrium, averaging along the line of sight over regions with different properties and interstellar chemistry. The first two factors are being addressed through laboratory measurements of the properties of molecules involved, and increases in angular resolution and sensitivity; the third factor may be the major obstacle to a complete understanding of molecular clouds. In part, this is because interstellar chemistry is an inherently non-equilibrium process. It may not be possible to measure all of the important influences affecting cloud chemistry. For example, which molecules are produced is the gas phase, and which on surfaces of dust grains? In this sense, the study of dust grains is a direct extension of molecular astronomy.

One indication of progress in molecular astrophysics over the last 25 years is the more than 4500 molecular spectral lines in the table of Recommended Rest Frequencies by F.J. Lovas (1992 J. Phys. Chem Ref. Data, 21, 181-272); before 1968 this table would have had just seven entries! Even more impressive have been the surprises in this field. The most recent, high-red-shifted CO, shows the connection between molecules and cosmology. Recently, it has been suggested that large numbers of molecular clouds with kinetic temperatures of ~3K exist, and may have a large influence on the dynamics of the outer parts of irregular and spiral galaxies.

Over the past 25 years, technically, there has been a more than hundred-fold improvement in sensitivity, and a steady movement toward higher frequencies. In 1968, the construction of a spectral line receiver for 1.3 cm was a challenge. Now, it is a spectral line receiver for 0.13 millimeters! In spite of the complexity of receivers and data reduction, it is clear that the era of the specialist radio astronomer has disappeared; today, most radio astronomers are astrophysicists.

A number of contributors have made predictions about the future. Hopefully, in a few years we can again come together and examine which predictions have been fulfilled.

We thank the A.-von-Humboldt Foundation, through the award of a Max-Planck-Forschungspreis to TLW and KJJ, and the Max-Planck-Geschellschaft for financial support. Ms. Angela Linhart, MPIfR, was an invaluable help with the preparation of this book. Mr. W. Fusshoeller and G. Hutschenreiter helped to produce a large portion of the visual materials. The photos were taken by Dr. R. Schwartz.

Bonn, Germany
Washington, D.C., USA
23 August, 1994

T. L. Wilson
K.J. Johnston

Contents

List of Participants

Dr. S. Ames
Astronomy Dept. Bonn University, Auf dem Hügel 71, 53121 Bonn, Germany

Prof. A. Baudry
Observatoire de Bordeaux, BP 89, F 33270 Floirac, France

Prof. S.V.W. Beckwith
Max-Planck-Institut für Astronomie, Königsstuhl 17, 69117 Heidelberg, Germany

Prof. M. Felli
Osservatorio Astrofisico de Arcetri, Largo E. Fermi 5, I 50125 Firenze, Italy

Dr. R.A. Gaume
Code 7213, Remote Sensing Division, Naval Research Laboratory, Washington, DC, 20375-5351, USA

Dr. J. Gomez-Gonzales
Centro Astronomico Yebes, Apartado 148, E 19080 Guadalajara, Spain

Dr. M. Guélin
IRAM, 300 Rue de la Piscine, Domaine Universitaire de Grenoble, F 38460 St. Martin d'Hères, France

Dr. R. Güsten
Max-Planck-Institut für Radioastronomie, Auf dem Hügel 69, 53121 Bonn, Germany

Dr. A. Harris
Max-Planck-Institut für Extraterrestrische Physik, K. Schwarzschild-Str. 2, 85740 Garching, Germany

Dr. L. d'Hendecourt
Institut d'Astrophysique Spatiale, 91405 Orsay, France

Dr. C. Henkel
Max-Planck-Institut für Radioastronomie, Auf dem Hügel 69, 53121 Bonn, Germany

Prof. E. Herbst
Departments of Physics and Astronomy, The Ohio State University, Columbus, OH, 43210, USA

Prof. R. Hills
Mullard Radio Astronomy Observatory, Cavendish Laboratory, Cambridge University, Madingley Road, Cambridge, CB3 OHE, England

Dr. K.J. Johnston
United States Naval Observatory, 3450 Massachusetts Ave., NW, Washington, D.C.,
20392-5420

Prof. R.B. Larson
Yale Astronomy Department Box 6666, New Haven, Ct., 06511-6666, U.S.A

Dr. J. Martin-Pintado
Centro Astronomico Yebes, Apartado 148, E 19080 Guadalajara, Spain

Dr. R. Mauersberger
Max-Planck-Institut für Radioastronomie, Auf dem Hügel 69, 53121 Bonn, Germany

Dr. S.T. Mageath
Max-Planck-Institut für Radioastronomie, Auf dem Hügel 69, 53121 Bonn, Germany

Dr. K.M. Menten
Harvard-Smithsonian Center for Astrophysics, 60 Garden Street, Cambridge, MA, 02138, USA

Dr. P.G. Mezger
Max-Planck-Institut für Radioastronomie, Auf dem Hügel 69, 53121 Bonn, Germany

Prof. J.M. Moran
Harvard-Smithsonian Center for Astrophysics, 60 Garden St., Cambridge, MA, 02138, USA

Mr. D. Muders
Max-Planck-Institut für Radioastronomie, Auf dem Hügel 69, 53121 Bonn, Germany

Dr. P.C. Myers
Harvard-Smithsonian Center for Astrophysics, MS 42, 60 Garden St., Cambridge, MA, 02138, USA

Dr. L. Pagani
DEMIRM, URA 336 du CNRS, Observatoire de Paris, Section de Meudon, F 92190 Meudon, France

Prof. D. Sanders
Inst. of Astronomy Univ. of Hawaii, 2680 Woodlawn Dr., Honolulu, Hawaii, 96822, USA

Prof. J. Schmid-Burgk
Max-Planck-Institut für Radioastronomie, Auf dem Hügel 69, 53121 Bonn, Germany

Dr. R. Schwartz
Max-Planck-Institut für Radioastronomie, Auf dem Hügel 69, 53121 Bonn, Germany

Prof. L.E. Snyder
Astronomy Department, University of Illinois, 1002 West Green Street, Urbana, IL, 61801, USA

Prof. P.M. Solomon
Astronomy Program, State Univ. of New York, Stony Brook, N.Y., 11794, USA

Prof. P. Thaddeus
Harvard-Smithsonian Center for Astrophysics, 60 Garden St., Cambridge, MA, 02138, USA

Prof. C.H. Townes
Physics Dept. Univ. of California, 557 Birge Hall, Berkeley, CA, 94720, USA

Prof. W. Tscharnuter
Institut f. Theoretische Astrophysik der Univ. Heidelberg, Im Neuenheimer Feld 561, Germany

Dr. P.A. Vanden Bout
National Radio Astronomy Observatory, 520 Edgemont Road, Charlottesville, VA, 22903-2475, USA

Prof. W.D. Watson
Deptartment of Physics, University of Illinois at Urbana-Champaign, 1110 West Green Street Urbana, IL 61801-3080, USA

Prof. W.J. Welch
Radio Astronomy Laboratory Univ. of California, Berkeley, CA, 94720, USA

Dr. T.L. Wilson
Max-Planck-Institut für Radioastronomie, Auf dem Hügel 69, D 53121 Bonn, Germany

Dr. H. Zinnecker
Inst. f. Astronomie und Astrophysik, Würzburg Universität, Am Hubland, 8700 Würzburg, Germany

Key to the photograph: 1. P. Myers, 2. W.J. Welch, 3. R.A. Gaume, 4. A. Baudry, 5. L.E. Snyder, 6. P.G. Mezger, 7. S. Beckwith, 8. J.M. Moran, 9. M. Felli, 10. K.-M. Menten, 11. T. Megeath, 12. M. Guelin, 13. E. Herbst, 14. P. Thaddeus, 15. R. Larson, 16. C. Henkel, 17. R. Mauersberger, 18. C.H. Townes, 19. P.M. Solomon, 20. J. Gomez-Gonzales, 21. T.L. Wilson, 22. S. Ames, 23. L. Pagani, 24. D.B. Sanders, 25. P.A. vanden Bout, 26. K.J. Johnston, 27. J. Schmid-Burgk, 28. R. Güsten, 29. J. Martin-Pintado, 30. D. Muders

Introduction: Radio and IR Studies of Molecular Clouds

Charles H. Townes

University of California, Berkeley

Abstract: An overview of the development of molecular astronomy, together with future prospects, is given.

It is difficult to make even an introduction to the subject of radio and IR studies of molecular clouds which is brief because these studies, although they are all rather recent, have already produced a rich and large field of astronomy.

In considering how to introduce the field, I thought it might be of interest to compare its initiation and growth with that of high resolution microwave spectroscopy, a field which began in 1946, 22 years before the events we celebrate at this meeting. It laid the ground work for microwave studies of molecular clouds, and its development has some similarities with that of microwave work on interstellar molecules. We celebrate at this meeting the discovery of ammonia, water, and formaldehyde in interstellar clouds 25 years ago. It is not accidental that ammonia and water were also the first two molecules which were studied by the new high resolution microwave spectroscopy shortly after World War II, with the aid of the then newly developed radar components which could conveniently produce and detect microwaves. Figure 1 shows the growth in the discovery of the microwave spectra of molecules by late 1951, five years later. There was some small decrease in this rate for a time because the job was becoming somewhat more difficult as scientists moved towards refractory molecules or very complex ones. However, I also made a summary in 1967 of the number studied by then, which turned out to be almost 300, so the rate had continued at about 14 new ones per year. As in the astronomical case, the simple molecules continued to provide great interest as rare isotopic species, hyperfine structure, additional excited states and new phenomena were probed.

Although ammonia and water were the first stable or complex molecules found in interstellar space, the study of interstellar molecules actually had a somewhat different beginning and considerably earlier. Optical astronomers had detected the free radicals CH, CH^+, and CN by about 1940, but after that nothing further was turned up for some time. In 1963 Alan Barrett, as a result of his great persistence and faith that it was there, detected the microwave spectrum of OH. This was a notable discovery. However, it unfortunately did not lead astronomers to look much further, perhaps because they felt that OH was

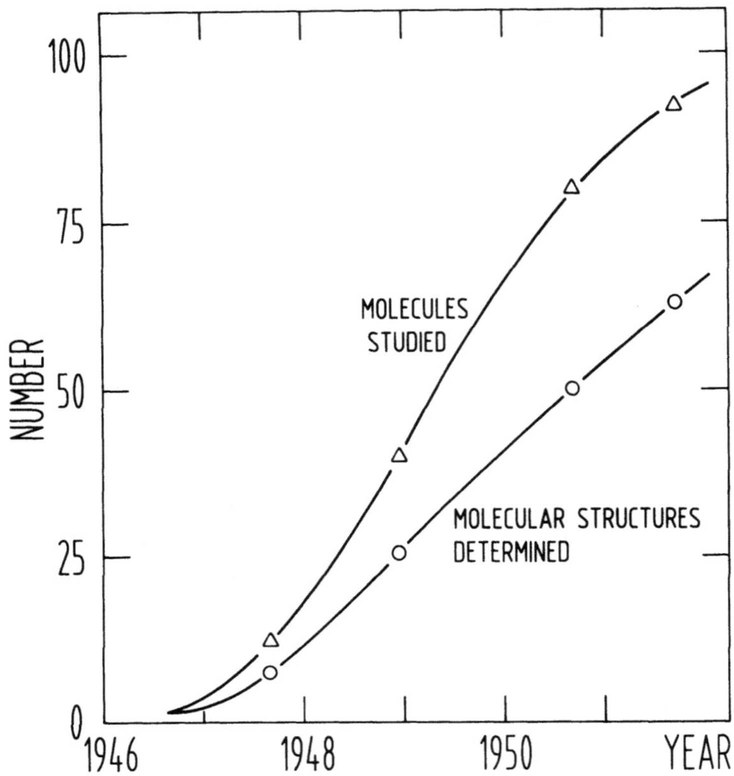

Fig. 1. The number of molecules studied and structures determined by laboratory microwave spectroscopy before a given date (Townes 1952).

a free radical which more or less completed the class of those already found, and that other types of molecules would not be present. In any case, there were no further fruitful searches carried out until 1968, when a group at Berkeley discovered ammonia, which was chosen for the search because it has a particularly strong spectrum, lying in a region where the University had an available antenna and where receivers could be easily built. At that time water seemed an unlikely candidate because its microwave line involved a rather highly excited rotational state. However, since after finding ammonia they had a receiver and antenna available, the Berkeley group also looked for water and found it in surprising strength. This strength was rather soon proven to be because it involved maser action. The discovery of formaldehyde by a group from Harvard and NRAO followed quickly thereafter. After this detection of water, ammonia, and formaldehyde, the field obviously attracted the interest of many astronomers, and the rate of discovery shot upwards. Figure 2 shows the rate of discovery of new molecules in interstellar space, beginning with the ones discovered by optical spectroscopy in the early 1940s. As in the case of the earlier microwave spectroscopy after ammonia and water came along, with minor variations the

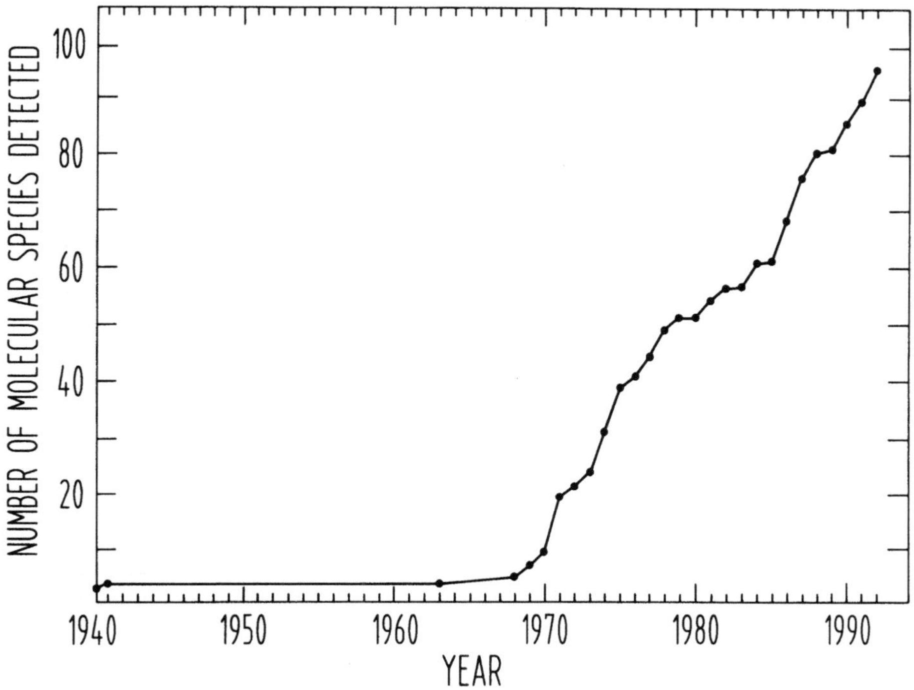

Fig. 2. The number of interstellar molecules identified by their spectra before a given date. (Data kindly provided by Patrick Thaddeus.)

rate of new discoveries has kept an approximately constant pace. Total numbers have now reached nearly 100, involving a rate of about four per year. Most of these have been detected in the microwave region, but some, such as hydrogen which has no microwave spectrum, have relied on infrared for discovery. In still other cases, infrared has been very fruitful in the examination of excited states, such as the higher rotational states of CO which tend to lie in the far infrared. Table 1 lists the molecules which have been found so far. The number of organic species is clearly striking; the list particularly abounds in linear chains of carbon, including molecules which were previously unknown in terrestrial laboratories. Whether this list will continue to grow at its past rapid rate is somewhat uncertain. This will depend in part on the introduction of new techniques and more sensitive instruments, but it will also depend on the eagerness of astronomers to push towards discovery of further molecules by comparison with their interest in using the known molecules as probes for the many types of astronomical phenomena. Figure 3 shows a histogram of the numbers of new molecules discovered each year since the beginning of the field. This figure and Table 1 have been kindly provided to me by Patrick Thaddeus who also makes the observation

Table 1. Interstellar molecules identified by March 1993. (Patrick Thaddeus, unpublished.)

March 1993

NUMBER OF ATOMS

2	3	4	5	6	7	8	9
H_2	H_2O	NH_3	SiH_4	CH_3OH	CH_3CHO	$CHOOCH_3$	CH_3CH_2OH
OH	H_2S	H_3O^+ ?	CH_4	NH_2CHO	CH_3NH_2	CH_3CCCN	$CH_3)_2O$
SO	SO_2	H_2CO	CHOOH	CH_3CN	CH_3CCH		CH_3CH_2CN
SO^+	HN_2^+	H_2CS	$HC\equiv CCN$	CH_3NC	CH_2CHCN		$H(C\equiv C)_3CN$
SiO	HNO	HNCO	CH_2NH	CH_3SH	$H(C\equiv C)_2CN$		$H(C\equiv C)_2CH_3$
SiS	SiH_2 ?	HNCS	NH_2CN	C_5H	C_6H		
NO	HCN	CCCN	H_2CCO	HC_2CHO			
NS	HNC	HCO_2^+	C_4H	$CH_2{=}CH_2$			**11**
HCl	HCO	CCCH	C_3H_2	H_2CCCC			
NaCl	HCO^+	c-CCCH	CH_2CN				$H(C\equiv C)_4CN$
KCl	OCS	CCCO	C_5				
AlCl	CCH	CCCS	SiC_4				
AlF	HCS^+	HCCH	H_2CCC				**13**
PN	SiCC	$HCNH^+$	HCCNC				
SiN	CCO	HCCN	HNCCC				$H(C\equiv C)_5CN$?
NH	CCS						
CH	C_3						
CH^+	MgNC						
CN							
CO							
CS							Total: 96
C_2							
SiC							
CP							

that the lull in rate of discovery about 1980 was probably due to the necessity for more intensive laboratory work on finding the spectra of new molecules for which laboratory spectra were not already available, and also a short delay in the development of good equipment at the shorter wavelengths.

One of the great boons which the discovery of such spectra provides is the possibility of using them to probe conditions within clouds, such as the cloud density, its temperature, the presence of shock waves, and of course the Doppler velocity. Fortunately, the present collection of molecules provides good probes for a very wide variety of conditions. Figure 4 indicates molecules typically used as probes for various ranges of densities and temperatures. From this figure, it can be seen that most of the normal density and temperature ranges are covered with molecular spectra we now know well, except for the region which combines very low temperature and very high density. Good test probes in this region are missing for fundamental physical reasons. To measure density, the rate of collision must not be very different than the rate of radiation of the molecule; that is, conditions must be somewhere near the so-called critical density. For a given dipole matrix element, the rate of radiation is proportional to the cube of the

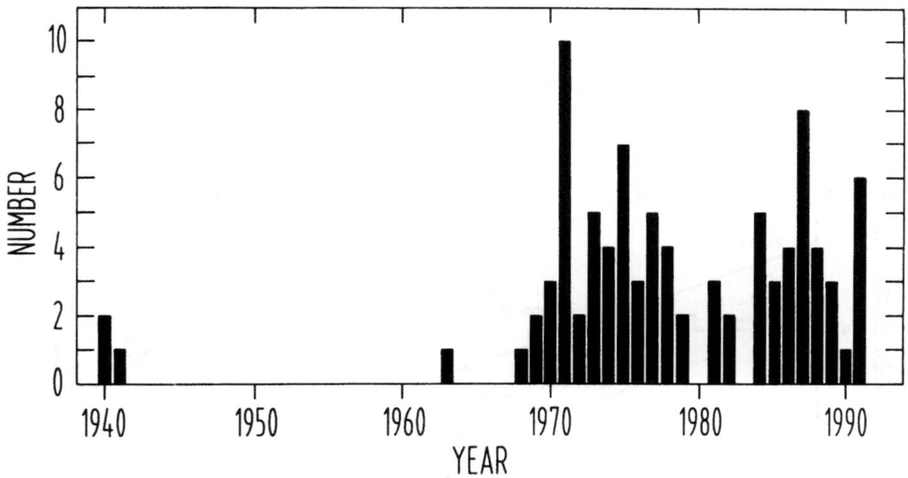

Fig. 3. Histogram of the number of interstellar molecules identified each year prior to 1993. (Patrick Thaddeus, unpublished.)

frequency. However, high frequencies require high temperatures for excitation. Hence, at low temperatures where high frequencies are not excited, one cannot expect to have cases with spontaneous rates of radiation fast enough to give a good measure of the very high densities which produce very rapid collisions. Perhaps one possible mode of getting at these high densities is the study of astronomical masers. This is because the electromagnetic fields in masers can be much more intense than the zero point fluctuations responsible for spontaneous emission, and hence the stimulated rate of molecules is much faster, giving an opportunity to probe very high rates of collision.

The far infrared as well as the microwave region is a useful medium for a study of molecular lines and their interpretation in terms of astronomical phenomena. It is particularly useful for studying the more excited molecular states and hence relatively warm and dense clouds. The far infrared also gives us access to the detection of fine structure lines of common atoms or ions such as C, N, and O in the ground state, which lie in the far infrared. Detection of ionic fine structure gives information both on gas characteristics such as densities, temperatures, and velocities, and also on the UV photon environment which produces the ionization. Figure 5 provides an overview of the densities and the excitation potentials thus tested by various atoms and ions. To the right of an excitation potential of 13.6 eV, the ionization potential of H, the densities measured are those of electrons, which are primarily responsible for the fine structure excitation. To the left of an ionization potential of 13.6 unless otherwise indicated the densities referred to are those of H or H_2, which in that case are the primary agents of excitation through collision. Atoms such as OI or ions such

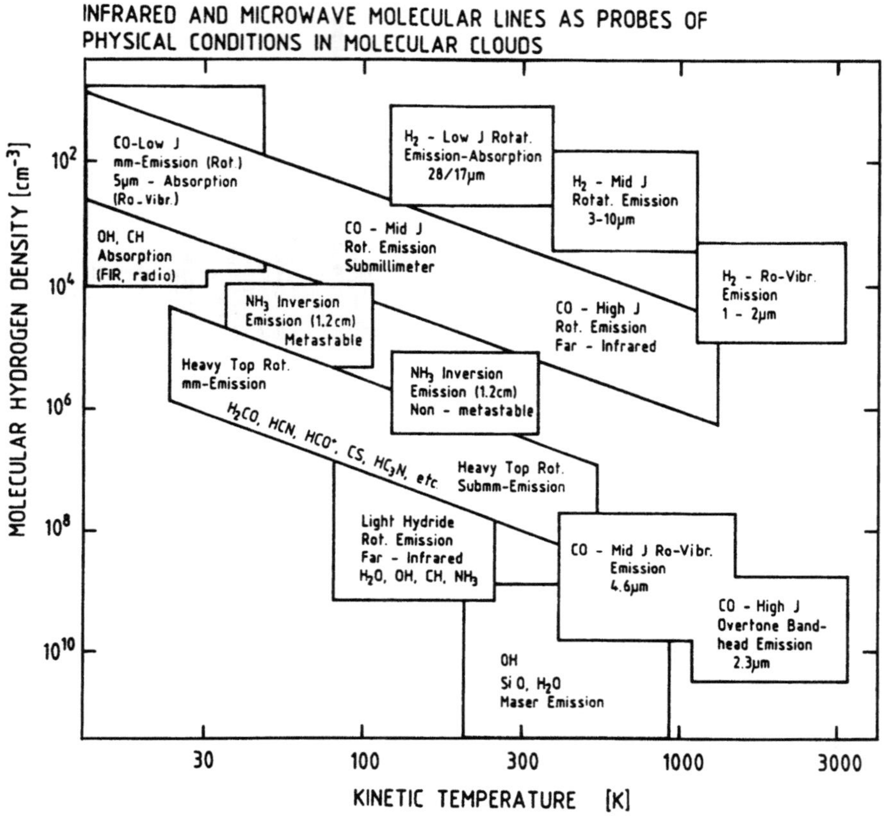

Fig. 4. Diagram showing the range of gas densities and temperatures which can be effectively determined by a variety of specific molecular spectra (Genzel 1991).

as OIII, which have two fine structure transitions in the far IR, are especially useful in that measurement of both transitions can allow determination of both density and kinetic temperature. Figure 6 illustrates the spectrum from the star burst galaxy M82, with both the continuum radiation and the more prominent line spectra shown in the same figure. Towards the extreme left one sees the microwave spectra, and towards the center and right are infrared spectra. With this varied collection of spectra and species, a very wide variety of conditions can be studied within the galaxy. A general correlation between the infrared intensity of the CII transition and that of the $J = 1 \rightarrow 0$ transition of CO has been proposed, either one giving a measure of the total activity in typical star burst galaxies. One can see from Figure 7 that in fact the intensity of the CII line is directly proportional to that of the $J = 1 \rightarrow 0$ CO transition for star burst galaxies or any galactic star-forming region, as indicated by the straight lines

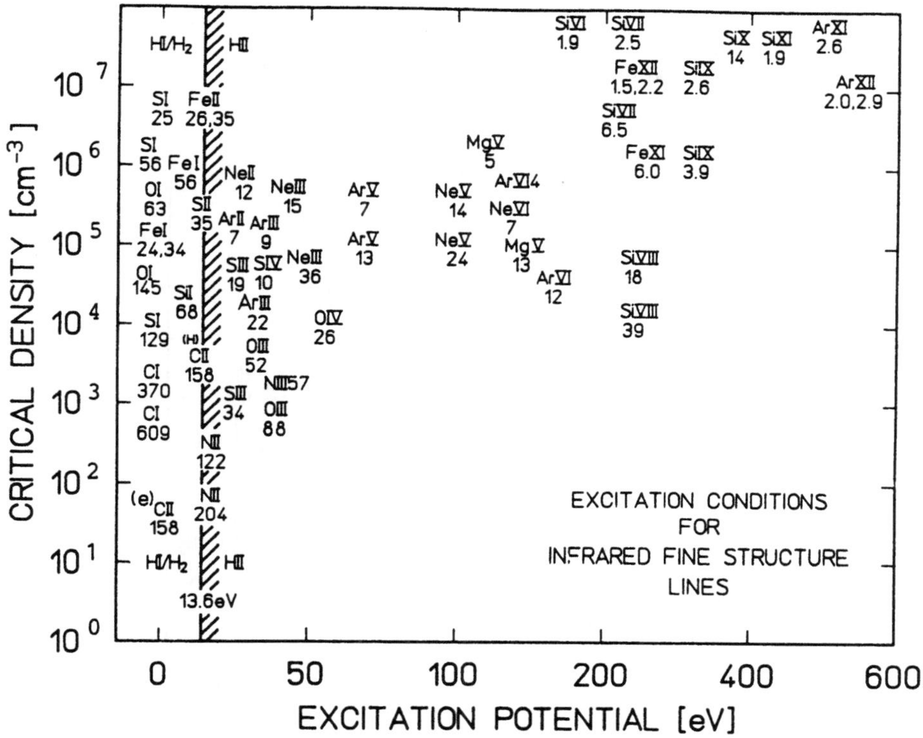

Fig. 5. Diagram showing densities, which can be probed by ground state fine structure transitions of a variety of atomic and ionic species. The ions also give a measure of the ultraviolet spectra, which produce the ionization indicated in the abscissae (Genzel 1991).

and the points near it. On the other hand, for galactic molecular clouds where not a great deal of stellar activity is occurring and for many spiral galaxies, the intensity of the CII line is very much lower than that indicated by the straight line. This is not unexpected for those regions that are less energetically excited. The probing of shock waves is illustrated in Figure 8, which shows the results of far infrared measurements of highly excited rotational lines of CO and also of the vibration excitation of molecular hydrogen. In the case of CO, the observed relative intensities of various rotational lines predict a temperature of about 750° and a density of a few times 10^6 per cubic centimeter, assuming that the material examined is in fact reasonably uniform. Molecular hydrogen measures quite different conditions, showing temperatures of about 2000° at densities of about 10^6 per cubic centimeter.

The origin and chemistry of molecules observed in interstellar clouds is often very difficult to pin down. This is partly because molecular formation can be quite complex, involving excitation by photons, electrons, and molecular formation on surfaces whose characteristics are somewhat uncertain. But it is also

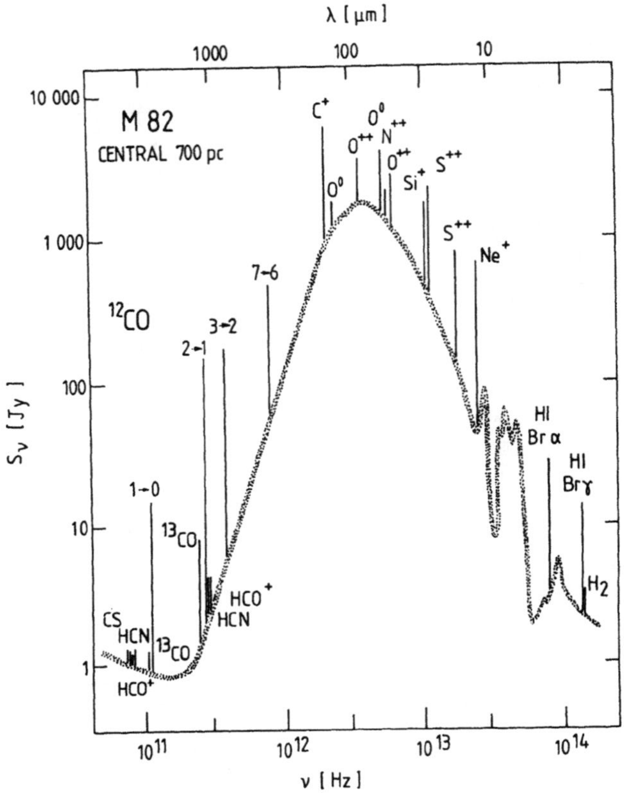

Fig. 6. The infrared and radiospectrum of the starburst galaxy M82. The diagram shows the continuum spectrum with some absorption due to dust and sharp emission line due to molecules, atoms, and ions (Genzel 1991).

because we generally do not know very well the past history of the cloud, and the molecules we observe now may have been formed under conditions which are quite different from those which presently exist. Perhaps one method of examining molecular chemistry of this type under better known conditions is to study molecules being formed in dust shells around stars. We know that they were originally emitted as atoms from the stars and we know something of the temperature and density conditions which they have undergone as the gas moves radially away from the star. This may allow better tests than we have had otherwise of theoretical models for formation.

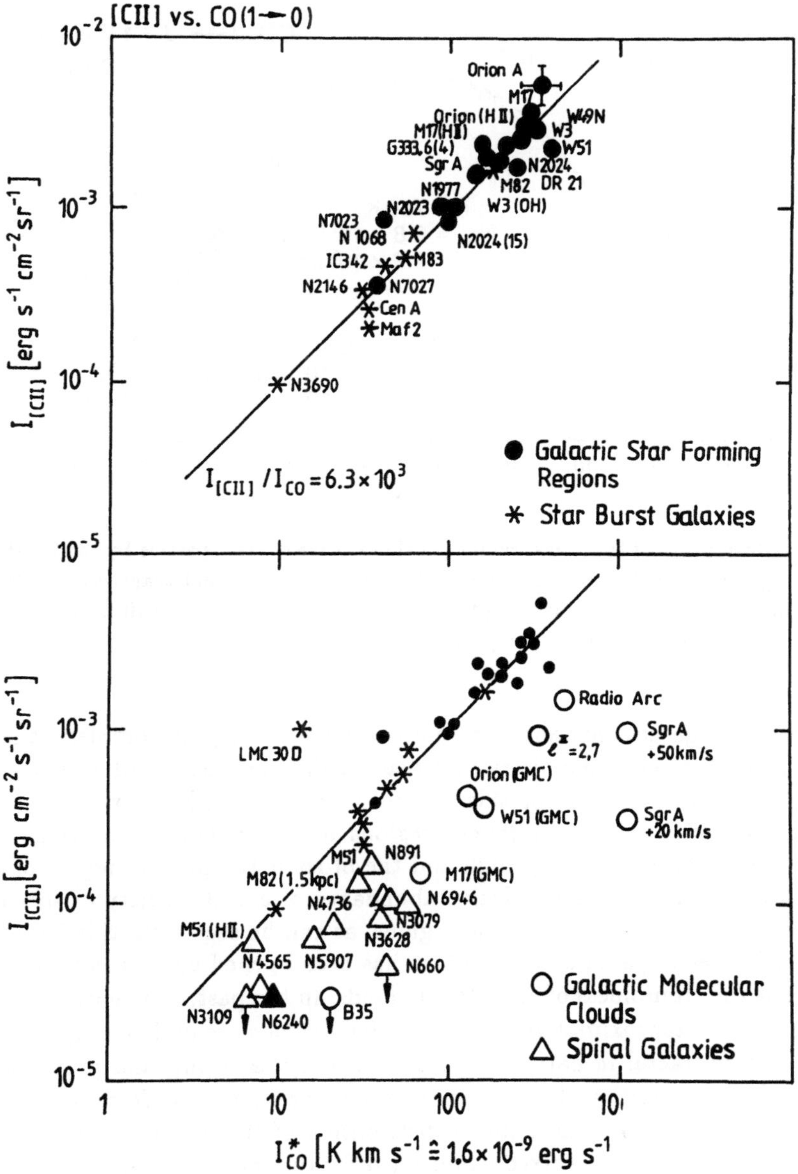

Fig. 7. The relation between intensities of the CII fine structure line and the CO (J=1 − 0) rotational line. For star forming regions, there is an approximate proportionality between the two, shown by the straight line. For regions not actively forming stars, the CII intensity is relatively weaker and is below the line (Genzel 1991).

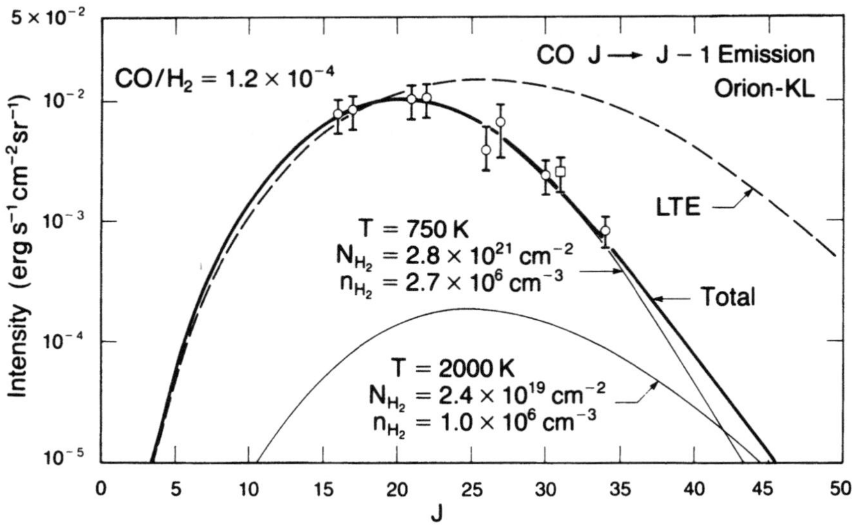

Fig. 8. Illustration of the measurement of shock wave characteristics by the far IR spectra of high rotational states of CO and the rotation-vibrational transistion of H_2. The curves indicate relative intensities of transitions measured. The resulting temperatures and densities deduced are shown (Watson et al. 1985).

Even though circumstances in molecular clouds may be complex, it is clear that different types of chemistry take place in different areas and we understand something of the reasons for this. Figure 9 shows a striking example of quite different chemistry in two different regions in Orion. One is the ridge region where excitation is relatively high and the other is the plateau region of somewhat lower excitation. The relative abundance of a wide variety of species are systematically different in the two regions, as can be seen from this figure. Still another interesting area of study which has been opened up by recent molecular studies is the clumpiness of interstellar clouds. In the past, clouds were generally assumed to be approximately uniform. In fact, clouds are quite clumpy and even appear to be fractal in nature. Figure 10 demonstrates this and provides some remarkable generalities about their fractal nature over a wide range of scales. This figure shows that the number distribution of the masses of the clumps is exponential with a rather fixed exponent α for the relation between the number ΔN of clouds of a given mass and the mass M, $M = M_0(\Delta N)^{-\alpha}$.

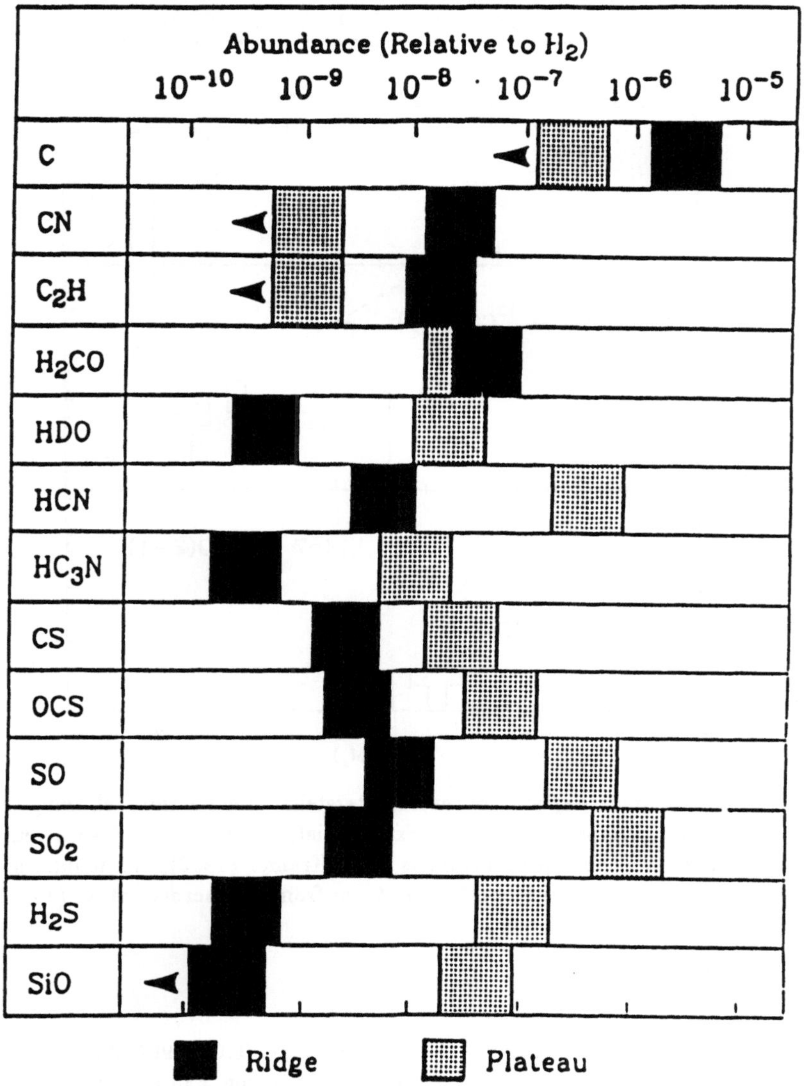

Fig. 9. Relative abundances of various molecular species in two different regions of Orion. The ridge region corresponds to one of rather high excitation; the plateau region involves lower excitation (Blake et al. 1987).

This summary can only briefly outline the rich field of molecular cloud studies now available through microwave and infrared techniques, most of which have been developed during the last 25 years which we celebrate. Clearly the field continues to grow in size and potency. We can look forward in this conference to many new and interesting results.

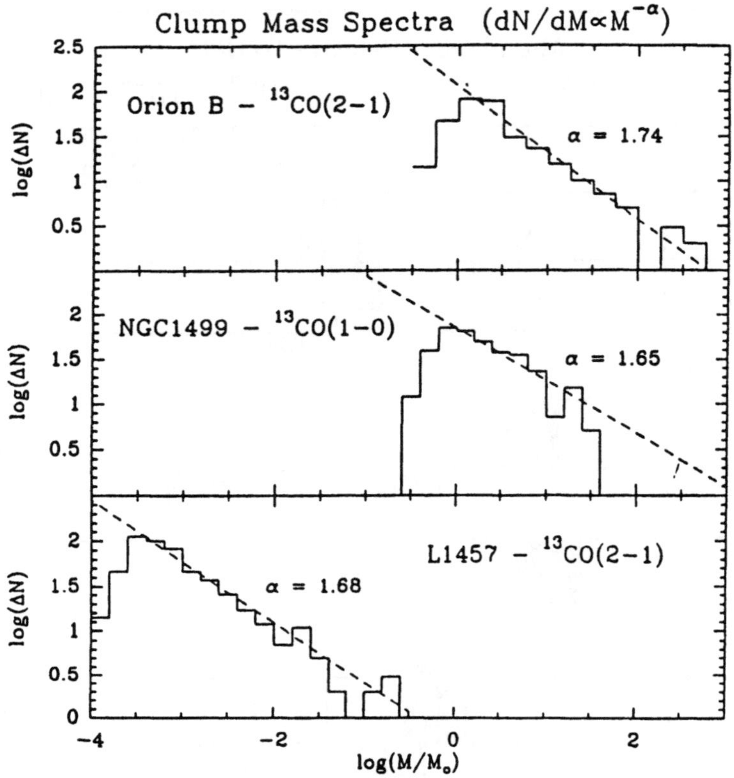

Fig. 10. The clump mass spectra of several interstellar cloud regions, showing the probability of a given clumps mass to vary exponentially with mass and with remarkably constant exponent. The top spectrum is from Kramer (1993), the middle spectrum from Herbertz (1993) and the bottom spectrum from Zimmerman et al. (1993).

References

Blake, G.A., Sutton, E.C., Masson, C.R., and Phillips, T.G., 1987, Ap. J. 315, 621.
Genzel, R. 1991, Saas Fee Lectures, Ed. Burton, Elmegreen, and Genzel, Springer 1992.
Herbertz, R., 1993, Ph.D. Thesis, Univ. of Köln
Kramer, C., 1993, Ph.D. Thesis, Univ. of Köln
Townes, C.H., 1952, Ann. N.Y. Acad. of Sciences, 55, 745.
Watson, D.M., Genzel, R., Townes, C.H., and Storey, J.W.V., 1985, Ap. J. 298, 316.
Zimmerman, T. et al. , 1993, in preparation

The Evolution of Molecular Clouds

Richard B. Larson

Yale Astronomy Department, Box 6666, New Haven, CT 06511, U.S.A.

Abstract: Molecular cloud birth, lifetime and destruction are discussed. Cloud collapse, fragmentation and star formation are also presented.

1 Introduction

Molecular clouds constitute the densest parts of the interstellar medium in galaxies, and ever since their discovery 25 years ago they have been of prime interest as the sites of star formation. Molecular clouds also play an important galactic role in that massive stars recently formed in them provide the main energy sources for the interstellar medium, partly by destroying their birth clouds and recycling their matter back into more diffuse forms. Thus an understanding of the life cycle of molecular clouds is of central importance not only for understanding how stars form, but also for understanding the dynamics of the interstellar medium and ultimately the evolution of galaxies as a whole. Many complex processes are clearly involved, most of which are not yet well understood; however, the increasingly detailed observational information that is becoming available about molecular clouds and star formation is beginning to allows us at least to identify some of the most import ant processes at work. In this contribution, I shall review briefly some relevant basic characteristics of molecular clouds and some of the many processes that are likely to play a role in their formation, evolution, and destruction.

2 Basic Cloud Characteristics

One fundamental characteristic of molecular clouds is that they are not, as has sometimes been assumed, isolated 'billiard balls' moving about independently in space, but instead are just dense *condensations* in more widely distributed, mostly atomic gas. Although molecular clouds may often appear to have sharp boundaries, these boundaries do not represent the edge of the matter distribution but just rapid transitions from the molecular gas to the surrounding atomic gas, which is distributed in extended envelopes that typically have comparable mass (Blitz 1988, 1991). The molecular composition of interstellar clouds depends on the local gas density n and on the column density N of material providing shielding from dissociating ultraviolet radiation, and the gas is predicted to be predominantly molecular wherever the product $nN^{2/3}$ exceeds a critical value that depends on the ultraviolet radiation flux and the dust abundance (Elmegreen 1989, 1993). Since the condition required for the gas in an interstellar cloud to be mainly molecular is independent of the condition for self-gravity to be important, there is no reason why molecular clouds should necessarily be gravitationally bound, or vice versa; the molecular content of a cloud or region could even change rapidly with time because of its sensitivity to local conditions such as the radiation field that can vary with time (Elmegreen 1993). Thus, there need be no close correspondence between the amount of molecular gas in a region and the amount of star formation occurring there, and indeed it is found that star formation rates in galaxies correlate better with the total gas content than with the content of molecular gas alone (Buat, Deharveng, & Donas 1989; Kennicutt 1989, 1990; Evans 1991). The Small Magellanic Cloud is an example of an actively star-forming system in which only a small fraction of the interstellar medium is molecular, owing to a high ultraviolet flux and a low dust abundance (Rubio, Lequeux, & Boulanger 1993).

Another important characteristic of molecular clouds is that they are *transient* structures and do not survive without major changes for more than a few times 10^7 years (Blitz & Shu 1980; Larson 1981). The short lifetimes of molecular clouds are directly indicated by the fact that the range in age of the young stars associated with them is only about 10 to 20 Myr, comparable to the internal dynamical timescales of large molecular clouds (Larson 1981); age spans of this order are found both for the T Tauri stars in dark clouds (Cohen & Kuhi 1979) and for the subgroups of OB stars in more massive star-forming complexes (Blaauw 1964, 1991). Subgroups of OB stars whose ages are greater than 10 Myr no longer contain significant amounts of gas or dust, so the gas from which such groups form is evidently cleared away in a time of this order (Blaauw 1991); those clouds or parts of clouds that form star clusters are also dispersed in a similar time, since little or no molecular gas is seen within 25 pc of clusters older than 10 Myr (Bash, Green, & Peters 1977; Leisawitz, Bash, & Thaddeus 1989). There cannot be any long 'dead time' between the formation of a massive molecular cloud and the onset of star formation in it, since very few of the largest molecular clouds are not forming stars (Blitz 1991), and the number of molecular clouds of all masses that are not forming stars is only about comparable to the number of star-forming clouds (Mooney & Solomon 1988; Solomon, this conference). Therefore the total lifetimes of molecular clouds, or at least of those parts of them that form systems of stars, cannot exceed twice the period of active star formation in them, or perhaps ~ 20 Myr (see also Elmegreen 1991a, who estimates a cloud lifetime of ~ 40 Myr from similar arguments). Thus there is no need, as has often been supposed, for molecular clouds to be supported against gravity in a quasi-equilibrium configuration for many dynamical times. While it is true that virial balance is often roughly satisfied, at least in the larger clouds (Larson 1981), this does not rule out the possibility of rapid dynamical evolution of these clouds, or of those parts of them that form groups or clusters of stars.

Additional evidence that molecular clouds are short-lived, or at least that their structures change rapidly with time, is provided by the fact that they appear to be chemically relatively unevolved; that is, the abundances of various molecules are are often far from those expected to prevail in chemical equilibrium, and resemble instead those predicted to occur at an early stage of chemical processing less than 1 Myr after cloud formation (Prasad et al. 1987; Herbst & Leung 1989; Herbst, this conference). It has also been a long-standing problem to understand why most of the molecules in dense cold molecular clouds have not frozen out on dust grains, as would be expected if these clouds are stable long-lived objects. A number of authors have therefore suggested that molecular clouds are actually relatively young (Stahler 1984), or that they experience rapid dynamical evolution that leads to frequent chemical reprocessing of their gas, for example by turbulent cycling between the interiors and the surfaces of clouds (Chièze & Pineau des Forêts 1990) or by the frequent formation and dispersal of dense clumps in them (Prasad 1991).

A third notable property of molecular clouds is that they are highly *irregular* structures and have complex shapes that do not at all resemble equilibrium configurations. Many of them have wispy or filamentary shapes resembling those of atmospheric clouds, as is well illustrated in the IRAS 100-micron maps of molecular clouds (Scalo 1990; Boulanger 1991). The atmospheric clouds with wispy structures are transient and are typically dispersing rapidly, and this may be true also of many molecular clouds, especially the smaller and more diffuse ones which are mostly not gravitationally bound (Herbertz, Ungerechts, & Winnewisser 1991; Falgarone, Puget, & Pérault 1992). The irregular boundaries of molecular clouds on contour maps have been found by a number of authors to have fractal properties (Scalo 1990; Dickman, Horvath, & Margulis 1990; Falgarone, Phillips, & Walker 1991; Falgarone 1992; Zimmermann & Stutzki 1992), and the fractal dimension inferred for the surfaces of these clouds is intriguingly similar to the fractal dimension that has been found for various interfaces in turbulent flows (Falgarone & Phillips 1991; Sreenivasan 1991). Although the significance of this result is not presently clear, it seems at least consistent with the possibility that molecular clouds are condensations in a turbulent medium and are structured partly by turbulence (Larson 1979, 1981; Falgarone & Phillips 1991). The possibility that turbulence plays an important role in determining cloud properties is supported by the fact that the basic scaling relations satisfied by molecular clouds, such as the size-linewidth relation, are similar for both bound and unbound clouds, and are similar also to those expected for turbulent flows (Falgarone et al. 1992; Falgarone 1992). It is clear in any case that simple models cannot provide an adequate description of the structure or dynamics of molecular clouds. Models incorporating a realistic degree of structural complexity, such as fractal models, may for some purposes provide a more useful description (e.g. Hetem & Lépine 1993).

3 Cloud Formation

Since molecular clouds are transient features, it follows that they are constantly being formed and destroyed. Therefore, it is necessary to understand the processes by which they are continually being reassembled from more dispersed gas. The rate at which interstellar gas is presently being collected into star-forming molecular clouds in our Galaxy is related to the star formation rate, and it can be estimated empirically from the observed star formation rate and the efficiency of star formation in molecular clouds. The total rate of star formation in our Galaxy is of the order of 3 $M_\odot$ per year, and since only about 2 percent or less of the mass of a typical molecular cloud is converted into stars (Myers et al. 1986; Leisawitz et al. 1989; Evans & Lada 1991), this implies that at least 150 $M_\odot$ of gas per year is being turned into star-forming molecular clouds. Since the total amount of gas in our Galaxy is about 5×10^9 $M_\odot$, the average time required to collect gas into giant molecular clouds must then be about 30 Myr. A similar estimate for the solar neighborhood, where the timescales are somewhat longer than the Galactic average, yields a formation timescale for molecular clouds of about 50 Myr (Larson 1991a, 1992b). Note that this estimated formation time is not much longer than the cloud lifetime of $\sim$ 20 Myr (Section 2); thus the formation of molecular clouds must itself be a rather rapid process, and cannot take many dynamical timescales. Since the timescales for the formation, internal evolution, and destruction of molecular clouds are all of the same order, these processes probably cannot be clearly separated in time, and they may all go on simultaneously in different parts of a star-forming complex.

Two possible formation mechanisms for molecular clouds that have been considered in some detail are (1) cloud growth by random collisions and coalescence, and (2) gravitational instability or 'swing amplification' effects in the Galactic gas layer; these and other mechanisms have been discussed by Larson (1987, 1988, 1992b), and in more detail by Elmegreen (1990a,b, 1991a,b). The first possibility, i.e. the building of large clouds from smaller ones by random collisions and coalescence, predicts formation times of at least 100 Myr for giant molecular clouds, and therefore probably cannot be the primary formation mechanism because this is longer than the cloud formation timescale estimated empirically; in any case, most collisions between smaller clouds are probably disruptive and so are not likely to result in coalescence. The second possible mechanism, i.e. large-scale gravitational instability and amplification effects in the Galactic gas layer, is almost certainly more important because it can collect gas into large complexes on a timescale that is only about 40 Myr in the solar neighborhood, in good agreement with the estimated cloud formation time. Evidence that gravitational instability effects are indeed primarily responsible for both molecular cloud formation and star formation in galaxies is provided by the fact that star formation is observed to occur only where the surface density of gas in galactic disks exceeds a threshold which is close to the critical value predicted for the onset of gravitational instability (Kennicutt 1989, 1990).

A true instability with unlimited growth of density perturbations probably does not occur in most galaxies, since the stability parameter Q is typically somewhat larger than unity and so is in the stable regime; nevertheless, limited 'swing amplification' of shearing density perturbations can still occur in this situation (Toomre 1981), and this is probably the effect mostly responsible for collecting interstellar gas into large star-forming cloud complexes and spiral arm segments in galaxies. The threshold gas surface density found by Kennicutt (1989, 1990) to be required for star formation actually corresponds to a value for Q of about 1.7, approximately the value below which swing amplification becomes important (Larson 1992b), thus supporting the relevance of this mechanism for star formation. Density waves may of course also play a role in driving cloud formation and star formation in galaxies, but their presence is not necessary, as is shown by the example of the irregular galaxies, and it may be that in many cases the role of density waves is largely just one of organizing star formation that would have occurred anyway (Elmegreen 1992).

The formation of massive molecular clouds by gravitational instability or swing amplification effects in a turbulent interstellar medium must also involve complex smaller-scale processes, including collisions between the small clouds that were present in the initial medium and the building up of larger clouds by accretion processes of some kind. The possible role of gas accumulation processes in the internal evolution of molecular clouds has been discussed by Larson (1982) and Myers (1991), and simulations of cloud collisions and coalescence in a contracting cloud complex have been carried out by Monaghan & Varnas (1988). While collisions almost certainly play a role in the building up of large molecular clouds, it nevertheless seems clear that purely *random* collisions cannot build them fast enough and that more ordered large-scale motions are therefore required, such as those that are involved in the formation of large cloud complexes or spiral arm segments in galaxies. This conclusion is, of course, consistent with the fact that most of the star formation in galaxies is observed to occur in large complexes or spiral arm segments (Efremov 1989).

4 Cloud Collapse and Fragmentation

As has been noted, a giant molecular cloud must begin forming stars soon after the cloud itself has formed, since relatively few of the largest molecular clouds are not forming stars. Even if as many as half of all molecular clouds are not forming stars (Solomon, this conference), the time delay between the formation of a molecular cloud and the onset of star formation in it cannot exceed the subsequent duration of the star formation activity, which is of the order of 10 Myr and comparable to the internal dynamical timescale (Section 2). Since it takes somewhat longer than this to build large molecular clouds, it is likely that star formation begins already in molecular clouds while they are still being assembled; moreover, star formation must begin within a time not much longer than the dynamical or free-fall time of such a cloud. Collapse and star formation can occur in the densest part of a cloud even if the cloud as a whole is not collapsing, and this must in fact be what usually happens because there is no evidence that most star-forming clouds are undergoing any rapid overall collapse, and there is even evidence that many of them are being dispersed, as will be discussed in Sections 5 and 6.

According to the classical picture, star formation involves the collapse of a cloud or part of a cloud under gravity and the associated fragmentation of the cloud into smaller and smaller bound clumps; this is expected to occur because molecular clouds typically contain many times the 'Jeans mass', which is the minimum mass for gravitationally bound fragments. It is possible, however, that much of the small-scale structure that eventually develops into stars and groups of stars is present from the beginning because star-forming clouds are assembled from gas that already has much small-scale structure. There is as yet little direct evidence bearing on this question, but one possible piece of evidence that fragmentation does occur in star-forming clouds, at least in the sense that small-scale density fluctuations are strongly amplified with time, is provided by the fact that one of the few large molecular clouds that is *not* presently forming stars contains relatively few small clumps compared with clouds that are forming stars (Williams & Blitz 1993); if this cloud is at an early stage of evolution and will later evolve to a star-forming stage, this suggests that many more small clumps will be formed in this process.

For a variety of reasons, including the effects of initial asymmetries, magnetic fields, and turbulence, the collapse of the densest parts of molecular clouds will almost certainly not be spherical but will tend to produce flattened or filamentary structures (Larson 1985; Miyama, Narita, & Hayashi 1987a,b). Even if flattened structures are the first to be formed, for example by collapse along magnetic field lines, they may tend to break up into filaments before finally fragmenting into clumps (Miyama et al. 1987a,b); there is in any case abundant evidence that molecular clouds often have elongated or filamentary shapes (Schneider & Elmegreen 1979; Scalo 1990; Myers et al. 1991; Bally et al. 1991). Most numerical simulations of collapse and fragmentation have modeled the idealized case of an initially nearly uniform, spherical, rapidly rotating cloud that collapses first to a disk and then fragments into clumps (Larson 1978; Miyama, Hayashi, & Narita 1984; Monaghan & Lattanzio 1991). The typical masses of the clumps that form are comparable to the Jeans mass calculated for a fragmenting gas layer, and with typical molecular cloud temperatures and column densities, this mass is of the order of one solar mass (Larson 1985). The numerical studies quoted above also illustrate in increasing detail the tendency for fragmenting disks to develop filamentary and even spiral small-scale structures before most of the gas condenses into clumps; such spiral structure plays an important role in the redistribution of angular momentum by gravitational torques that must occur if most of the mass is to end up in condensed objects (Larson 1984).

The above calculations may not, however, be very relevant to real molecular clouds, most of which are far from spherical and are rotating only rather slowly (Goodman et al. 1993). The probably more realistic case of the collapse of an elongated cylindrical cloud has been studied numerically by Bastien et al. (1991) and Arcoragi et al. (1991), and the effect of including a small amount of rotation has been studied by Bonnell et al. (1991, 1992) and Bonnell & Bastien (1992). A cylindrical cloud tends to collapse first toward its axis to form a thin filament, and it then fragments along its length into two or more dense clumps (Larson 1972). The masses of the clumps are similar in order of magnitude to the Jeans mass for a fragmenting cylinder, but the tendency of the cloud to collapse first to a thin filament results in somewhat smaller fragments which may have masses as small as one-quarter of the initial Jeans mass (Bastien et al. 1991), possibly accounting for the formation of stars with masses as small as ~ 0.1 solar masses (Larson 1992a).

Qualitatively new results are found when the initially cylindrical cloud is given a slow rotation about an arbitrary axis: the fragments then develop surrounding disks, and these disks may fragment further and may also interact with each other as the fragments fall closer together, leading eventually to the formation of binary and multiple systems with a variety of properties (Bonnell et al. 1991, 1992). Somewhat similar results have been found by Boss (1993) and by Burkert & Bodenheimer (1993), who have calculated the collapse of initially spherical rotating clouds with bar-like density perturbations; in some cases these clouds collapse to elongated filaments that fragment along their lengths into two or more objects. Such phenomena may well account for the origin of many of the observed binary and multiple systems. Protostellar interactions involving disks can also have a number of interesting consequences, including the tilting of the disks and the stripping of mass from them, as well as the formation of additional binary systems by capture and the modification of binary orbits by gravitational drag effects (Herbig & Terndrup 1986; Larson 1990; Pringle 1991; Heller 1993a,b,c). Such effects could be especially important if stars form in dense clusters, as appears to be the case for most stars (see Section 5).

Direct observational evidence for collapse and fragmentation processes has proven difficult to find. One reason for this is the well-known fact that stars generate vigorous outflows as soon as they are formed, and these outflows tend to dominate the observed properties of star-forming regions and to obscure the relatively modest expected infall motions. The collapse of a dense protostellar clump into a single star or small group of stars would in any case be difficult to observe because measurable infall velocities are expected only in a very small region. Collapse might be easier to observe if a massive molecular cloud core collapses to form an entire cluster of stars, since larger infall velocities could then be produced over a larger region of space. If a newly formed cluster is observed at a sufficiently early time, the effects of the young stars on their environment might still remain confined to a small region and infall motions might still be observable outside this region. A possible example of this is provided by the core of the W49A molecular cloud, which contains a compact group of newly formed O stars surrounded by ultracompact H ±II regions (Welch et al. 1987). The molecular line profiles observed toward these H ±II regions have redshifted absorption components attributable to infalling gas on the near side of the cluster, and the inferred infall velocity of $\sim 10\ \mathrm{km\,s^{-1}}$ is comparable to the predicted free-fall velocity of the massive W49A cloud core. Further possible examples of the infall of residual cloud material toward newly formed clusters of massive stars have been described by Welch (1993; this conference). In all of these cases, as with the nearer and more extensively studied examples of cluster formation discussed below, the timescale for the collapse of the cloud core and the formation of the associated cluster appears to be quite short, i.e. of the order of 1 Myr or less, suggesting that the evolution of the densest part of a molecular cloud can be quite rapid and can approach a free-fall collapse.

5 Formation of Star Clusters

The numerical simulations discussed above suggest that stars may generally form in multiple systems or groups of some sort, possibly hierarchical in structure, but simulations of the formation of multiple stellar systems are still at an early stage, and simulations of the formation of an entire cluster of stars are not yet feasible owing to the complexity of the processes involved. Thus, theoretical work cannot yet address many of the details of how systems of stars form. However, observational studies of star formation have made great strides in recent years, thanks largely to new instrumentation such as infrared cameras and improved millimeter-wavelength telescopes and interferometers, and the resulting data are beginning to provide the basis for a phenomenological approach to understanding of the formation of systems of stars. Infrared cameras have, in particular, made it possible to survey large areas of star-forming clouds for newborn stars still embedded in them, and a major discovery has been the finding that most of the recently formed stars in nearby molecular clouds are located in compact clusters associated with massive dense cloud cores; it thus appears that the formation of compact clusters of stars is the dominant mode of star formation, at least in well-studied nearby star-forming regions (Lada & Lada 1991; Lada 1992; Carpenter et al. 1993; McCaughrean 1993; Lada, Strom, & Myers 1993; Zinnecker, McCaughrean, & Wilking 1993). The isolated mode of star formation that has so far received the most theoretical attention may therefore not be relevant for most stars, and strongly clustered star formation and processes such as the protostellar interactions mentioned above may be more important than has previously been recognized.

Even in the Taurus-Auriga dark clouds, which have usually been regarded as exemplifying an isolated mode of star formation (e.g., Lada et al. 1993), most of the young stars are actually not completely isolated but are concentrated in small groupings associated with the most prominent dark clouds (Herbig 1977; Larson 1982; Myers 1987). Gomez et al. (1993) have analyzed the spatial distribution of all of the known young stars in the Taurus-Auriga region, and have found evidence for clustering on all scales from 0.005 pc to 5 pc; in particular, they have found that the two-point angular correlation function for these stars is approximately a power-law function of separation over this range of distances, suggesting that completely isolated star formation is rare even in the Taurus-Auriga clouds, and also that similar mechanisms may be operating over a wide range of length scales to produce the observed approximately scale-free clustering. One possible way of producing such scale-free clustering might be that gravitational drag or 'dynamical friction' effects, which have no preferred length scale, act on dense protostellar clumps or forming stars in a collapsing and fragmenting cloud and cause them to become increasingly clustered, possibly in a hierarchical fashion (Larson 1990, 1991b).

The nearest newly formed star cluster is the compact grouping of more than 100 infrared stellar objects embedded in the dense core of the ρ Ophiuchi molecular cloud (Wilking 1992; Greene & Young 1992; Lada et al. 1993). This cluster contains most of the young stars known to be associated with the ρ Oph cloud, and it is also the most active current site of star formation in the entire Ophiuchus-Scorpius-Centaurus region, in which several earlier episodes of star formation have produced OB associations or subgroups (Blaauw 1991). It is thus of considerable interest to examine the circumstances in which the ρ Oph cluster has formed, since they may yield clues to the processes involved. The CO survey of this region by de Geus, Bronfman, & Thaddeus (1990) shows that the ρ Oph cloud core and its embedded cluster are located in the head of a comet-shaped cloud closely adjacent to or partly within the Upper Scorpius OB association, and that this cloud and several others in its vicinity have long tails that point radially away from the center of this association. The windswept appearance of these clouds strongly suggests that they have been shaped by outflows from the OB association, perhaps in somewhat the same way that comets are shaped by the solar wind. The explosive effects of outflows or pressures generated by the OB association on the surrounding interstellar medium are clearly demonstrated by the presence of a large cavity in the distribution of the atomic gas in this region, bounded by a shell that appears to be expanding away from the OB association (de Geus 1991; Blaauw 1991). The molecular clouds in this region mostly appear to have been formed, or at least compressed, where this shell interacts with the denser atomic gas closer to the Galactic plane.

Similar phenomena appear to be occurring on a larger scale in the Orion region, especially in the Orion A cloud which contains the prominent and highly concentrated Trapezium cluster of more than 500 young stars partially embedded in the dense OMC1 core region of the cloud (Herbig & Terndrup 1986; McCaughrean 1989; McCaughrean, Rayner, & Zinnecker 1991; Zinnecker et al. 1993). The Trapezium cluster is closely associated with, and appears to have formed from, a dense clumpy molecular filament that contains three currently active sites of star formation, including the luminous and intensively studied IRc2 region (Hasegawa 1987; Sargent & Mundy 1988; Genzel & Stutzki 1989). This dense $\int$-shaped filament or 'ridge' of molecular gas is also the dominant feature in the head of the strikingly comet-like and filamentary Orion A cloud (Bally et al. 1987, 1991). As is the case with the ρ Oph cloud, the Orion A cloud is located closely adjacent to an OB association, and its long tail points away from the center of the association; again, the structure of this cloud strongly suggests that it has been shaped and compressed by outflows or pressures from the OB association (Bally et al. 1987, 1991). The cometary shape of the Orion A cloud, and even the dense filament in its head, are reminiscent of the results of Woodward's (1978) simulation of the deformation of an interstellar cloud swept over by a shock front in the intercloud medium; most of the material in this cloud is ablated away into a long 'comet tail', but a small part of it is compressed into a dense elongated clump in the head of the comet.

The adjacent Orion B cloud also borders on the Orion OB association, and it also shows evidence of being compressed by outflows from the association: the edge of this cloud nearest the center of the association is relatively sharply bounded and contains several dense clumps (Blaauw 1991; Bally et al. 1991), most of which again contain embedded clusters of young stars similar to but smaller than the Trapezium cluster (Lada et al. 1991a,b; Lada 1992). The compact young clusters that have been found in both the Orion A and B clouds contain most of the young stars known to be associated with these clouds (Lada et al. 1993); thus, in both the Orion region and the Ophiuchus-Scorpius-Centaurus region, most of the recent star formation has taken place in clusters embedded in dense molecular clumps located at the edges of OB associations. These examples and others, such as the dense cloud cores adjacent to the W4 H ±II region which also contain embedded young clusters of stars (Lada et al. 1978; Elmegreen 1992), suggest that clusters of stars may often form in massive clumps of molecular gas that have been compressed to particularly high densities by the effects of prior episodes of star formation in their vicinity. The formation of dense clusters may thus represent a type of 'triggered' star formation (Elmegreen & Lada 1977; Elmegreen 1992), at least in the sense that external compression has caused the stars to form in an exceptionally compact configuration; however, the processes involved are evidently quite complex, and many of the details are not yet clear. The formation of a bound cluster of stars might even be the culminating event in a region of star formation, occurring after several previous episodes of star formation have compressed a small part of the residual gas to a sufficiently high density (Larson 1993).

Another possibly general feature of star formation in clusters is that most of the cluster stars may form in dense filaments like the OMC1 filament associated with the Trapezium cluster. It may be significant, for example, that the second most prominent newly formed cluster in Orion, the NGC 2024 cluster in the Orion B cloud, also has an associated molecular filament containing a string of dense clumps which appear to be sites of ongoing star formation (Mezger et al. 1992; Mezger, this conference). The NGC 2024 clumps are extremely dense and are also very cold, not yet having been noticeably heated by embedded stars; moreover, they are observed mainly as dust continuum sources and show relatively little molecular line emission, apparently because most of their molecules are frozen out onto dust grains (Mauersberger et al. 1992). These clumps therefore probably represent a very early stage of star formation, possibly even an isothermal phase preceding the appearance of a central stellar object (Mezger et al. 1992); however, at least one of these clumps has already produced a bipolar outflow, indicating that it has already formed or begun to form a central star. Even the ρ Oph cloud contains in the dense ρ Oph A core region a similar chain of very dense and cold clumps which are prominent only in millimeter continuum maps and show little molecular line emission, suggesting that they represent a very early stage of star formation (André, Ward-Thompson, & Barsony 1993), although once again one of them has already produced a bipolar outflow. The fact that the sites of ongoing star formation in all three of the clusters mentioned are dense clumpy filaments supports the possibility that the fragmentation of filaments is a generally important mechanism of star formation, as discussed in Section 4.

The origin of the filaments that seem to be such ubiquitous features in star-forming regions is not presently understood, but it may be relevant that filamentary structures are a rather general result of gravitational collapse; it was noted in Section 4, for example, that filaments are often produced even in simulations of the collapse of initially spherical clouds. The general tendency for collapsing clouds to become increasingly elongated had in fact been noted much earlier in analytical and numerical studies showing that any initial elongation of a non-spherical cloud tends to be amplified when the cloud collapses gravitationally (Mestel 1965; Lin, Mestel, & Shu 1965; Larson 1972). However, other effects could also be responsible for the formation of the observed filaments; for example, purely gas-dynamical phenomena such as shear and ablation could play a role, as could helical magnetic fields (Bally et al. 1991; Fukui & Mizuno 1991). In any case, it is again clear that complex dynamical phenomena must be involved.

6 Cloud Destruction and Recycling

Clearly, star formation is highly destructive to molecular clouds; as was emphasized in Section 2, these clouds typically survive for only a few times 10^7 years before being destroyed, or at least restructured in a major way by the effects of star formation. The most effective cloud destruction mechanism may simply be ionization by massive hot stars, which can evaporate away from a star-forming cloud many times the amount of mass that goes into stars; for example, if stars form with a normal mass function, it is only necessary to convert about 4 percent of the mass of a cloud into stars to obtain enough hot stars and enough ionizing photons to completely ionize the rest of the cloud (Whitworth 1979; Larson 1987, 1988; Franco 1992). If star formation proceeds continuously in molecular clouds, this implies that its efficiency will only be of the order of 4 percent, since once this fraction of a cloud's mass has been turned into stars, the remainder will soon be ionized and dispersed. The fact that the observed efficiency of star formation in molecular clouds is actually somewhat smaller than this, typically only $\sim$ 2 percent or less, may be explainable if stellar winds also contribute importantly to the destruction of these clouds (Leisawitz et al. 1989). Both ionization and winds operate over a period of $\sim$ 10 Myr or less that is determined by the lifetimes of the most massive stars; this is comparable to the inferred lifetimes of molecular clouds after the onset of star formation in them.

The gas ionized and evaporated from molecular clouds is dispersed in outflows that may often take the form of 'champagne flows' emerging from ionized cavities (Tenorio-Tagle 1979). The total mass flux in these ionized flows is very large, perhaps accounting for as much as half of the mass cycled through molecular clouds in our Galaxy, which was estimated in Section 3 to be at least 150 $M_\odot$/yr. Ionized outflows from regions of star formation will therefore play an important role in replenishing the warm intercloud component of the interstellar medium. A comparable amount of molecular cloud material may also be dispersed in neutral form by dynamical effects including the pressure exerted by ionized gas, stellar winds, and eventually supernova explosions, and the resulting cloud debris may be swept up into large expanding shells of mostly atomic gas. Such shells are in fact often observed around regions of star formation, including both the Ophiuchus-Scorpius-Centaurus and the Orion regions; a well-known example is Barnard's Loop, which partially surrounds the Orion star-forming region (Blaauw 1991; Bally et al. 1991). The eventual disruption of such shells will replenish the cold neutral component of the interstellar medium, and may in fact be the major source of new interstellar clouds, as was originally suggested by Oort (1954).

The fact that both the ionized and the neutral debris from molecular clouds are predicted to be dispersed with velocities of the order of 10 $\mathrm{km\,s^{-1}}$ and the fact that gas is observed to be receding from some young clusters with velocities of this order (Leisawitz et al. 1989), strongly suggest that the destruction of star-forming clouds is also a major contributor to the observed turbulent motions in the interstellar medium, and that it may largely account for the typical cloud velocity dispersion of $\sim$ 5 to 10 $\mathrm{km\,s^{-1}}$ (Larson 1987; Tenorio-Tagle & Bodenheimer 1988; Bally et al. 1991). This is possible because the interstellar medium is continually being cycled through star-forming clouds on a timescale that was estimated in Section 3 to be about 50 Myr locally; this cycling time, which is also the time required to regenerate cloud velocities of the order of 10 $\mathrm{km\,s^{-1}}$, is not much longer than the timescale of $\sim$ 20 Myr estimated for the damping of interstellar cloud motions (Spitzer 1978), or the time interval of $\sim$ 30 Myr between the collisions of large shells (Bally et al. 1991). Supernova explosions will of course generate additional interstellar cloud motions, but they need not be the dominant source of such motions, as has usually been assumed. Thus, the cycling of interstellar matter through star-forming clouds and the subsequent disruption of these clouds by the various effects of star formation may account for many of the observed properties of the interstellar medium.

The cycle of cloud formation and destruction is finally closed when the debris from old star-forming clouds becomes the raw material for building new molecular clouds by the large-scale gravitational instability effects discussed in Section 3. Clearly, many forms of interstellar matter and many types of processes will play a role in the continual cycling of interstellar matter between diffuse and condensed forms that has been discussed. These processes are important not only because of their implications for molecular cloud evolution and star formation, but also because of their broader significance for galactic evolution. One example of the implications of cloud formation and destruction processes for galactic evolution is that they play an important role in the dispersal through the interstellar medium of the heavy elements produced in stars, and hence in the chemical evolution of galaxies (Bateman & Larson 1993). Thus, the cycle of cloud formation and destruction is a central aspect not only of the evolution of molecular clouds but also of the evolution of galaxies as a whole, and one that has many fundamental implications.

References

André, P., Ward-Thompson, D., & Barsony, M. 1993, ApJ, 406, 122
Arcoragi, J.-P., Bonnell, I., Martel, H., Benz, W., & Bastien, P. 1991, ApJ, 380, 476
Bally, J., Langer, W. D., Stark, A. A., & Wilson, R. W. 1987, ApJ, 312, L45
Bally, J., Langer, W. D., Wilson, R. W., Stark, A. A., & Pound, M. W. 1991, in Fragmentation of Molecular Clouds and Star Formation, IAU Symposium No. 147, eds. E. Falgarone, F. Boulanger, & G. Duvert (Kluwer, Dordrecht), p. 11
Bash, F. N., Green, E., & Peters, W. L. 1977, ApJ, 217, 464
Bastien, P., Arcoragi, J.-P., Benz, W., Bonnell, I., & Martel, H. 1991, ApJ, 378, 255
Bateman, N. P. T., & Larson, R. B. 1993, ApJ, 407, 634
Blaauw, A. 1964, ARA&A, 2, 213
Blaauw, A. 1991, in The Physics of Star Formation and Early Stellar Evolution, eds. C. J. Lada & N. D. Kylafis (Kluwer, Dordrecht), p. 125
Blitz, L. 1988, in Millimetre and Submillimetre Astronomy, eds. R. D. Wolstencroft & W. B. Burton (Kluwer, Dordrecht), p. 269
Blitz, L. 1991, in The Physics of Star Formation and Early Stellar Evolution, eds. C. J. Lada & N. D. Kylafis (Kluwer, Dordrecht), p. 3
Blitz, L., & Shu, F. H. 1980, ApJ, 238, 148
Bonnell, I., & Bastien, P. 1992, ApJ, 401, 654
Bonnell, I., Martel, H., Bastien, P., Arcoragi, J.-P., & Benz, W. 1991, ApJ, 377, 553
Bonnell, I., Arcoragi, J.-P., Martel, H., & Bastien, P. 1992, ApJ, 400, 579
Boss, A. P. 1993, ApJ, 410, 157
Boulanger, F. 1991, in Fragmentation of Molecular Clouds and Star Formation, IAU Symposium No. 147, eds. E. Falgarone, F. Boulanger, & G. Duvert (Kluwer, Dordrecht), cover photograph
Buat, V., Deharveng, J. M., & Donas, J. 1989, A&A, 223, 42
Burkert, A., & Bodenheimer, P. 1993, MNRAS, in press
Carpenter, J. M., Snell, R. L., Schloerb, F. P., & Skrutskie, M. F. 1993, ApJ, 407, 657
Chièze, J. P., & Pineau des Forêts, G. 1990, in Physical Processes in Fragmentation and Star Formation, eds. R. Capuzzo-Dolcetta, C. Chiosi, & A. Di Fazio (Kluwer, Dordrecht), p. 17
Cohen, M., & Kuhi, L. V. 1979, ApJS, 41, 743
de Geus, E. J. 1991, in The Formation and Evolution of Star Clusters, ed. K. Janes (ASP, San Francisco), p. 40
de Geus, E. J., Bronfman, L., & Thaddeus, P. 1990, A&A, 231, 137
Dickman, R. L., Horvath, M. A., & Margulis, M. 1990, ApJ, 365, 586
Efremov, Y. N. 1989, Soviet Scientific Reviews E: Astrophysics and Space Science Reviews, 7, 105
Elmegreen, B. G. 1989, ApJ, 338, 178
Elmegreen, B. G. 1990a, in The Evolution of the Interstellar Medium, ed. L. Blitz (ASP, San Francisco), p. 247
Elmegreen, B. G. 1990b, ApJ, 357, 125
Elmegreen, B. G. 1991a, in The Physics of Star Formation and Early Stellar Evolution, eds. C. J. Lada & N. D. Kylafis (Kluwer, Dordrecht), p. 35

Elmegreen, B. G. 1991b, ApJ, 378, 139
Elmegreen, B. G. 1992, in Star Formation in Stellar Systems, eds. G. Tenorio-Tagle, M. Prieto, & F. Sánchez (Cambridge University Press, Cambridge), p. 381
Elmegreen, B. G. 1993, ApJ, 411, in press
Elmegreen, B. G., & Lada, C. J. 1977, ApJ, 214, 725
Evans, N. J. 1991, in Frontiers of Stellar Evolution, ed. D. L. Lambert (ASP, San Francisco), p. 45
Evans, N. J., & Lada, E. A. 1991, in Fragmentation of Molecular Clouds and Star Formation, IAU Symposium No. 147, eds. E. Falgarone, F. Boulanger, & G. Duvert (Kluwer, Dordrecht), p. 293
Falgarone, E. 1992, in Astrochemistry of Cosmic Phenomena, IAU Symposium No. 150, ed. P. D. Singh (Kluwer, Dordrecht), p. 159
Falgarone, E., & Phillips, T. G. 1991, in Fragmentation of Molecular Clouds and Star Formation, IAU Symposium No. 147, eds. E. Falgarone, F. Boulanger, & G. Duvert (Kluwer, Dordrecht), p. 119
Falgarone, E., Phillips, T. G., & Walker, C. K. 1991, ApJ, 378, 186
Falgarone, E., Puget, J.-L., & Pérault, M. 1992, A&A, 257, 715
Franco, J. 1992, in Star Formation in Stellar Systems, eds. G. Tenorio-Tagle, M. Prieto, & F. Sánchez (Cambridge University Press, Cambridge), p. 515
Fukui, Y., & Mizuno, A. 1991, in Fragmentation of Molecular Clouds and Star Formation, IAU Symposium No. 147, eds. E. Falgarone, F. Boulanger, & G. Duvert (Kluwer, Dordrecht), p. 275
Genzel, R., & Stutzki, J. 1989, ARA&A, 27, 41
Goodman, A. A., Benson, P. J., Fuller, G. A., & Myers, P. C. 1993, ApJ, 406, 528
Gomez, M., Hartmann, L., Kenyon, S. J., & Hewett, R. 1993, AJ, 105, 1927
Greene, T. P., & Young, E. T. 1992, ApJ, 395, 516
Hasegawa, T. 1987, in Star Forming Regions, IAU Symposium No. 115, eds. M. Peimbert & J. Jugaku (Reidel, Dordrecht), p. 123
Heller, C. H. 1993a, ApJ, 408, 337
Heller, C. H. 1993b,c, ApJ, in press
Herbertz, R., Ungerechts, H., & Winnewisser, G. 1991, A&A, 249, 483
Herbig, G. H. 1977, ApJ, 214, 747
Herbig, G. H., & Terndrup, D. M. 1986, ApJ, 307, 609
Herbst, E., & Leung, C. M. 1989, ApJS, 69, 271
Hetem, A., & Lépine, J. R. D. 1993, A&A, 270, 451
Kennicutt, R. C. 1989, ApJ, 344, 685
Kennicutt, R. C. 1990, in The Interstellar Medium in Galaxies, eds. H. A. Thronson & J. M. Shull (Kluwer, Dordrecht), p. 405
Lada, C. J., & Lada, E. A. 1991, in The Formation and Evolution of Star Clusters, ed. K. Janes (ASP, San Francisco), p. 3
Lada, C. J., Elmegreen, B. G., Cong, H.-I., & Thaddeus, P. 1978, ApJ, 226, L39
Lada, E. A. 1992, ApJ, 393, L25
Lada, E. A., Bally, J., & Stark, A. A. 1991a, ApJ, 368, 432
Lada, E. A., DePoy, D. L., Evans, N. J., & Gatley, I. 1991b, ApJ, 371, 171
Lada, E. A., Strom, K. M., & Myers, P. C. 1993, in Protostars & Planets III, eds. E. H. Levy, J. I. Lunine, & M. S. Matthews (University of Arizona Press, Tucson), in press
Larson, R. B. 1972, MNRAS, 156, 437
Larson, R. B. 1978, MNRAS, 184, 69
Larson, R. B. 1979, MNRAS, 186, 479

Larson, R. B. 1981, MNRAS, 194, 809
Larson, R. B. 1982, MNRAS, 200, 159
Larson, R. B. 1984, MNRAS, 206, 197
Larson, R. B. 1985, MNRAS, 214, 379
Larson, R. B. 1987, in Starbursts and Galaxy Evolution, eds. T. X. Thuan, T. Montmerle, & J. Tran Thanh Van (Editions Frontières, Gif Sur Yvette), p. 467
Larson, R. B. 1988, in Galactic and Extragalactic Star Formation, eds. R. E. Pudritz & M. Fich (Kluwer, Dordrecht), p. 459
Larson, R. B. 1990, in Physical Processes in Fragmentation and Star Formation, eds. R. Capuzzo-Dolcetta, C. Chiosi, & A. Di Fazio (Kluwer, Dordrecht), p. 389
Larson, R. B. 1991a, in Frontiers of Stellar Evolution, ed. D. L. Lambert (ASP, San Francisco), p. 571
Larson, R. B. 1991b, in Fragmentation of Molecular Clouds and Star Formation, IAU Symposium No. 147, eds. E. Falgarone, F. Boulanger, & G. Duvert (Kluwer, Dordrecht), p. 261
Larson, R. B. 1992a, MNRAS, 256, 641
Larson, R. B. 1992b, in Star Formation in Stellar Systems, eds. G. Tenorio-Tagle, M. Prieto, & F. Sánchez (Cambridge University Press, Cambridge), p. 125
Larson, R. B. 1993, in The Globular Cluster – Galaxy Connection, eds. J. P. Brodie & G. H. Smith (ASP, San Francisco), in press
Lin, C. C., Mestel, L., & Shu, F. H. 1965, ApJ, 142, 1431
Leisawitz, D., Bash, F. N., & Thaddeus, P. 1989, ApJS, 70, 731
Mauersberger, R., Wilson, T. L., Mezger, P. G., Gaume, R., & Johnston, K. J. 1992, A&A, 256, 640
McCaughrean, M. J. 1989, BAAS, 21, 712; see also Sky and Telescope, 77, 352
McCaughrean, M. 1993, in Massive Stars: Their Lives in the Interstellar Medium, eds. J. P. Cassinelli & E. B. Churchwell (ASP, San Francisco), p. 80
McCaughrean, M., Rayner, J., & Zinnecker, H. 1991, Mem. Soc. Astr. Ital., 62, 715
Mestel, L. 1965, QJRAS, 6, 161
Mezger, P. G., Sievers, A. W., Haslam, C. G. T., Kreysa, E., Lemke, R., Mauersberger, R., & Wilson, T. L. 1992, A&A, 256, 631
Miyama, S. M., Hayashi, C., & Narita, S. 1984, ApJ, 279, 621
Miyama, S. M., Narita, S., & Hayashi, C. 1987a, Prog. Theor. Phys., 78, 1051
Miyama, S. M., Narita, S., & Hayashi, C. 1987b, Prog. Theor. Phys., 78, 1273
Monaghan, J. J., & Lattanzio, J. C. 1991, ApJ, 375, 177
Monaghan, J. J., & Varnas, S. R. 1988, MNRAS, 231, 515
Mooney, T. J., & Solomon, P. M. 1988, ApJ, 334, L51
Myers, P. C. 1987, in Star Forming Regions, IAU Symposium No. 115, eds. M. Peimbert & J. Jugaku (Reidel, Dordrecht), p. 33
Myers, P. C. 1991, in The Formation and Evolution of Star Clusters, ed. K. Janes (ASP, San Francisco), p. 73
Myers, P. C., Dame, T. M., Thaddeus, P., Cohen, R. S., Silverberg, R. F., Dwek, E., & Hauser, M. G. 1986, ApJ, 301, 398
Myers, P. C., Fuller, G. A., Goodman, A. A., & Benson, P. J. 1991, ApJ, 376, 561
Oort, J. H. 1954, BAN, 12, 177
Prasad, S. S. 1991, in Fragmentation of Molecular Clouds and Star Formation, IAU Symposium No. 147, eds. E. Falgarone, F. Boulanger, & G. Duvert (Kluwer, Dordrecht), p. 93
Prasad, S. S., Tarafdar, S. P., Villere, K. R., & Huntress, W. T. 1987, in Interstellar Processes, eds. D. J. Hollenbach & H. A. Thronson (Reidel, Dordrecht), p. 631

Pringle, J. E. 1991, in The Physics of Star Formation and Early Stellar Evolution, eds. C. J. Lada & N. D. Kylafis (Kluwer, Dordrecht), p. 437
Rubio, M., Lequeux, J., & Boulanger, F. 1993, A&A, 271, 9
Sargent, A. I., & Mundy, L. G. 1988, in Galactic and Extragalactic Star Formation, eds. R. E. Pudritz & M. Fich (Kluwer, Dordrecht), p. 261
Scalo, J. 1990, in Physical Processes in Fragmentation and Star Formation, eds. R. Capuzzo-Dolcetta, C. Chiosi, & A. Di Fazio (Kluwer, Dordrecht), p. 151
Schneider, S., & Elmegreen, B. G. 1979, ApJS, 41, 87
Spitzer, L. 1978, Physical Processes in the Interstellar Medium (Wiley, New York)
Sreenivasan, K. 1991, Ann. Rev. Fluid Mech., 23, 539
Stahler, S. W. 1984, ApJ, 281, 209
Tenorio-Tagle, G. 1979, A&A, 71, 59
Tenorio-Tagle, G., & Bodenheimer, P. 1988, ARA&A, 26, 145
Toomre, A. 1981, in The Structure and Evolution of Normal Galaxies, eds. S. M. Fall & D. Lynden-Bell (Cambridge University Press, Cambridge), p. 111
Welch, Wm. J. 1993, in Massive Stars: Their Lives in the Interstellar Medium, eds. J. P. Cassinelli & E. B. Churchwell (ASP, San Francisco), p. 15
Welch, Wm. J., Dreher, J. W., Jackson, J. M., Terebey, S., & Vogel, S. N. 1987, Science, 238, 1550
Whitworth, A., 1979, MNRAS, 186, 59
Wilking, B. A. 1992, in Low Mass Star Formation in Southern Molecular Clouds, ESO Scientific Report No. 11, ed. B. Reipurth (ESO, Garching), p. 159
Woodward, P. R. 1978, ARA&A, 16, 555
Williams, J. P., & Blitz, L. 1993, ApJ, 405, L75
Zimmermann, T., & Stutzki, J. 1992, Physica A, 191, 79
Zinnecker, H., McCaughrean, M. J., & Wilking, B. A. 1993, in Protostars & Planets III, eds. E. H. Levy, J. I. Lunine, & M. S. Matthews (University of Arizona Press, Tucson), in press

Interstellar Chemistry in the Last Two Decades

Eric Herbst

Departments of Physics and Astronomy, The Ohio State University, Columbus, OH 43210, USA

Abstract: Since the discovery of polyatomic molecules in interstellar space more than two decades ago, much progress has been made in understanding how these molecules are both produced and destroyed. In this review, the history of the progress in interstellar chemistry is discussed.

1 Introduction

Interstellar molecules represent no recent discovery; by the 1940's the diatomic species CH, CH^+, and CN had been discovered by optical spectroscopy to be present in what are now known as diffuse interstellar clouds [1-4]. The diatomic radical OH was discovered via radio techniques in 1963 [5] but it was not until the end of the 1960's that the crucial discoveries of polyatomic molecules in dense interstellar clouds began. By now, almost 100 different gaseous molecules, plus numerous isotopic variants, have been discovered in interstellar space and circumstellar shells of evolved stars [6-7]. A list is shown in Table 1. These molecules have been detected mainly in emission in the radio region of the spectrum although infra-red absorption spectra seen along the line of sight to protostellar or stellar sources embedded in dense interstellar or circumstellar clouds have also been identified. The quantized quantum mechanical motions of gas phase molecules that lead to high resolution radio and infra-red spectra are rotation and vibration, respectively. Both motions have been studied extensively in the laboratory, so that the rotational and vibrational transitions of most stable and many unstable simple molecules are well known.

Gaseous interstellar molecules range in complexity from diatomic species to a thirteen-atom, unsaturated (hydrogen-poor), linear organo-nitrogen species $HC_{11}N$. As can be seen in Table 1, the molecules are mainly organic in nature, and are evenly divided between standard molecules, which can be purchased, and highly unusual neutral and ionic species, which have only the most ephemeral existence in the terrestrial laboratory. No molecules with more than two atoms have been seen in diffuse interstellar clouds. These regions, characterized by low gas densities and high penetration by dissociative, external ultra-violet radiation from stars, do not appear to be conducive to molecular development. Indeed, the gas in these regions is dominated by atoms except for the case of H_2, which can have an abundance comparable to that of atomic hydrogen. Dense interstellar clouds, on the other hand, possess enough dust to shield molecules to a great extent from photodissociation by external ultra-violet radiation. The most abundant molecule by far in dense clouds is H_2; the second most abundant molecule, CO, possesses a fractional abundance of 10^{-4}, while all other molecules can legimately be regarded as trace species. The largest organic molecules detected have fractional abundances on the order of 10^{-10} [8-9]. Despite their small fractional abundances, the absolute abundances of these species are much larger than that of corresponding species on earth. Although chemistry clearly plays an important role in producing the fractional abundances, the dominance of hydrogen owes much to the uniquely high elemental abundance of this element in all but the smallest, gravitationally weakest sources, such as our own planet. Typical elemental abundances (by number with respect to hydrogen) are 0.10 for He, and $\approx 10^{-4}$ for C, N, and O, with O being in excess of C [10].

Molecular abundances can differ from source to source, especially for the more complex species. For example, the cyanopolyynes ($HC_{2n+1}N$) are most abundant in a small source known as TMC-1 in the large Taurus complex of our own galaxy. Other well-known sources in the Milky Way include Sgr B2 (near the galactic center) and the Orion Nebula. Molecules are also seen in clouds of external galaxies. Some of the molecules listed in Table 1 have only been detected in circumstellar envelopes, in particular the nearby carbon-rich (elemental C $\geq$ elemental O) circumstellar envelope IRC+10216.

Spectral characterization of the dust particles, which accompany the gas in interstellar and circumstellar sources and comprise perhaps 1 % of their mass, is difficult. Scattering and absorption of background starlight indicate a size distribution peaking at perhaps 0.1 μ. In addition, broad infra-red spectral features can be assigned with some degree of ambiguity to the vibrations of molecules in dust particles [11-13]. Strong infra-red features originating from circumstellar dust particles indicate the presence of silicates. In dense clouds, the additional observations of ices of CO, H_2O, CH_3OH, and other species indicate that dust particles probably contain both cores of refractory-type materials (e.g., silicates) and mantles of ices. The observation of strong infra-red emission from regions surrounding stars and stellar-type objects has been interpreted in terms of very small dust particles or particularly large gas phase molecules; spectral analysis of particular emission features suggests strongly that the species can be thought of as small particles of graphitic or amorphous carbonaceous material, or as an ensemble of differing polycyclic aromatic hydrocarbons (PAH's) [14-18]. Despite the carbonaceous nature of the material, there is no evidence that fullerenes are present in observable quantities. Whatever their detailed nature, the small dust particles contain perhaps 0.01% of the mass and 10^{-7} of the number density of the localized sources in which they are found. Since the observation of interstellar molecules began, scientists have been trying to understand where and how they are formed. The possibility that molecules are not synthesized where they are found arises because of the seemingly hostile conditions for chemical

Table 1. List of Gaseous Interstellar Molecules

H_2	AlF	NaCl	H_2O	CCC	NH_3	HCCH
CH^+	CP	KCl	H_2S	CCO	H_3O^+	HCCN
CH	CSi	AlCl	HCN	CCSi	H_2CO	CCCN
NH	CS		HNC	CCS	H_2CS	CCCO
OH	NS		HCO^+	SO_2	CCCH	CCCS
HCl	SO		HCO	OCS	c-C_3H	
CC	SiO		HN_2^+	MgNC	HNCO	
CN	SO^+		HNO		HNCS	
CO	PN		CCH		$HOCO^+$	
NO	SiS		HCS^+		$HCNH^+$	
CH_4	HNCCC	CH_3OH	CH_3NH_2	CH_3OCH_3	HC_9N	
SiH_4	C_4H	CH_3SH	CH_3CCH	C_2H_5OH	$HC_{11}N$	
H_2CCC	C_5	CH_3CN	CH_3CHO	C_2H_5CN		
c-C_3H_2	C_4Si	CH_3NC	C_2H_3CN	CH_3C_4H		
H_2CNH		C_2H_4	HC_5N	HC_7N		
HCOOH		H_2CCCC	C_6H			
CH_2CN		$HCONH_2$				
NH_2CN		HC_2CHO	$HCOOCH_3$	CH_3C_5N		
HCCCN		C_5H	CH_3CCCN	CH_3COCH_3		
HCCNC						

development in dense interstellar clouds. At typical gas densities of 10^4 cm^{-3}, collisions between molecules occur quite rarely. Moreover, most of these collisions cannot be expected to be reactive, given the low prevailing temperatures of 10-70 K. Evolved atmospheres of old stars might constitute a more favorable atmosphere as far as density and temperature are concerned. Indeed, molecules and dust particle cores are observed to form in such sources. As the molecules are blown out into interstellar space, however, ultra-violet radiation from other stars dissociates them almost immediately (within 100-1000 years) [19]. Dust particles and possibly PAH's are more fortunate and survive [20]. Thus, the material blown out of older generations of stars that congregates together to form new generations of interstellar clouds is probably composed of dust particles and atoms; the formation of small molecules must occur, albeit quite slowly, in the clouds themselves.

The in situ formation of molecules in interstellar clouds from precursor atoms can occur either in the gas or on the surfaces of dust particles. Perhaps the most important role of surface chemistry is the conversion of atomic into molecular hydrogen. In the gas, this conversion can occur via the sticking of two hydrogen atoms, a process only possible at low densities if the system emits a photon during the collision to radiate away its excess energy. Although this process, labelled radiative association, is now known to occur for reactions involving more than two atoms, it is too inefficient for two hydrogen atoms to explain the interstellar formation of H_2 [21]. Rather, the dominant process appears to involve the sticking of hydrogen atoms to the surfaces of dust particles; a sticking that occurs efficiently on cold particles because they are thermodynamic entities with many degrees of freedom. Once the H atoms stick, they can migrate rapidly if, as occurs on most surfaces, they are bound loosely by van der Waals'-type forces ("physisorption") [22-23]. Although all weakly-adsorbed species can migrate to some degree, H atoms, due to their lightness, can tunnel efficiently under the potential barriers that exist between binding sites on grain surfaces, and so migrate far more rapidly than heaver species. Once two hydrogen atoms encounter one another, they can form H_2 by releasing the excess chemical energy to the many degrees of freedom of the dust particle. The surface H_2 formed in this manner can then be returned to the gas by simple thermal evaporation or by non-thermal desorption, perhaps by using some of the excess chemical energy [24].

The formation of other molecules on dust surfaces should occur as well, although the laboratory understanding of surface chemistry is still poor. As was recognized quite soon after the discovery of polyatomic interstellar molecules, saturation of adsorbed heavy atoms by successive H atom addition should occur readily [25-27]. For example, the hydrogenation of C atoms into methane (CH_4) can occur sequentially: $C \rightarrow CH \rightarrow CH_2 \rightarrow CH_3 \rightarrow CH_4$ since the intermediate CH_n radicals are thought to be quite reactive down to low temperatures. The formation of saturated species such as H_2O, NH_3, H_2S, and CH_4 is a salient feature of grain chemistry as long as the supply of H atoms lasts. Hydrogenation via reactions with molecular hydrogen is much slower. General grain reactions will be considered in more detail in Section III. Unlike the case of H_2, it is far from apparent how the heavy molecules formed on dust surfaces can evaporate into the gas. A variety of non-thermal desorption mechanisms have been suggested over the years, but none has been shown conclusively to be efficient [28]. Assuming that desorption is indeed generally inefficient at low temperature, regions of star formation are an obvious place to look for chemical signatures of dust syntheses since the rising temperatures are certain to result in significant desorption [29]. The production of most gaseous molecules in cool regions appears to require a gas phase route.

Once H_2 is formed efficiently on dust particle surfaces and released into the gas, gas phase chemistry can be initiated [6-7]. The low temperature dictates that only exothermic chemical processes can occur. In addition, the majority of exothermic chemical reactions between neutral species are known to possess "activation energy," which is a potential energy barrier caused by the breaking of old chemical bonds before new ones can be formed. The rate coefficient k(T) for gas phase reactions that possess activation energy E_a is given by the formula

$$k(T) = A(T)exp(-E_a/kT) \tag{1}$$

where T is the temperature in K, k is the Boltzmann constant, and A(T) is a weakly temperature-dependent factor which depends on the long-range nature of reactive collisions. A "typical" activation energy of 1-5 eV leads to a negligibly small rate coefficient for cool interstellar temperatures. Neutral radicals are thought to react with zero activation energy, but do not seem to be sufficiently abundant to produce the variety of molecules detected. Reactions involving atomic or molecular ions are a different story. As was realized by astrophysicists soon after the discovery of interstellar polyatomic molecules and before the confirmation that ions actually existed in clouds, exothermic reactions between ions and neutral molecules can synthesize polyatomic molecules in a relatively efficient manner [24,30-32]. Ion-molecule reactions tend not to possess activation energy barriers because of strong long-range attractive forces. For non-polar neutrals, the simple Langevin model predicts, moreover, that A(T) is completely independent of temperature [30]. In addition, it is virtually the same for all reactants. For polar neutrals, the situation appears to be even better, since theories indicate an inverse temperature dependence for A(T) [33]. In the last twenty years, numerous laboratory studies of ion-molecule reactions at low temperatures have more than confirmed these simple views [34].

Ionization in dense interstellar clouds stems mainly from cosmic ray bombardment of H_2 [31]:

$$H_2 + CR \longrightarrow H_2^+ + e^- + CR, \tag{2}$$

where CR stands for cosmic ray, although cosmic ray ionization of abundant helium atoms to produce He^+ is also important. Cosmic rays are relativistic nuclei, mainly protons. The flux of cosmic rays that impinges upon and penetrates interstellar clouds is small, yet sufficient to power a complex ion-molecule chemistry. According to models discussed below, the fractional ionization is only $10^{-6} - 10^{-8}$. In diffuse interstellar clouds, the penetration of external ultra-violet radiation leads to a far greater degree of ionization. Essentially all of the elemental abundance of carbon is in the form of C^+; the resulting fractional ionization is approximately 10^{-4}. Cosmic ray penetration and ionization leads indirectly to a small ultra-violet flux in even the densest clouds.

2 Ion-Molecule Chemistry

Ion-molecule syntheses of interstellar molecules can occur via the following sequences of reactions [6,7,35]. The H_2^+ formed by cosmic ray bombardment of H_2 reacts with H_2 to form the simplest polyatomic species H_3^+:

$$H_2^+ + H_2 \longrightarrow H_3^+ + H. \tag{3}$$

The H_3^+ ion then reacts with a heavy atom A (A = C, O, S etc.):

$$A + H_3^+ \longrightarrow AH^+ + H_2 \tag{4}$$

via proton transfer if the process is exothermic. The newly formed AH^+ ion reacts sequentially with ubiquitous H_2 via either standard atom-transfer reactions

$$AH_j^+ + H_2 \longrightarrow AH_{j+1}^+ + H, \tag{5}$$

or, if such a reaction is endothermic, by a much slower radiative association reaction

$$AH_j^+ + H_2 \longrightarrow AH_{j+2}^+ + h\nu. \tag{6}$$

Hydrogenation can also commence with the atomic ion A^+ via a standard ion-molecule reaction

$$A^+ + H_2 \longrightarrow AH^+ + H, \tag{7}$$

or by a radiative association process. In this manner, the saturated ion AH_n^+ can eventually be synthesized. Production of the neutral species then occurs via dissociative recombination reactions with electrons. Although these processes are known to be rapid, the assorted neutral product channels have not been fully characterized for any important system [36-37], and simplistic theoretical estimates are still necessary. It is normally assumed that the dominant products are

$$AH_n^+ + e^- \longrightarrow AH_{n-1} + H; AH_{n-2} + 2H(H_2). \tag{8}$$

The interstellar syntheses of gaseous H_2O, OH, NH_3, NH_2, CH_4, CH_3, etc. can occur in this manner. In dense interstellar clouds, synthesized gaseous molecules are depleted by gas phase reactions and by adsorption onto dust particles. In diffuse clouds, ultra-violet induced photodestruction is more potent for neutral species. Once simple molecules are produced in dense interstellar clouds, more complex species can be formed via several types of reactions. Let us consider the synthesis of hydrocarbons. An important synthetic reaction for hydrocarbons is labelled carbon insertion; in this process C^+ reacts with a neutral species by inserting itself and expelling a hydrogen atom; viz.:

$$C^+ + C_nH_m \longrightarrow C_{n+1}H_{m-1}^+ + H. \tag{9}$$

Another important type of synthetic process is the so-called condensation reaction, in which a hydrocarbon ion and a hydrocarbon neutral react to form a much larger hydrocarbon ion as well as a hydrogen atom or atoms; viz.:

$$C_nH_m^+ + C_kH_l \longrightarrow C_{n+k}H_{m+l-1}^+ + H. \tag{10}$$

Finally, radiative association processes; viz.:

$$C_nH_m^+ + C_kH_l \longrightarrow C_{n+k}H_{m+l}^+ + h\nu \tag{11}$$

become efficient for systems with perhaps 10 total atoms. All of the standard ion-molecule reactions discussed here are well studied processes in the laboratory. Radiative association reactions have only been so studied recently [38]; for many years theoretical treatments of the rates of these processes or estimates based on experimental measurements of three-body association rates at high density were all that was available. The theoretical treatments have in the main been proven to be quite reliable [39].

As more complex hydrocarbon ions are synthesized by the above reactions, competitions ensue between hydrogenation via reaction with H_2 and dissociative recombination with electrons. In general, reactions between multi-carbon-atom hydrocarbon ions and H_2 become endothermic when more than two hydrogen atoms are added to the carbon skeleton. Even radiative association reactions appear to have activation energy barriers when more than two hydrogen atoms are present [40]. The result is that saturated complex molecules are not produced efficiently in the gas, unlike the saturation that is achieved for heavy atoms.

Synthetic routes via ion-molecule reactions also provide a natural explanation for the wide variety of unusual neutral molecules detected in dense interstellar clouds. These species are as likely to be produced by dissociative recombination reactions as are "normal" species, and ion-molecule reactions destroy both 'normal' and unusual molecules with equal rate coefficients. As an example, consider the formation of the interstellar molecules c-C_3H_2, H_2CCC, c-C_3H, and HCCC, where "c" stands for "cyclic." All of these structures can be formed from the precursor ion $C_3H_3^+$, which is itself formed with sufficient energy to undergo rapid isomerization between a cyclic and a non-cyclic (propargyl) form H_2CCCH^+ [41]. The cyclic (cyclopropenyl) ion undergoes dissociative recombination to form cyclic products:

$$c - C_3H_3^+ + e^- \longrightarrow c - C_3H_2 + H; c - C_3H + H_2 \tag{12}$$

while the non-cyclic ion undergoes dissociative recombination to form non-cyclic products:

$$H_2CCCH^+ + e^- \longrightarrow H_2CCC + H; HCCC + H_2. \tag{13}$$

Ion-molecule reactions also provide a ready explanation for much of the strong isotopic fractionation seen in interstellar chemistry. Just as H_2 is the major repository of hydrogen in dense clouds, HD is the major repository of deuterium. Although the D/H elemental abundance ratio is normally $\approx 3 \times 10^{-5}$, trace molecules can possess singly deuterated "isotopomers" with abundances up to 1-10 % of the normal species. The explanation for this effect has been understood for some time [42-43]. At very low temperatures, small differences in zero-point energy become very significant in determining the exothermic or endothermic character of isotopic exchange reactions. Consider the reactions

$$H_3^+ + HD \rightleftharpoons H_2D^+ + H_2. \tag{14}$$

The left-to-right direction is the exothermic one; its exothermicity (divided by the Boltzmann constant) is only 230 K including the fact that the lowest rotational level of the H_3^+ ion is forbidden by Pauli Principle considerations. At temperatures much lower than 230 K, the backwards reaction is seriously hindered and the abundance of H_2D^+ can reach a sizable fraction of that of H_3^+. The effect is clearly a function of temperature, with the lowest temperature clouds predicted to have the strongest fractionation effect. A variety of reactions between molecular ions and HD such as the one above serve to produce large abundances of singly deuterated ions [44-45]. These ions then undergo ion-molecule reactions, and the net effect is that large abundances of singly deuterated species - both ionic and neutral- are produced.

Despite the appealing nature of the gas phase hypothesis, there are several sources of uncertainty in our understanding of the low temperature gas phase chemistry of interstellar clouds. First, in the realm of ion-molecule reactions, not enough reactions have been measured to confirm the hypothesis that these reactions can produce the more complex interstellar molecules detected. Even the reactions that have been studied have typically been measured at room temperature, and extrapolations from 300 K to 10 K can in certain instances be quite dangerous [34]. Secondly, the extent of neutral-neutral chemistry involving atoms and radicals is still unknown since until recently there were esssentially no measurements at temperatures under 100 K. Recent low temperature measurements indicate that a variety of neutral-neutral reactions may be considerably more rapid than was heretofore believed [46]. Thirdly, the neutral products formed in the dissociative recombination of molecular ions with electrons have not been fully characterized for all but the simplest reactions [36-37]. It is entirely possible that the detailed results of gas phase models of cold interstellar clouds, discussed below, will be changed dramatically in the future.

The discussion of gas phase reactions has focused on low temperature. The interstellar medium is rather heterogeneous, and some of the material in dense interstellar clouds is warm for a variety of reasons such as ultra-violet photon penetration, star formation, and shock waves. As the temperature increases, the restrictions on which types of chemical processes in the gas can occur decrease. Exothermic neutral-neutral reactions turn on first, especially those involving one stable species and one radical. At still higher temperatures, weakly endothermic processes become competitive in rate, and eventually even strongly endothermic processes assume some importance. As an example, consider the reaction

$$C^+ + H_2 \longrightarrow CH^+ + H, \tag{15}$$

which is endothermic by 0.39 eV. Completely unimportant in cool interstellar clouds (where the radiative association between the reactants is significant), this endothermic reaction has been postulated to produce the observed abundance of the CH^+ species in diffuse interstellar clouds at temperatures temporarily elevated to several thousands of degrees via shock waves [47].

As more types of chemical reactions become possible, thermodynamic equilibrium in a chemical sense becomes more attainable. In protostellar regions, the increasing density and temperature will drive the system to chemical equilibrium, whereas the approximation of chemical equilibrium in cool interstellar clouds leads to ludicrous results.

Although the discussion has centered on reactive collisions, it is important to note that non-reactive collisions play a significant role in the thermal balance and excitation of the gas in dense clouds. Strong theoretical programs by Green and Flower have given astrophysicists many necessary inelastic cross sections [48-49]; experimental programs that measure inelastic collision rates directly or that study pressure broadening of rotational lines at low temperatures are also quite useful.

3 Dust Surface Chemistry

Dust surface chemistry is initiated by the sticking of neutral atoms and molecules onto cold surfaces. The dust surfaces tend to be negatively charged, because electrons move more rapidly in a thermal system than do the heavier positive ions. The sticking of gas phase positive ions is thought to be inefficient due to acceleration. Binding energies E_D/k for "physisorbed" surface species range from a few hundred K for light, small species, to a few thousand K for heavier molecules, and are somewhat dependent on the nature of the surface [24,50]. Based on pioneering work by Salpeter and associates [21, 23-25] and by Williams [22], it is generally thought that the chemistry on large dust particles occurs via what surface chemists refer to as a Langmuir-Hinshelwood mechanism, in which migration of weakly adsorbed particles from binding site to site on a dust particle surface is necessary for reaction. The rate of a particular reaction then depends on several factors, including the surface migration rates of the reacting species, and, once the molecules are close enough to react with one another, the chemical (activation energy) barrier for a particular reaction. Surface migration can occur via thermal hopping over the potential barriers E_b between binding sites, or via tunneling under these barriers. A simple expression for the rate t_{hop}^{-1} of thermal hopping is given by the equation [50-52]

$$t_{hop}^{-1} = \nu_0 exp(-E_b/kT) \tag{16}$$

where ν_0 is the adsorbate vibrational frequency, which is typically in the range $10^{12} - 10^{13} s^{-1}$. The potential barrier is 25-30 % of the binding energy E_D. For dust particles at 10 K, the exponential dependence of the hopping rate on the barrier height means that only first-row atoms and light molecules can migrate appreciably. For the lightest atoms and molecules (H, He, H_2), tunneling under these barriers is more rapid by several orders of magnitude [23]. These species migrate much more rapidly than somewhat heavier atoms such as C,N, and O and light hydrides. The result is that the migration factor in dust chemistry is dominated by the superior migratory ability of H and H_2 as long as the dust particle can be regarded as a thermal system.

The activation energy barrier factor discriminates in favor of atoms and radicals as opposed to stable species, since exponential barriers, as appear in eq. (1), slow the rates of surface reactions as well as gas phase reactions. Reactions in which both reactant partners are radicals most probably proceed as simple associations in which the dust particle takes up the excess chemical energy:

$$A + B \longrightarrow AB, \tag{17}$$

because formation of a strong chemical bond between radicals is known to dominate in gas phase systems as long as there is a mechanism (normally collisional) for stabilization of the product species. Reactions in which one partner is a stable species, and for which a potential barrier exists, probably occur via non-associative routes since these should normally possess lower barriers. Such barriers range from 0.1 eV - 0.5 eV or (roughly) 1000 K - 5000 K. Unlike the gas, the possibility of tunneling under chemical barriers must be taken seriously on surfaces, since reactant molecules can remain in the vicinity of one another for longer periods of time [51,53]. Tunneling occurs for light reactants such as H and H_2; for H, tunneling can occur for reactions involving stable molecules, whereas for H_2, tunneling is competitive only for reactions with radicals. The second type of reaction is particularly crucial in determining whether or not complex molecules can be formed on the surfaces of dust particles given the large abundance of H_2 for most of the cloud lifetime. If the (unknown) activation energies for surface processes such as

$$C_n + H_2 \longrightarrow C_nH + H, \tag{18}$$

and

$$C_nH + H_2 \longrightarrow C_nH_2 + H, \tag{19}$$

are small enough ($E_a/k \leq 2000K$), then hydrogenation of radicals into stable (saturated) species occurs on grain surfaces efficiently even if the major form of surface hydrogen is molecular. The result is that the saturated molecules remain unreactive and do not lead to the production of more complex species via clustering reactions such as

$$C + C_n \longrightarrow C_{n+1}. \tag{20}$$

The desorption of surface molecules can occur thermally or non-thermally. Thermal desorption occurs with a rate given by the equation

$$t_{evap}^{-1} = \nu_0 exp(-E_D/kT). \tag{21}$$

It is too slow to be of importance for all but the lightest species (H, He, H_2) on 10 K grains; at higher temperatures, volatile molecules such as CO begin to desorb at meaningful rates. Non-thermal desorption can occur via a number of poorly studied mechanisms. Cosmic rays can induce desorption, either by heating the whole grain following ionization, or via spot heating of portions of the grain [28]. Ultra-violet radiation can induce desorption via a direct quantum mechanical process akin to the photoelectric effect, via spot heating caused by photodissociation of surface species, or via actual grain explosions induced by the formation and subsequent runaway reactions of radicals [54-55]. Exothermic chemical reactions can induce desorption if sufficient energy can be transferred to the product adsorbate-surface bond [24, 56]. The energy of exothermic reactions might even be convertible into translation of the products, allowing heavy species to migrate more efficiently before they are thermalized [57]. Even grain-grain collisions can result in desorption if they are sufficiently inelastic [58]. Although some of these processes have been studied in the laboratory or theoretically, no consistent set of rates or quantum yields is available. It appears likely that in cold clouds, these non-thermal processes can only delay the eventual condensation of gas phase matter involving heavy atoms, not prevent it [57,59].

4 Models of Cool Homogeneous Sources

Although dense interstellar clouds are quite heterogeneous, it is probably useful to consider ideal cases of homogeneous regions that are not penetrated by external ultra-violet radiation. Such models need not pertain to entire clouds, but to central regions of the clouds not proximate to star formation activity. Such regions are labelled "quiescent." The original models of quiescent sources were steady-state in nature. In 1973, Herbst and Klemperer [31] published the first gas phase model of dense clouds. An earlier work by Solomon and Klemperer [60] considered the gas phase chemistry of diffuse clouds, where the role of external radiation is paramount. Consisting of only 100 reactions involving 35 species of up to five constituent atoms, and including only the elements H, He, C, N, and O, the Herbst-Klemperer model was able to predict fractional abundances (concentrations divided by the gas density) as functions of overall gas density and temperature in reasonable agreement with observation. Grain chemistry as well as adsorption and desorption were ignored except for the efficient conversion of atomic into molecular hydrogen. The apparent success of the Herbst-Klemperer model prompted others to improve it by enlarging the number of molecules and elements considered. Mitchell, Ginsburg, and Kuntz [61] utilized 454 gas phase reactions involving 100 species of up to 4 atoms in size, and enjoyed a similar degree of success. A similarly complex model of diffuse clouds was formulated by Black and Dalgarno [62]. More recently, Black and van Dishoeck, Federman et al., and other groups have constructed more detailed models of diffuse clouds with special consideration of photodestruction processes [63-68]. So-called high latitude clouds, the characteristics of which are in between diffuse and dense clouds, have also been studied by a variety of authors.

In parallel with steady-state gas phase models, steady-state models by Watson and Salpeter [25] and by Pickles and Williams [27] employing dust chemistry were undertaken. In these models, the production of molecules occurs totally or mainly on dust surfaces, desorption is efficient, and the resulting gaseous species are either destroyed in the gas, or depleted by adsorption onto dust particles. Given the uncertainties in dust chemistry, these models could not really be considered as quantitative in nature. The state of dust chemistry at the time was reviewed in some detail by Watson [24].

By the mid 1970's, chemical time dependence began appearing in both gas phase and dust surface models of dense clouds. Langer and Glassgold [69] realized that the time necessary for simple molecules such as CO and H_2O to reach steady state via gas phase processes is quite long compared with simple gravitational collapse times. Allen and Robinson [56] studied the time-dependent dust chemistry occurring on interstellar grains small enough that (a) typical exothermicities of chemical reactions yield enough energy to heat the grains and desorb mantles, and (b) migration is not a significant bar to reactivity of heavy species. Utilizing approximately 600 surface association reactions between radicals, they built species up to ten atoms in size. The newly synthesized species are desorbed into the gas, where no reactions were allowed to take place, and then depleted from the gas via adsorption. Thus, the gas phase existence of most molecules is transitory since, as unreactive molecules are formed, the desorption process winds down.

The 1980's saw the full flowering of the so-called pseudo-time-dependent gas phase approach. In this method, the gas phase chemistry is followed from given initial abudances under the conditions of fixed density and temperature. The initial abundances can be either atomic or analogous to what is found in diffuse clouds, where the hydrogen is found in both atomic and molecular forms. Steady-state is reached by 10^7 yr. The original objection that such a long time scale exceeds the free-fall gravitational collapse time is probably not serious, because it is now understood that clouds collapse at a much slower rate than this except in localized areas. The time scale is long, however, compared with the time necessary for a gas phase species to strike and stick to a cool dust particle. The time scale for adsorption is $\approx 3 \times 10^5$ yr for a cloud of gas density 10^4 cm^{-3}. Thus, performing a pseudo-time-dependent calculation and comparing the calculated final steady-state abundances with observation necessitates the ignoring of adsorption onto dust particles.

The first large-scale pseudo-time dependent calculations were performed by Prasad and Huntress [70-71] in 1980 and by Graedel et al. [10] in 1982. Prasad and Huntress first noticed that at intermediate times, before steady-state is reached, certain molecules have peak abundances. As ion-molecule laboratory work by people such as Smith and Adams, Bohme and Schiff, McEwan, and Anicich and Huntress flourished [34], astrochemists began to include syntheses of larger molecules in their codes, despite the fact that many of the reactions along the synthetic pathways were still unstudied in the laboratory. The major pathways for synthesis (discussed briefly in Section II) were suggested by Mitchell and Huntress, Suzuki, ourselves, and others [72-77]. The role of radiative association became better appreciated, and detailed theories of reaction rates for this process became available [39]. By the mid-80's, work by ourselves [78] and by Millar and collaborators [79-80] had discovered that calculated abundances of organic molecules all reach peaks at so-called "early times" of $10^5 - 10^6$ yr, after which they strongly decline. Agreement with observation for organic molecules in dense clouds such as TMC-1, Sgr B2, and Orion, where these molecules are most abundant, is only possible at early times. Thus, the apparent agreement between steady-state models and observation achieved in the 1970's does not extend to more complex molecules. Indeed, the overall agreement at early times for both small and large molecules is reasonably good, and early-time gas phase model results can now be regarded as the so-called standard model used by observational astronomers in comparing their results with theory. The reason that organic abundances peak at early time has to do with the transitory abundance of the neutral carbon atom. In models in which the elemental oxygen abundance exceeds the elemental carbon abundance, atomic carbon eventually is oxidized to form CO. Before this occurs, atomic carbon is active in molecular synthesis via carbon-insertion reactions with hydrocarbon ions. The early-time results are not affected much by the inclusion of adsorption onto dust since the gas phase becomes heavily depleted only at later times. Nor, within reason, are the peak early-time abundances strongly dependent on initial abundances. If one chooses to start with atomic initial conditions, the peak early-time abundances are achieved somewhat later and, for complex molecules, are somewhat smaller than if one starts with significant amounts of molecular hydrogen [52]. If one is interested in the abundances of small molecules, steady-state treatments are adequate, and many investigators have continued to use such treatments. Gerin et al. [81] have shown that the chemisty of NO is well represented by a steady-state model and Schilke et al. [82] have shown the same for HCN and HNC in Orion.

By the end of the 1980's and early 1990's, the pseudo-time-dependent approach had been been extended to include deuterium and other isotopic fractionation [44-45], molecules as large as 10 or more atoms [83-84], and molecules including the elements phosphorus, sulfur and silicon [85-87]. Current models contain perhaps 3000 - 4000 gas phase reactions involving $\approx$ 300 species. Most recently, we and others have begun to include significant numbers of possibly rapid neutral-neutral reactions; initial results indicate that inclusion of these reactions can strongly modify early-time abundances [88].

Most pseudo-time-dependent calculations either ignore photons or consider the effects of external photons as a simple exponential function of the so-called visual extinction. It was originally realized by Prasad and Tarafdar [89] that cosmic rays can cause significant ultra-violet photon fluxes deep inside dense clouds via a process in which electrons, produced in primary ionization events, excite molecular hydrogen to emitting states. The effect of such internally generated radiation was studied by Dalgarno and co-workers and found to be most important at times well past "early time." [90-92]

Grain chemistry was not neglected during the 1980's. In 1982, Tielens and Hagen [51] developed a model for the chemical evolution of grain mantles on large grains. These authors utilized the gas phase pseudo-time-dependent model of Prasad and Huntress [70]. The steady-state abundances of the gas phase model were allowed to adsorb onto grain mantles, and the evolution was followed using reactions of the types discussed in Section III. The dominant surface species was found to be water ice. The first time-dependent model in which gas phase and grain surface chemistry were considered simultaneously was presented in 1985 by d'Hendecourt et al. [93]. These authors started with an atomic gas and bare grain cores. Their gas phase chemistry consists of 551 reactions whereas their grain chemistry consists of only 18 reactions, chiefly involving atoms and small radicals. Efficient non-thermal desorption was included in most of the calculations, permitting a steady-state or quasi-steady-state to be reached both in the gas and on grain surfaces. Once again, the dominant surface molecule was found to be water ice. Tielens and Allamandola reviewed the subject of dust chemistry in 1987 [50]. In 1990, Brown and Charnley [94] added some grain reactions involving atomic hydrogen to a gas phase pseudo-time-dependent calculation. In 1990, Brown [57] showed that complex molecules can be synthesized on standard grains if and only if atomic hydrogen reactions, leading to saturated and unreactive species, can be curtailed. One promising mechanism for such curtailment is a reduced migration ability due to surface traps, as studied by Smoluchowski

[95]. Brown did not consider hydrogenation via surface reactions involving H_2. He assumed that a portion of the exothermicity of chemical reactions could be utilized to desorb some of the reaction products. Another mechanism for complex molecule formation on grains - photoprocessing - has been amply demonstrated by Greenberg and co-workers in the laboratory although this mechanism has not been incorporated into detailed models to the best of our knowledge [96].

In the later 1980's, the role of PAH's in gas phase models of both diffuse and dense clouds was considered by Omont [97] and by Dalgarno and co-workers. Lepp and Dalgarno [98] argued that in dense clouds, the dominant carrier of negative charge should be negatively-charged PAH-type molecules. The transformation from neutral PAH's occurs via electron-sticking reactions of the type

$$PAH + e^- \longrightarrow PAH^- + h\nu. \tag{22}$$

The consequence of a large fractional abundance of negatively-charged PAH molecules is a larger abundance of neutral atomic carbon at times approaching steady state, leading to somewhat greater abundances of complex molecules because of the synthetic power of C. The distinction between steady-state and early-time complex molecule abundances is still present, if somewhat reduced. Later authors have challenged the view that negatively-charged PAH's are more abundant than electrons. The issue is related to the size of the electron affinities of the interstellar PAH's and even to the existence of PAH's in regions far from where they are observed.

The increasing evidence for the existence of icy mantles on grains in dense interstellar clouds led us to incorporate grain chemistry into our hitherto pristine gas phase chemical models in the last few years. We are especially interested in both reproducing the mantle observations and in determining whether or not complex molecules can be formed on grain surfaces. The grain chemistry followed pertains to large grains which are considered to be thermal to the extent that migration and tunneling rates are determined by temperature. A temperature of 10 K was assumed for both the gas and the dust. Added to the large file of gas phase reactions was a set of a few hundred surface reactions with individually determined rates, mainly involving atoms and radicals, but including less reactive species if the activation energy is small enough to permit efficient tunneling. In the initial paper [52], only thermal desorption was allowed, and hydrogenation reactions involving molecular hydrogen were not included. Models were run with hydrogen initially in the atomic form as well as in the molecular form. The neglect of non-thermal desorption means that heavy molecules are depleted strongly from the gas at times after 10^6 yr. At times before this, both an active gas phase and surface chemistry occur. Complex molecules are synthesized in the gas, with peak abundances occurring at so-called "early time." Grain surface chemistry produces a variety of smaller molecules, with water ice dominant, as in previous models. The production of complex molecules occurs on grain surfaces only if the initial form of hydrogen is H_2; otherwise, atomic hydrogen rapidly converts surface molecules into saturated, stable forms. At times past 10^6 yr, complex molecules from the gas adsorb onto the dust particles and are slowly hydrogenated by small amounts of surface atomic hydrogen. The final result is that the dust particles contain all of the neutral heavy matter, and are best described as organic snowballs. The widespread observation of surface CO in dense clouds is only reproduced by models with hydrogen initially in its molecular form; otherwise CO is hydrogenated too rapidly into methanol. Even the successful models can only reproduce the surface CO abundance at times well before steady-state conditions are achieved.

In the second paper of the series [53], C_n and C_nH radical-H_2 reactions and non-thermal desorption via cosmic ray heating of the entire grain were added to the surface reaction set. The radical - H_2 reactions hinder the synthesis of complex molecules on grain surfaces even if H_2 is the initial form of hydrogen, if the activation energy for such reactions is 2100 K. It would appear that, in the absence of photoprocesses [96], complex molecules are not produced on large grains. Cosmic-ray-induced desorption prevents total condensation of gaseous heavy molecules at very long times; however, only the more volatile species such as CO have non-negligible gas phase abundances. Although all models produce copious amounts of water ice, once again only models in which the initial form of hydrogen is H_2 form sufficient amounts of solid carbon monoxide to agree with observations.

In the final paper of the series [99], the simple physical picture of grains as consisting of cores and surfaces was replaced by a more realistic three-phase picture in which grains consist of cores, mantles, and surfaces. The adsorbed material was considered to be part of the mantle if additional adsorbed material lies atop it. In addition, mantle species were assumed to be unreactive, so that unsaturated molecules can be protected to some extent against surface hydrogenation, as long as additional species are being adsorbed atop them. Thus, as opposed to earlier model results, significant condensed phase abundances of unsaturated species can be maintained indefinitely. The surface + mantle abundance of CO is too low at any time to explain observations because of a change in the surface reaction network, reflecting new calculations, which has the effect of increasing the surface H abundance. Work in progress suggests that the H abundance can be reduced and the CO surface abundance increased by changing the elemental abundances used in the model. Normally, the elemental abundances for the gas-surface-mantle system are set to the so-called "depleted metal" values, in which metals and heavy elements such as S and Si are reduced by two orders of magnitude from their more standard values. This is done to account for dust particle cores. If a higher abundance is used for S, the CO surface abundance can be made to agree with observation. However, the production of complex molecules in the gas suffers, unless the high elemental sulfur abundance is placed initially on dust mantles.

Willacy and Williams [59] have recently published a model calculation in which several mechanisms for non-thermal desorption (via direct cosmic ray heating and photon heating initiated by cosmic rays) are included in a gas-grain chemical model of dense interstellar clouds. The only surface process considered is hydrogenation of atoms and radicals via H atoms. Complex molecules peak in the gas at the standard "early time." Typically lower, quasi-steady state gaseous abundances are found to exist for periods between 1×10^6 and 5×10^7 yr, during which adsorption and desorption are roughly in balance. The agreement with observation during this time is satisfactory for smaller species. Thereafter most gaseous abundances drop sharply. Surface abundances are not followed. The non-thermal desorption in the main delays rather than halts condensation of the gaseous matter.

5 Towards Physical Heterogeneity

The homogeneous models described in Section IV do not begin to come to grips with the heterogeneous and time-dependent physical conditions in dense interstellar clouds. A conceptually relatively simple improvement is to consider the collapse of the cloud or, more realistically, a fragment of the cloud, either by assuming some sort of free-fall or delayed free-fall collapse, or by solving hydrodynamic equations. In 1978, Gerola and Glassgold [100] developed a one-dimensional hydrodynamic picture of self-gravitating interstellar clouds with gas phase chemistry. More complete gas phase chemical networks were included in the general hydrodynamic collapse model of Tarafdar et al. [101], published seven years later. These authors calculated both column densities (concentrations multiplied by path length) of atoms and diatomic molecules in collapsing diffuse clouds as well as fractional abundances of moderately complex molecules in the cores of dense clouds. Early-time abundances in dense cloud cores stem naturally from this treatment because as a diffuse cloud collapses, the amount of time spent in the density range characteristic of dense clouds is finite, so that steady-state abundances can never be reached. A review of hydrodynamic models was later presented by Prasad et al. [102]. Most recently, the possibility that hydrodynamics with turbulence and magnetic fields can lead to dense cores which can evolve into star forming regions or return to regions of low density was considered by Prasad et al. [103] in a model with gas phase chemistry comparable in complexity to our recent gas phase homogeneous models. Once again, emphasis was placed on obtaining a natural reason for the evolution of and importance of early-time abundances. A simpler picture of chemistry during collapse was given by Flower and Pineau des Forêts [104-105] who considered an isobaric system with hot and diffuse initial conditions.

Even the picture of a cloud or portion of a cloud undergoing hydrodynamic motion cannot represent all the dynamic motion in a large molecular cloud complex. An early view of the influence of large scale dynamic motions on the cloud chemistry was promoted by Boland and De Jong [106] who suggested that such motions could cycle dense material from the interiors of clouds towards cloud edges, where the material would be subjected to ultra-violet radiation. The ultraviolet radiation could break down complex molecules and perhaps photodesorb material from dust surfaces, maintaining a supply of carbon ions and atoms needed to recreate large abundances of complex molecules and explain the observation of large amounts of neutral atomic carbon by Phillips and Huggins [107]. A cycling time of 10^5 yr or so could also explain why early-time abundances make physical sense. However, this view of large scale dynamic motions was attacked on physical grounds by Williams and Hartquist [108] who proposed a cycle of carbon adsorption and desorption based on shock waves (see below).

The effect of external radiation on the outer edges (envelopes, halos) of clouds, both dense and diffuse, has been well studied, with the dominant form of carbon as a function of depth into the cloud (e.g., C^+, C, CO) a salient feature of the studies. A general model for dense cloud edges was published by Tielens and Hollenbach [109] and, for clouds in external galaxies in which the heavy elements are less abundant, by Maloney and Black [110]. Abgrall et al. [111] have studied the balance between H and H_2 in diffuse clouds with great attention paid to the radiative transfer problem. An older model produced to study this balance is that of Federman et al. [112]. Van Dishoeck and Black's recent work on ultra-violet penetration into diffuse clouds focuses on the CO problem [63]. Dense cloud envelopes have been investigated theoretically by Federman and Allen [113], who showed that observed abundances of H_2CO (formaldehyde) can be best explained as (photo?)desorbed material from icy grain mantles. Wannier et al. [114] have observationally studied warm neutral halos around molecular clouds using lines of HI (atomic hydrogen), CO, and OH.

There is now ample observational evidence that cloud interiors are very heterogeneous, with denser regions embedded in more diffuse material. How does such a clumpy medium affect the chemical processes? Chièze and Pineau des Forêts have developed models in which diffuse and dense gas coexist and mix with each other [115-116]. The mixing of diffuse gas into dense gas injects neutral and ionized carbon atoms into the dense gas, perpetuating the conditions that exist at early time in homogeneous pseudo-time-dependent models. An extension of their chemical model to include complex organic molecules [117] shows that under certain circumstances (rapid mixing time, high neutral carbon abundance) it is indeed possible to maintain large steady-state abundances of complex molecules in dense clumps comparable to the early-time abundances of homogeneous pseudo-time-dependent models. This two-phase model does not consider the radiative transfer problem in any detail, however. Clearly, the influx of external ultra-violet radiation or radiation from newly born stars into a clumpy medium will have a significant effect on the chemistry of much of the material. Radiation transfer in a clumpy medium has now been considered by Boissé [118].

Observations by Genzel, Stutzki et al. of C^+, C, and CO have shown that much of the clumpy, dense material in selected molecular cloud complexes is warm and beset by radiation [119-120]. Known now as photodissociation regions or PDR's, these hot portions of dense or diffuse clouds are characterized by a gas phase that is predominantly neutral and a gas phase chemistry that is strongly influenced by heating caused by ultra-violet radiation in the range 6-13.6 eV. PDR's resemble cloud envelopes, discussed above, except that they often lie next to intense ultra-violet sources, so that the radiation influx is up to five orders of magnitude greater than the normal value. General chemical-thermal models of PDR's have been undertaken by Tielens and Hollenbach [109] and by Hollenbach et al. [121]. These models predict chemical abundances as a function of optical depth, and intensities of atomic fine structure transitions. Sternberg and Dalgarno [122] have studied the infrared response of H_2 to the environment of a PDR and Escalante et al. [123] have concentrated on the near IR emission of neutral atomic carbon. LeBourlet et al. [124] have studied dark cloud envelopes with ultra-violet radiation up to 1000 times the norm.

6 Shocks and Star Formation

The influence of shock waves emanating from events such as high mass star formation on both diffuse and dense interstellar clouds has been studied by a large number of scientists [125-126]. The physical properties of shocked material depend on the velocity of the shock front and the size of the perpendicular magnetic field. If the magnetic field can be ignored, the shock is termed a J-shock and is characterized by a dramatic rise in temperature followed by slow cooling. The initial temperature rise is a strong function of the shock velocity; high enough velocities produce what are known as dissociative shocks, in which the molecules in the gas tend to be destroyed and have to be reformed as the material cools. These have been studied by Neufeld and Dalgarno [127-128]. Lower velocities lead to temperatures in excess of a few thousand degrees, permitting endothermic reactions such as (15). Once the magnetic field exceeds a certain critical value, the characteristics of the shock change, and it is referred to as a C-type shock. The neutral gas now shows a smooth variation in temperature rather than a sharp boundary, but there is a net velocity between the neutral and ionic material which can power endothermic ion-molecule reactions.

Much of the work on shock chemistry has dealt with the CH^+ problem in diffuse interstellar clouds [47]. Although this review does not consider models of diffuse clouds in any detail, it is safe to say that these models cannot produce sufficient amounts of the molecules CH^+ and NH. The latter species may well be produced on grains or as a photoproduct of ammonia which is produced on grains and subsequently desorbed [129]. Although similar suggestions have been made for CH^+ over the years, the dominant view for the last decade has been that the molecule is produced in shocks via reaction (15). The most recent shock models, however, do not appear to handle the problem adequately [130]. In addition, the shock picture may not be in accord with recent high resolution optical studies of diffuse clouds, which permit detailed dynamical understanding of the gas for the first time. Duley et al. have recently considered other possibilities [131]. C-type shock models of dense clouds have been considered also; a recent one pertaining to Orion has been undertaken by Schilke et al. [82].

A well-studied region with chemistry thought by many to be indicative of shocks is the so-called Plateau source in Orion KL, a hot, outflow region from the infrared source IRc2, which betrays itself by large spectral linewidths and a gaseous chemistry featuring heavy element-containing molecules such as sulfur-bearing species and silicon monoxide [9,132]. Differing shock models yield quite different abundances for these species [133-134]. Moreover, quiescent models with elevated temperatures can also reproduce some of the observations. For example, the model of Langer and Glassgold accounts for SiO production [135]. The high temperature chemistry of protostellar winds has itself been investigated by Glassgold et al. [136].

Nearby the Plateau source in Orion are other distinct sources. Called the Hot Core and the Compact Ridge, respectively, they are both more quiescent than the rather turbulent Plateau source, but differ strongly in their chemical signatures from the surrounding cooler and less dense material, known as the Extended Ridge, and from each other [8-9]. The Hot Core has a gas density of perhaps 10^7 cm^{-3} and a temperature of 200 K. It is a compact region of material slightly to the southeast of the strong infrared source IRc2. The Compact Ridge is somewhat cooler, with a temperature of 100 K, and less dense, with a gas density of 10^6 cm^{-3}. It is somewhat larger in size than the Hot Core and to the southeast of it. Other condensations in the same region have also been detected [137]. The chemistry of the Hot Core is characterized by a higher degree of saturation than is normally observed in quiescent sources, with the complex nitrogen-containing molecules vinyl cyanide (CH_2CHCN) and ethyl cyanide (CH_3CH_2CN) salient components. The Hot Core was originally detected as a region of enhanced ammonia (NH_3) abundance. The Compact Ridge, on the other hand, is characterized by large abundances of methanol and the more complex oxygen-containing organic molecules methyl formate ($HCOOCH_3$) and dimethyl ether (CH_3OCH_3). Methanol is enhanced in the Compact Ridge relative to its abundance in both the quiescent gas and the Hot Core; methyl formate and dimethyl ether are not seen in either of the other sources.

Based on the idea of Pauls et al. [29] that the enhanced ammonia abundance in the Hot Core arose largely as a result of mantle evaporation, Brown et al. developed a model for the chemistry of the Hot Core [138]. In their model, three distinct temporal phases are considered. Initially, the gas and dust are quiescent and possess a temperature and density characteristic of quiescent material. During the first phase, chemistry occurs in the gas and dust as the material undergoes isothermal, free fall collapse. Eventually, all of the material with heavy atoms is on grain mantles. The grain surface chemistry considered is chiefly hydrogenation to more saturated forms via H-atom association reactions. During the second phase, the temperature rises suddenly due to heat from the protostar IRc2. The increased temperature results in sudden evaporation from the grains of all of the material adsorbed and chemically processed during the first phase. Finally, in the third phase, the gas phase chemistry occurs in a high temperature, high density medium. Essentially very little happens for the first 10^4 yr, so that the imprint of surface chemistry (a high degree of saturation) is not lost rapidly by subsequent gas phase chemical processing. This model is also useful in explaining the unexpectedly high degree of deuteration seen in the Hot Core [139-140]. In the initial stage, gas phase chemistry produces a large abundance of deuterium atoms, principally via the reaction sequence

$$H_2D^+ + CO \longrightarrow DCO^+ + H_2, \tag{23}$$

$$DCO^+ + e^- \longrightarrow D + CO, \tag{24}$$

as suggested by Dalgarno and Lepp [141]. The deuterium atoms adsorb onto the dust grains and produce enhanced abundances of singly deuterated isotopomers such as NH_2D and HDO. Even doubly deuterated isotopomers such as D_2CO can be produced in significant abundance [142].

The chemistry of the Compact Ridge must be different in some manner. Blake et al. [9] suggested that an inflow of water from IRc2 might explain the enhanced abundances of oxygen-containing species via a gas phase chemistry initiated by the ion-molecule radiative association reaction

$$CH_3^+ + H_2O \longrightarrow CH_3OH_2^+ + h\nu. \tag{25}$$

Millar et al. [143] tested this hypothesis via a pseudo-time-dependent gas phase model in which external water is admitted. They found that they could not reproduce the observed enhanced abundances with any reasonable rate of water injection. They did find, however, that the enhanced abundances of methyl formate and dimethyl ether can be reproduced by injection of methanol, and suggested that the source of the methanol is evaporation from grain mantles. Charnley et al. [144] developed this line of thought by building a model consisting of the second and third phases of the Hot Core model of Brown et al. Specifically, they started with grain mantles rich in methanol and some other species, and followed the gas phase chemistry after desorption due to rising temperatures. Rapid gas phase conversion of some of the methanol into methyl formate and dimethyl ether then occurs, in agreement with observation. But why should the grain mantles following the cold phase of the Compact Ridge source be so different from those of the Hot Core, which are rich in saturated nitrogen-containing species? One possibility, of course, is that the elemental abundances are different in the two sources. A more reasonable possibility is that the chemistry is different.

Caselli et al. [145] have very recently developed a consistent model of both the Hot Core and the Compact Ridge in which the chemical differences arise mainly from temperature differences in the initial (cold) phase. Using a large gas-grain chemical code, they followed the gaseous and surface chemistry in the three-phase scenario of Brown et al. The Hot Core starts from a temperature of 40 K whereas the Compact Ridge starts from a temperature of 20 K. At the higher temperature, carbon monoxide is too volatile to reside on grain surfaces long enough to be hydrogenated to methanol. Thus, the grain mantles in the two sources become quite distinct chemically, as is required when the material is desorbed.

Even TMC-1, the almost classic example of a quiescent source, may not be that at all. Rawlings et al. [146] have developed a model for collapsing envelopes of low mass protostars, of which TMC-1 may be an example. In this model, collapse occurs isothermally and some material is depleted by adsorption onto dust particles. The observed shapes of spectral lines are looked at closely, and used to constrain the model. Spectral wings indicative of collapsing material are hidden to some extent by the adsorption of gaseous material.

7 The Future

In the next few years, progress in interstellar chemistry should proceed along two paths. The first path is directed towards a greater understanding of the myriad chemical processes used in models. The two greatest needs at present in gas phase chemistry are for studies of the neutral products of dissociative recombination reactions and for rates and products of neutral-neutral reactions at low temperatures. Since ion-molecule syntheses proceed through ion precursors to neutral species via recombination reactions, a knowledge of the products of the recombination reactions is critical. As of now, insufficient experimental studies have been undertaken to characterize fully any interstellar recombination of importance. There is hope that within a few years, the situation will improve markedly for recombination reactions with electrons. If PAH's are a major carrier of the negative charge in dense interstellar clouds, recombination reactions between positive molecular ions and negatively charged PAH^- ions are of importance. Although it is thought that such reactions result in less fragmentation of the products than do dissociative recombination reactions involving electrons, there is little or no experimental evidence concerning possible neutral products. More work on the products of positive ion-negative ion recombinations is needed. There is some hope that experimentalists will be able to study such systems shortly.

The role of neutral neutral reactions in low temperature chemistry remains poorly understood. In the last year, a number of reactions involving the radical CN have been studied at very low temperatures and found to be surprisingly rapid [147]. How general is this phenomenon? We still have little idea. Recent room temperature measurements involving neutral atomic carbon and a host of reaction partners reveal rapid reactions, but do not tell us what the products are. Moreover, the rapidity of the reactions may not extend to low temperature, where a previous theory suggests that the symmetry of the lowest fine structure state of C, which is the only one populated at low temperature, is too high for the atom to be reactive. Preliminary model studies by ourselves and by Howe and Millar indicate that neutral-neutral reactions involving the atoms C and O will have the greatest effect on gas phase model results. The effect of O atom reactions has already been studied by Bettens and Brown [88]. One can expect that theoreticians will begin to study these systems shortly, with experiments at low temperatures possible as well.

Laboratory investigations of gas phase reactions involving larger species than heretofore studied are a must if modellers are to extend their models to include complex hydrocarbons and perhaps even small PAH's. In particular, the efficiency of radiative association reactions at low temperatures needs to be investigated for reactive systems involving more than 5-10 atoms. It may well turn out to be the case that such reactions have very high efficiencies, so that the synthesis of complex molecules from simpler species occurs rather quickly. Recent ion cyclotron resonance experiments by groups at JPL [148] and at Case Western [149] give hope that relevant laboratory work will soon be undertaken.

Although there remain large gaps in our understanding of gas phase chemical processes, the situation is far worse for dust surface processes. Basic theoretical and especially experimental studies of reactions and relevant physical processes such as desorption and surface migration are needed. Such studies are beginning to be undertaken in the laboratory by investigators such as Allamandola and Ewing. Bohme [150] has recently studied PAH ion-molecule reactions.

All of these laboratory studies should help theoreticians to stay abreast of the latest observations of new and unusual gas phase molecules. One can of course expect the flood of molecular detections witnessed over the last twenty five years to continue unabated.

As our knowledge of the relevant chemical processes improves, modellers will also be attempting to incorporate spatial and temporal heterogeneities to a greater extent. Stimulated by interferometric observations and aided by ever more powerful computers, modellers should find it possible soon to include large scale gas-grain chemical codes along with hydrodynamics, clumping, and radiative transfer. Eventually, the chemistry of smaller and smaller regions such as protostars will be studied in some detail. Nor will dense interstellar clouds - in our own galaxy and a myriad of others - be the only sources studied in chemical detail. Both young stellar objects and evolved objects will be looked at closely. Building upon past work in carbon-rich and oxygen-rich circumstellar envelopes, not reviewed here, modellers are already turning their attention to molecular development in young stellar objects, planetary nebulae, and protoplanetary nebulae. In the evolved objects, the physics and chemistry of grain and PAH formation are receiving increasing attention [151-153], as kinetic models become advanced enough to replace the simple condensation models used in the past.

In the twenty-five years since polyatomic interstellar molecules were first detected, a large amount of progress has occurred in our knowledge of the extent of interstellar chemistry and the basic chemical processes occurring in interstellar clouds. The next twenty-five years promise to be a period in which a much greater understanding and synthesis of the field will be achieved.

8 Acknowledgments

I am pleased to acknowledge the support of the National Science Foundation (USA) for my research program in astrochemistry.

References

1. T. Dunham, Jr., *Pub. A. S. P.* 49, 26 (1937).
2. P. Swings and L. Rosenfeld, *ApJ* 86, 483 (1937).
3. A. McKellar, *Pub. A. S. P.* 52, 187 (1941).
4. A. E. Douglas and G. Herzberg, *ApJ* 94, 381 (1941).
5. S. Weinreb, A. H. Barrett, L. Meeks, and J. C. Henry, *Nature* 200, 829 (1963).
6. G. Winnewisser, E. Herbst, and H. Ungerechts, in *Spectroscopy of the Earth's Atmosphere and Interstellar Medium*, ed. K. N. Rao and A. Weber (Academic, NY, 1992), p. 423.
7. E. Herbst, *Ang. Chemie (Int. Ed.)* 29, 595 (1990).
8. W. M. Irvine, P. F. Goldsmith, and Å. Hjalmarson, in *Interstellar Processes*, ed. D. J. Hollenbach and H. A. Thronson, Jr. (Reidel, Dordrecht, 1987), p. 561.
9. G. A. Blake, E. C. Sutton, C. R. Masson, and T. G. Phillips, *ApJ* 315, 621 (1987).
10. T. E. Graedel, W. D. Langer, and M. A. Frerking, *ApJS* 48, 321 (1982).
11. A. G. G. M. Tielens, in *Interstellar Dust*, ed. L. J. Allamandola and A. G. G. M. Tielens (Kluwer, Dordrecht, 1989), p. 239.
12. D. C. B. Whittet, in *Dust and Chemistry in Astronomy*, ed. T. J. Millar and D. A. Williams (IOP Publ., 1993) in press.
13. L. J. Allamandola, S. A. Sandford, A. G. G. M. Tielens, and T. M. Herbst, *ApJ* 399, 134 (1992).
14. W. W. Duley and D. A. Williams, *MNRAS* 196, 269 (1981).
15. L. J. Allamandola, A. G. G. M. Tielens, and J. R. Barker, *ApJ* 290, L25 (1985).
16. A. Léger and J. L. Puget, *A&A* 137, L5 (1984).
17. L. J. Allamandola, A. G. G. M. Tielens, and J. R. Barker, in *Interstellar Processes*, ed. D. J. Hollenbach and H. A. Thronson, Jr. (Reidel, Dordrecht, 1987), p. 471.
18. L. J. Allamandola, A. G. G. M. Tielens, and J. R. Barker, *ApJS* 71, 733 (1989).
19. L. J. Stief, B. Donn, S. Glicker, E. P. Gentieu, and J. E. Mentall, *ApJ* 171, 21 (1972).
20. C. B. Seab, in *Interstellar Processes*, ed. D. J. Hollenbach and H. A. Thronson, Jr. (Reidel, Dordrecht, 1987), p. 491.
21. R. J. Gould and E. E. Salpeter, *ApJ* 138, 393 (1963).
22. D. A. Williams, *ApJ* 151, 935 (1968).
23. D. J. Hollenbach and E. E. Salpeter, *J. Chem. Phys.* 53, 79 (1970).
24. W. D. Watson, *Rev. Mod. Phys.* 48, 513 (1976).
25. W. D.Watson and E. E. Salpeter, *ApJ* 174, 321 (1972).
26. P. Aannestad, *ApJS* 25, 505 (1973).
27. J. B. Pickles and D. A. Williams, *Astr. Spa. Sci.* 52, 443 (1977).
28. A. Léger, M. Jura, and A. Omont, *A&A* 144, 147 (1985).
29. T. A. Pauls, T. L. Wilson, J. H. Bieging, and R. N. Martin, *A&A* 124, 123 (1983).
30. E. Herbst, in *Interstellar Processes*, ed. D. J. Hollenbach and H. A. Thronson, Jr. (Reidel, Dordrecht, 1987), p. 611.
31. E. Herbst and W. Klemperer, *ApJ* 185, 505 (1973).
32. W. D. Watson, *ApJ* 183, L17 (1973).
33. D. C. Clary, in *Rate Coefficients in Astrochemistry*, ed. T. J. Millar and D. A.Williams (Kluwer, Dordrecht, 1988), p. 1.
34. V. G. Anicich and W. T. Huntress, Jr., *ApJS* 84, 215 (1993).
35. D. Smith, *Chem. Rev.* 92, 1473 (1992).
36. C. R. Herd, N. G. Adams, and D. Smith, *ApJ* 349, 388 (1990).

37. N. G. Adams, C. R. Herd, M. Geoghegan, D. Smith, A. Canosa, J. C. Gomet, B. R. Rowe, J. L. Queffelec, and M. Morlais, *J. Chem. Phys.* 94, 4852 (1991).
38. D. Gerlich and S. Horning, *Chem. Rev.* 92, 1509 (1992).
39. D. R. Bates and E. Herbst, in *Rate Coefficients in Astrochemistry*, ed. T. J. Millar and D. A. Williams (Kluwer, Dordrecht, 1988), p. 17.
40. K. Giles, N. G. Adams, and D. Smith, *Int. J. Mass Spectrom. Ion Proc.* 89, 303 (1989).
41. S. A. Maluendes, A. D. McLean, and E. Herbst, *ApJ*, in press.
42. M. Guélin, W. D. Langer, R. L. Snell, and A. Wootten, *ApJ* 217, L165 (1977).
43. L. E. Snyder, J. M. Hollis, D. Buhl, and W. D. Watson, *ApJ* 218, L61 (1977).
44. R. D. Brown and E. H. N. Rice, *MNRAS* 223, 429 (1986).
45. T. J. Millar, A. Bennett, and E. Herbst, *ApJ* 340, 906 (1989).
46. B. R. Rowe, A. Canosa, and I. Sims, *J. Chem. Soc. Faraday Trans.* 89, xxx (1993).
47. M. Elitzur and W. D. Watson, *ApJ* 222, L141 (1978).
48. D. R. Flower, in *Interstellar Processes*, ed. D. J. Hollenbach and H. A. Thronson, Jr. (Reidel, Dordrecht, 1987), p. 745.
49. S. Green, S. Maluendes, and A. D. McLean, *ApJS* 85, 181 (1993).
50. A. G. G. M. Tielens and L. J. Allamandola, in *Interstellar Processes*, ed. D. J. Hollenbach and H. A. Thronson, Jr. (Reidel, Dordrecht, 1987), p. 397.
51. A. G. G. M. Tielens and W. Hagen, *A&A* 114, 245 (1982).
52. T. I. Hasegawa, E. Herbst, and C. M. Leung, *ApJS* 82, 167 (1992).
53. T. I. Hasegawa and E. Herbst, *MNRAS* 261, 83 (1993).
54. T. W. Hartquist and D. A. Williams, *MNRAS* 247, 343 (1991).
55. L. B. d'Hendecourt, L. J. Allamandola, F. Baas, and J. M. Greenberg, *A&A* 109, L12 (1982).
56. M. Allen and G. W. Robinson, *ApJ* 212, 396 (1977).
57. P. D. Brown *MNRAS* 243, 65 (1990).
58. W. A. Schutte and J. M. Greenberg, *A&A* 244, 190 (1991).
59. K. Willacy and D. A. Williams, *MNRAS* 260, 635 (1993).
60. P. M. Solomon and W. Klemperer, *ApJ* 178, 389 (1972).
61. G. F. Mitchell, J. L. Ginsburg, and P. J. Kuntz, *ApJ* 212, 71 (1977).
62. J. H. Black and A. Dalgarno, *ApJS* 34, 405 (1977).
63. E. F. Van Dishoeck and J. H. Black, *ApJ* 334, 771 (1988).
64. E. F. Van Dishoeck and J. H. Black, *ApJS* 62, 109 (1986).
65. Y. P. Viala, E. Roueff, and H. Abgrall, *A&A* 190, 215 (1988).
66. S. R. Federman and A. E. Glassgold, *A&A* 89, 113 (1980).
67. B. T. Januzzi, J. H. Black, C. J. Lada, and E. F. Van Dishoeck, *ApJ* 332, 995 (1988).
68. S. R. Federman, C. J. Strom, D. L. Lambert, J. A. Cardelli, V. V. Smith, and C. L. Joseph, pre-print.
69. W. D. Langer and A. E. Glassgold, *A&A* 48, 395 (1976).
70. S. S. Prasad and W. T. Huntress, Jr., *ApJS* 43, 1 (1980).
71. S. S. Prasad and W. T. Huntress, Jr., *ApJ* 239, 151 (1980).
72. G. F. Mitchell and W. T. Huntress, Jr., *Nature* 278, 722 (1979).
73. W. T. Huntress, Jr. and G. F. Mitchell, *ApJ* 231, 456 (1979).
74. H. Suzuki, *ApJ* 272, 579 (1983).
75. E. Herbst, N. G. Adams, and D. Smith, *ApJ* 269, 329 (1983).
76. H. I. Schiff and D. K. Bohme, *ApJ* 252, 740 (1979).
77. G. Winnewisser and C. M. Walmsley, *Astr. Spa. Sci.* 65, 83 (1979).
78. C. M. Leung, E. Herbst, and W. F. Huebner, *ApJS* 56, 231 (1984).

79. T. J. Millar and L. A. M. Nejad, *MNRAS* 217, 507 (1985).
80. T. J. Millar, C. M. Leung, and E. Herbst, *A&A* 183, 109 (1987).
81. M. Gerin, Y. Viala, F. Pauzat, and Y. Ellinger, *A&A* 266, 463 (1992).
82. P. Schilke, C. M. Walmsley, G. Pineau des Forêts, E. Roueff, D. R. Flower, and S. Guilloteau, *A&A* 256, 595 (1992).
83. E. Herbst and C. M. Leung, *ApJS* 69, 271 (1989).
84. E. Herbst and C. M. leung, *A&A* 233, 177 (1990).
85. T. J. Millar and E. Herbst, *A&A* 231, 466 (1990).
86. E. Herbst, T. J. Millar, S. Wlodek, and D. K. Bohme, *A&A* 222, 205 (1989).
87. T. J. Millar, *A&A* 242, 241 (1991).
88. R. P. A. Bettens and R. D. Brown, *MNRAS* 258, 347 (1992).
89. S. S. Prasad and S. P. Tarafdar, *ApJ* 267, 603 (1983).
90. R. Gredel, S. Lepp, and A. Dalgarno, *ApJ* 323, L137 (1987).
91. A. Sternberg, A. Dalgarno, and S. Lepp, *ApJ* 321, 383 (1987).
92. R. Gredel, S. Lepp, A. Dalgarno, and E. Herbst, *ApJ* 347, 289 (1989).
93. L. B. d'Hendecourt, L. J. Allamandola, and J. M. Greenberg, *A&A* 152, 130 (1985).
94. P. D. Brown and S. B. Charnley, *MNRAS* 244, 432 (1990).
95. R. Smoluchowski, *J. Phys. Chem.* 87, 4229 (1983).
96. J. M. Greenberg, in *Astrochemistry*, ed. M. S. Vardya and S. P. Tarafdar (Reidel, Dordrecht, 1987), p. 501.
97. A. Omont, *A&A* 164, 159 (1986).
98. S. Lepp and A. Dalgarno, *ApJ* 324, 553 (1988).
99. T. I. Hasegawa and E. Herbst, *MNRAS*, in press.
100. H. Gerola and A. E. Glassgold, *ApJS* 37, 1 (1978).
101. S. P. Tarafdar, S. S. Prasad, W. T. Huntress, Jr., K. R. Villere, and D. C. Black, *ApJ* 289, 220 (1985).
102. S. S. Prasad, S. P. Tarafdar, K. R. Villere, and W. T. Huntress, Jr., in *Interstellar Processes*, ed. D. J. Hollenbach and H. A. Thronson, Jr. (Reidel, Dordrecht, 1987), p. 631.
103. S. S. Prasad, K. R. Heere, and S. P. Tarafdar, *ApJ* 373, 123 (1991).
104. D. R. Flower and G. Pineau des Forêts, *MNRAS* 247, 500 (1990); *erratum:* 249, 191 (1991).
105. G. Pineau des Forêts, D. R. Flower, and E. Herbst, *MNRAS* 253, 359 (1991).
106. W. Boland and T. de Jong, *ApJ* 261, 110 (1982).
107. T. G. Phillips and P. J. Huggins, *ApJ* 251, 533 (1981).
108. D. A. Williams and T. W. Hartquist, *MNRAS* 210, 141 (1984).
109. A. G. G. M. Tielens and D. J. Hollenbach, *ApJ* 291, 722 (1985).
110. P. Maloney and J. H. Black, *ApJ* 325, 389 (1988).
111. H. Abgrall, J. Le Bourlet, G. Pineau des Forêts, E. Roueff, D. R. Flower, and L. Heck, *A&A* 253, 525 (1992).
112. S. R. Federman and A. E. Glassgold, *ApJ* 227, 466 (1979).
113. S. R. Federman and M. Allen, *ApJ* 375, 157 (1991).
114. P. G. Wannier, B-G. Andersson, S. R. Federman, B. M. Lewis, Y. P. Viala, and E. Shaya, *ApJ*, in press.
115. J. P. Chièze and G. Pineau des Forêts, *A&A* 221, 89 (1989).
116. J. P. Chièze and G. Pineau des Forêts, in *Physical Processes in Fragmentation and Star Formation*, ed. R. Capuzzo-Dolceetta, C. Choisi, and A. Di Fazio (Kluwer, Dordrecht, 1990), p. 17.
117. J. P. Chièze, G. Pineau des Forêts, and E. Herbst, *ApJ* 373, 110 (1991).
118. P. Boissé, *A&A* 228, 483 (1990).

119. R. Genzel, A. I. Harris, D. T. Jaffe, and J. Stutzki, *ApJ* 332, 1049 (1988).
120. J. Stutzki, G. T. Stacey, R. Genzel, A. I. Harris, D. T. Jaffe, and J. B. Lugten, *ApJ* 332, 379 (1988).
121. D. J. Hollenbach, T. Takahashi, and A. G. G. M. Tielens, *ApJ* 377, 192 (1991).
122. A. Sternberg and A. Dalgarno, *ApJ* 338, 197 (1989).
123. V. Escalante, A. Sternberg, and A. Dalgarno, *ApJ* 375, 630 (1991).
124. J. Le Bourlot, G. Pineau des Forêts, E. Roueff, and D. R. Flower, *A&A* 267, 233 (1993).
125. J. M. Shull and B. T. Draine, in *Interstellar Processes*, ed. D. J. Hollenbach and H. A. Thronson, Jr. (Reidel, Dordrecht, 1987), p. 283.
126. G. F. Mitchell, in *Chemistry and Spectroscopy of Interstellar Molecules*, ed. D. K. Bohme, E. Herbst, N. Kaifu, and S. Saito (Univ. of Tokyo Press, 1992), p. 255.
127. D. A. Neufeld and A. Dalgarno, *ApJ* 340, 869 (1989).
128. D. A. Neufeld and A. Dalgarno, *ApJ* 344, 251 (1989).
129. R. Wagenblast, D. A. Williams, T. J. Millar, and L. A. M. Nejad, *MNRAS* 260, 420 (1993).
130. D. R. Flower, T. S. Monteiro, G. Pineau des Forêts, and E. Roueff, in *Rate Coefficients in Astrochemistry*, ed. T. J. Millar and D. A. Williams (Kluwer, Dordrecht, 1988), p. 271.
131. W. W. Duley, T. W. Hartquist, A. Sternberg, R. Wagenblast, and D. A. Williams, *MNRAS* 255, 463 (1992).
132. L. M. Ziurys, P. Friberg, and W. M. Irvine, *ApJ* 343, 201 (1989).
133. T. W. Hartquist, M. Oppenhimer, and A. Dalgarno, *ApJ* 236, 182 (1980).
134. T. M. Leen and M. M. Graff, *ApJ* 325, 411 (1988).
135. W. D. Langer and A. E. Glassgold, *ApJ* 352, 121 (1990).
136. A. E. Glassgold, G. A. Mamon, and P. J. Huggins, *ApJ* 373, 254 (1991).
137. Y. C. Minh, M. Ohishi, D. G. Roh, M. Ishiguro, and W. M. Irvine, preprint.
138. P. D. Brown, S. B. Charnley, and T. J. Millar, *MNRAS* 231, 409 (1988).
139. P. D. Brown and T. J. Millar, *MNRAS* 237, 661 (1989).
140. P. D. Brown and T. J. Millar, *MNRAS* 240, 25P (1989).
141. A. Dalgarno and S. Lepp, *ApJ* 287, L47 (1984).
142. B. E. Turner, *ApJ* 362, L29 (1990).
143. T. J. Millar, E. Herbst, and S. B. Charnley, *ApJ* 369, 147 (1991).
144. S. B. Charnley, A. G. G. M. Tielens, and T. J. Millar, *ApJ* 399, L71 (1992).
145. P. Caselli, T. I. Hasegawa, and E. Herbst, *ApJ* 408, xxx (1993).
146. J. M. C. Rawlings, T. W. Hartquist, K. M. Menten, and D. A. Williams, *MNRAS* 255, 471 (1992).
147. I. R. Sims, J.-L. Queffelec, A. Defrance, C. Rebrion-Rowe, D. Travers, B. R. Rowe, and I. W. M. Smith, in preparation.
148. A. D. Sen, V. G. Anicich, and M. J. McEwan, in *Chemistry and Spectroscopy of Interstellar Molecules*, ed. D. K. Bohme, E. Herbst, N. Kaifu, and S. Saito (Univ. of Tokyo Press, 1992), p. 211.
149. R. C. Dunbar, *Int. J. Mass Spectrom. Ion Proc.* 100, 423 (1990).
150. D. K. Bohme, *Chem. Rev.* 92, 1487 (1992).
151. M. Frenklach and F. D. Feigelson, *ApJ* 341, 372 (1989).
152. I. Cherchneff and J. R. Barker, *ApJ* 394, 703 (1992).
153. M. Frenklach, C. S. Carmer, and E. D. Feigelson, *Nature* 339, 196 (1989).

Hunting Oxygen in the Interstellar Medium

L. Pagani [1], *P.J. Encrenaz* [1,2]

[1]DEMIRM, URA 336 du CNRS, Observatoire de Paris, Section de Meudon, 92190 Meudon, France
[2]Radioastronomie Millimetrique, URA 336 du CNRS, ENS, 24 Rue Lhomond, 75231 Paris Cedex 05, France

Abstract: Oxygen is the third most abundant element in the Universe after hydrogen and helium. Despite this, we have no clear idea about the form in which it appears in the interstellar medium, especially in the dark, dense clouds where the chemistry is complex. Our atmosphere, which is rich in O_2, H_2O and CO_2, precludes most ground based observations. In interstellar clouds, CO, OH and water ice abundances have been measured, but until very recently no quantitative estimates of water vapor and of atomic oxygen (OI) could be made, and no trace of O_2 could be found. Thanks to the development of high performance receivers and antennas, we are starting to get a first estimate of the abundance of these species. However, 25 years after the first detection of H_2O we still have not yet found half of the oxygen atoms which should be present in the interstellar clouds. Two possibilities are atomic oxygen and grain cores. In the next decade, the launch of several balloons and satellites will provide better observational estimates.

1 Introduction

Since the first complex molecules observed in the interstellar medium at the end of the 60's, there have been spectacular gains in the frequency coverage and sensitivity of receivers. This has resulted in an ever growing number of detections of species, each of which is less abundant or more intricate than those previously discovered. However until very recently, little was known about the possible reservoirs of oxygen. This is mostly due to the influence of our atmosphere, which prevents direct observations from the ground. The first two complex molecules found, NH_3 and H_2O were discovered in interstellar space by Cheung et al. (1968, 1969). At the same time Heiles (1968) showed that OH was present in dark clouds, not only in the diffuse medium. Thus as of 25 years ago, two out of the first three molecules discovered were oxygen-bearers. Despite this good start, we have not been able to locate most of the oxygen. Why try so hard to identify the reservoirs? Firstly, oxygen is the second most abundant nucleosynthezised atom after Helium. Its cosmic abundance, as measured relative to Hydrogen, is to 8.5×10^{-4} (Anders & Grevesse 1989) while that of Carbon is only 4.0×10^{-4}. Oxygen is injected into the interstellar medium via stellar winds, envelope ejections, planetary nebula, supernova, etc. Thus the contribution of oxygen to the chemistry of the dense medium should be very important.

The next oxygen-bearing molecules to be identified after OH and H_2O were H_2CO (Snyder et al. 1969) whose abundance is low, and the very abundant CO (Wilson et al. 1970). The discovery of CO led to the first quantitative estimate of the abundance of a major oxygen-bearing molecule in the dense interstellar medium. However, more detailed estimates proved to be difficult and are still a matter of debate. From observations of the rare isotopomer ^{13}CO, it became clear that lines of CO were optically thick and this fact makes the radiative transport difficult to model. The discovery of HDO in 1973 (Phillips et al. 1973) and numerous other transitions thereafter might have been important for the H_2O abundance, but chemical fractionation makes the relation of HDO to H_2O very uncertain. Apart from two lines of H_2O which were discovered using the Kuiper Airborne Observatory at the end of the 70's, we have had to wait for the dedicated submillimeter telescopes on Mauna Kea, Hawaii and the development of sensitive SIS receivers to see a wealth of new lines being discovered. These were mostly H_2O lines but transitions of $H_3O^+$and $^{16}O^{18}O$ have also been found. In parallel to this radio work, near infrared spectroscopic observations have been carried out since the early 70's to detect icy grain mantles toward obscured stars. Water ice had been identified very early and CO ice has been known for the last ten years. Today, a reasonable amount of data has been collected, which enables an estimate of the abundance of several oxygen-bearing species. From this has emerged a clear picture of where oxygen is... not! This contribution is concerned with quest to find oxygen-bearing species in the dense interstellar medium (ISM). We will not discuss the diffuse ISM nor the extragalactic sources, for which less information is available. This paper is divided in 5 sections. The major chemical reactions involving oxygen will be presented in the next section along with the model estimates of the abundances of various oxygen-bearing species. Section 3 will present the methods used to search for these species and Section 4 will analyse the results of searches. We will conclude in Section 5 with an account of future prospects.

2 Chemistry of Oxygen

Fig. 1 shows a simplified chemical network involving the most abundant oxygen-bearing molecules (adapted from Prasad et al. 1987). Oxygen was first predicted to play a role in the interstellar medium in the model by Herbst & Klemperer (1973). Since then, the reaction schemes involving oxygen have not changed much, although the models have grown in complexity. Recently a key reaction has been partly measured in the laboratory, that is

$$\begin{aligned} H_3O^+ + e^- &\longrightarrow OH + H_2 \\ &\longrightarrow H_2O + H \\ &\longrightarrow OH + 2H \\ &\longrightarrow O + H + H_2 \end{aligned} \tag{26}$$

Herd et al. (1990) and Adams et al. (1992) have determined that the branching ratio for OH was 65 % but could not determine the abundance of the other products. As can be seen in Fig. 1, this reaction is the only path to form H_2O and an important one to form OH and subsequently O_2 SO, NO, etc. Numerous

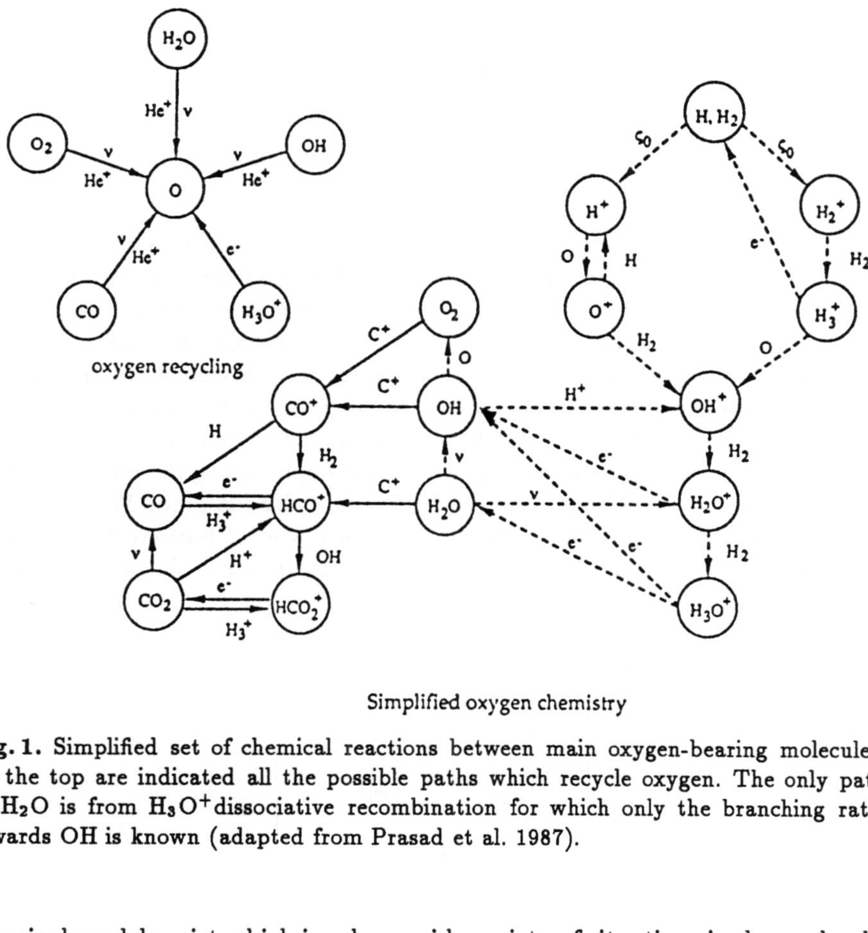

Fig. 1. Simplified set of chemical reactions between main oxygen-bearing molecules. At the top are indicated all the possible paths which recycle oxygen. The only path to H_2O is from H_3O^+dissociative recombination for which only the branching ratio towards OH is known (adapted from Prasad et al. 1987).

chemical models exist which involve a wide variety of situations in dense clouds, either cold and quiescent (for example, TMC1 or L134N) or warm (Orion KL, or Sgr B2). Typical models include e.g. gas-grain coupling, (Brown & Millar 1989; Charnley et al. 1992; Hasegawa et al. 1992), time-dependence (Langer & Graedel 1989; Millar et al. 1991), two-phase mixing (Pineau des Forets et al. 1992), or UV irradiation and depth dependence (Viala 1986). There also exists a series of models dealing with shocked gas (e.g. Draine 1991; Draine et al. 1983; Graff & Dalgarno 1987; Roueff & Pineau des Forets 1993) which are interesting because the oxygen chemistry is greatly affected in such models.

Considering these cloud models and choosing some representative solutions, we find that the average realative abundances for the four main species, OI, O_2 H_2O and CO, are in the range 10^{-5} to 10^{-4} in the gas phase (Table 1). All

Table 1. Typical abundances from chemical models, realative to H_2

Species	Abundance
O	4.70×10^{-5}
O_2	2.43×10^{-5}
OH	1.44×10^{-7}
CO	9.00×10^{-5}
H_2O	1.04×10^{-5}
H_3O^+	1.55×10^{-9}
HCO^+	4.00×10^{-9}

these models use depletion factors to represent the amount of atoms on grains. For the heavy elements like Fe, Mg, etc., the depletion factor is thought to be of the order of 10^{-3}, for C, N and O, the depletion factor is considered to be in the range 0.3 to 0.1. Thus the sum of the species gas-phase abundances of Table 1 is less than the cosmic abundance of these species. Obviously atomic oxygen, molecular oxygen and carbon monoxyde are the most abundant (in this Table, as an oxygen reservoir, O_2 is as important as OI because it contains 2 atoms) while H_2O is slightly lower, and OH is negligible. Although among the most abundant ions, the abundances of both H_3O^+ and $HCO^+$are also marginal. Water vapor is also abundant in many of these models, especially when grains are not included. For example, in the Viala model, which is displayed in Fig. 2. The Pineau des

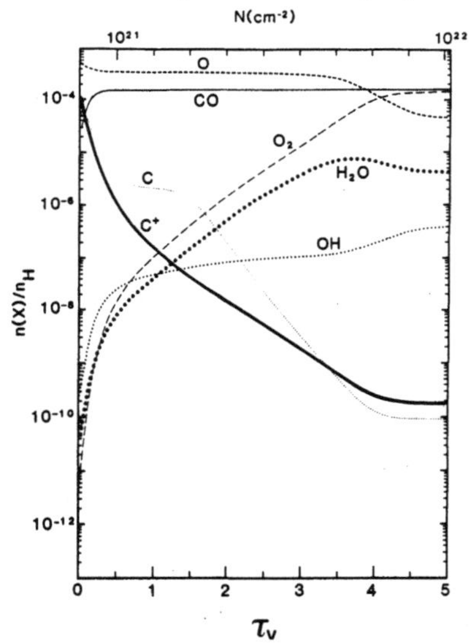

Fig. 2. UV-dependent chemical model as proposed by Viala (1986) as a function of optical depth.

Forets et al. (1992) models (Two-phase mixing with or without gas-grain interaction) are aimed at explaining the C/CO observed ratios. These models indicate low to very low CO and O_2 abundances in most cases. The latter, especially, would seem to be several orders of magnitude below the (tentatively) observed abundance. The observed value indicates that there is some efficient desorption mechanism from grains in the deep interior of clouds apart from recycling them in the UV illuminated diffuse part; even without this grain interaction, the model O_2 abundance is far too low compared to the recent measurement of this molecule (see below). This is due to the fact that O_2 is easily destroyed by ions brought in from the diffuse phase via any mixing mechanism. On the basis of observations, this mechanism is questionable. Another interesting model is the gas-grain pseudo time-dependent model of Hasegawa et al. (1992). In addition to giving the gas-phase abundances, this also predicts the solid-phase abundance (Table 2) and shows that for O_2 and CO the solid phase is only a small fraction of the gas phase. The reverse holds for H_2O (gas/ice = 0.01), making water ice a main reservoir of oxygen and even the most important in Model B of Table 2. Shock waves are an important aspect of high-mass star forming regions. The

Table 2. Grain mantle abundances of main oxygen-bearing neutrals as a fraction of H_2. Adapted from models A and B of Hasegawa et al. (1992) at epoch t = 3.2 $\times 10^5$ y.

Species	Fractional Abundance Model A	Model B
O	5.20×10^{-21}	8.70×10^{-16}
O_2	1.90×10^{-5}	3.40×10^{-11}
OH	2.00×10^{-15}	8.50×10^{-16}
CO	1.50×10^{-5}	1.30×10^{-10}
H_2O	2.90×10^{-5}	1.70×10^{-4}

chemistry under the extreme conditions involved by the passage of a shock wave is very different from that in a cold and quiescent medium. Fast endothermic reactions between neutral species which cannot take place in a cold medium are activated in the hot post shock region and determine the chemistry of oxygen. Some models (e.g. Draine et al. 1983; Graff & Dalgarno 1987; Flower & Pineau des Forets, private communication) show that OI and perhaps O_2 (in the model of Graff & Dalgarno but not in the model of Flower & Pineau des Forets) would disappear, being replaced by H_2O (which would also be evaporated from grains). This would lead to an abundance of H_2O which is comparable to or greater than that of CO, since the latter being hardly perturbed by the shock (Fig. 3).

3 Searching for Oxygen-bearing Species

Our own atmosphere is rich in oxygen-bearing species (O_2 H_2O, CO_2, and also a small quantity of CO in the mesosphere together with O_3). Thus, searching for these species in the interstellar medium from the ground is difficult except for

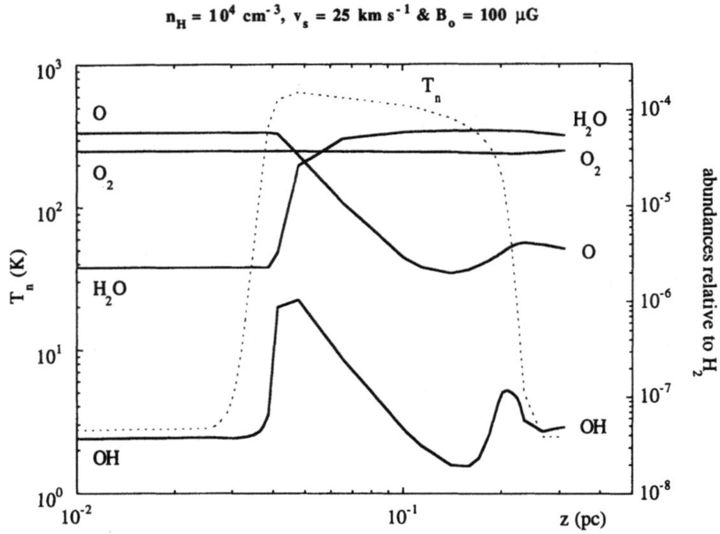

Fig. 3. MHD-shock driven chemical model. It can be noted that OI almost disappears being completely turned into H_2O. CO is not represented because its abundance does not vary in this non-dissociative shock (Flower & Pineau des Forets, private communication).

CO for which the mesospheric line is neither strong nor wide but interfere with astronomical observations of some galactic sources for several weeks per year. No O_2 lines are observable from the ground, as almost no H_2O lines with upper level energy is close to or below 300 K. Several ways exist to avoid this problem. First, observe high energy levels of H_2O: there exist in the millimeter and sub-millimeter bands several lines of H_2O which come from high energy levels. These levels are high enough not to be populated in the 250-300 K thermalized terrestrial atmosphere. All these millimeter lines exist because of a chance coincidence between pairs of levels from two consecutive J-ladders making the energy difference small enough to be in the millimeter domain (Fig. 4). The most well known is the ($J_{KK'} = 6_{16} - 5_{23}$), at 22 GHz. However all these lines proved to be masing and not directly usable as column density tracers. Still if many maser lines are observed, excitation models might be constrained well enough to derive important physical parameters such as column density (see Menten, this volume, for a detailed discussion). Second, observe isotopomers: ^{18}O is 500 times rarer than ^{16}O. Thus the atmosphere is much more transparent for rare isotopomer lines as shown by the observations of $H_2^{18}O$ by Jacq et al. (1988, 1990) and Wannier et al. (1991) and by the numerous attempts to detect $^{16}O^{18}O$ from the ground. The telluric isotopomers are still visible but can be usually avoided by selecting sources with large velocity corrections. In the case of water, HDO can also be searched for, being 5 orders of magnitude rarer in the terrestrial atmosphere. However, because the chemistry of Deuterium involves strong fractionation effects leading to unknown enhancements of the HDO abundance, it is commonly

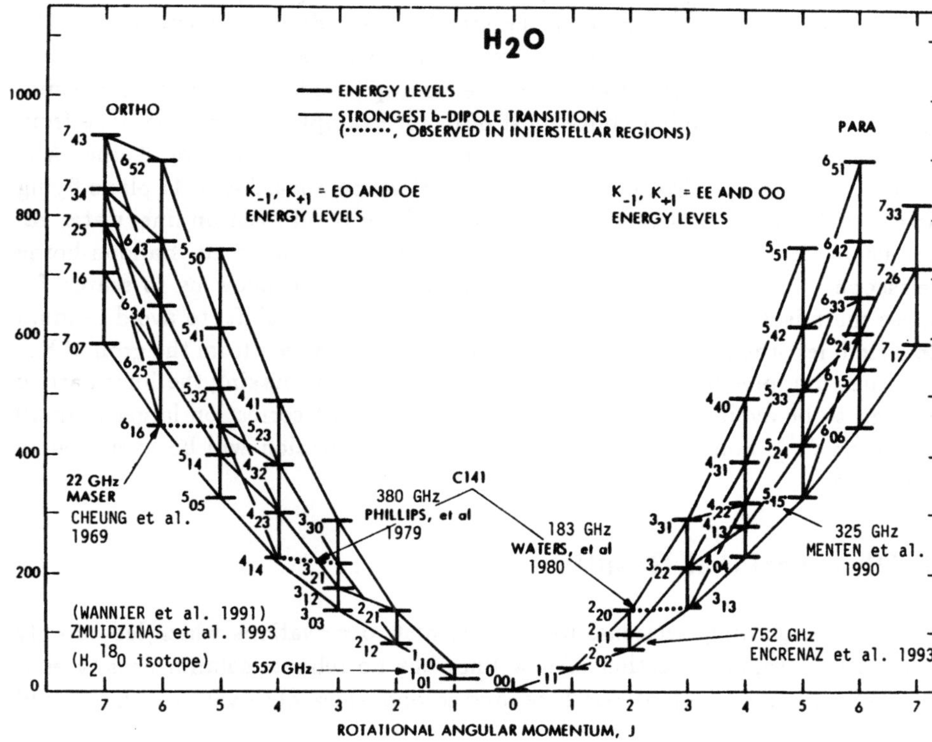

Fig. 4. H_2O energy level diagram. About half of the presently detected transitions are indicated (see Menten, this book., for a complete list). We have included the ($J_{KK'} = 1_{10} - 1_{01}$) 548 GHz $H_2^{18}O$ line detection because of its importance due to its position at the bottom of the ortho scale. 557 GHz is the corresponding $H_2^{16}O$ transition frequency.

believed that the HDO abundance cannot be related reliably to the H_2O abundance. No fractionation effect is presently known for ^{18}O; the terrestrial ratio of 500 (250 for $O_2/^{16}O^{18}O$) is usually assumed. The difficulty there lies in the fact that (1) the $^{16}O^{18}O$ line is extremely weak even for an abundance close to that of $C^{18}O$ and (2) the low frequency lines of $H_2^{18}O$ lie high in the rotational ladder, making the derivation of abundances highly model-dependent. Only the observation of the ($J_{KK'} = 1_{10} - 1_{01}$) ground rotational line of $H_2^{18}O$ at 548 GHz for the ortho form can be used confidently. Third, one could observe redshifted sources: those having significant velocity shifts can be used to search for lines of the main isotopomers if the redshift is large enough to bring the astrophysical line to a frequency where the atmosphere transparency is good. This has been attempted for moderately at low redshifted galaxies (Combes et al. 1991; Liszt 1985, 1992 and for the very high redshifted galaxy IRAS 10214+4724 at z = 2.286 (Encrenaz et al. 1993). Only the last named group could claim success (see below). However, for redshifted sources, one can search for the main isotopomers, which are stronger than the rare substitutions, but these galaxies are highly beam diluted, making the searches very difficult. Furthermore, the comparison of data from entire galaxies with local galactic cloud data is not simple. Fourth, observing from above the atmosphere: this eliminates the atmospheric problem by flying above most of it, for example This can be done from airplane-borne telescopes, satellites or stratospheric balloons. The Kuiper Airborne Observatory (KAO) carries a 91 cm telescope aboard a C141 plane flying at approximately 12 to 13 km with which $H_2^{18}O$ ground rotation line, water vapor near-infrared lines and infrared OI lines have been observed. Balloon-borne telescopes can fly up to 35 to 40 km altitude and will allow us to search for those main isotopomer lines which have a small velocity shift, to avoid residual telluric lines (the peak intensity is usually the same as seen from the ground but the line-wing pressure-broadening is much attenuated, making the lines narrow enough to be avoided for sources with a velocity shift correction larger than 10 to 20 $\mathrm{km\,s^{-1}}$). Some of the many projects in preparation will be presented in Section 5.

4 Observational Results

In the two following sections, we discuss, first, observations which led to only qualitative results (detection of new lines but no reliable column density estimates for the species of interest here) and second, those observations which could lead to column density estimates.

4.1 Qualitative Results

Until 1979, the only way to trace water was to observe the ($J_{KK'} = 6_{16} - 5_{23}$) ortho H_2O line at 22 GHz. This is a masering line, from energy levels lying at 450 K above the ground state. Then, two lines were discovered in the millimeter/submillimeter wavelength range (Phillips et al. 1979; Waters et al. 1980). These were the ($J_{KK'} = 4_{14} - 3_{21}$) transition of ortho-$H_2O$ at 380 GHz and the ($J_{KK'} = 3_{13} - 2_{20}$) transition of para-$H_2O$ at 183 GHz. Both were observed from the KAO. Unlike the 22 GHz line, the upper energy level of both lines lies below 300 K; these levels are thus reasonably populated in the Earth's atmosphere and prevent observations from the ground. With a water vapor content lower than at the best mountain sites and taking advantage of the maximum Doppler velocity shift the Orion-KL source can attain, the these two lines could be observed. However, the lines proved to be masing and thus useless to derive abundances. Cernicharo (1990) detected the 183 GHz line from a high mountain-site under exceptional meteorological conditions with the IRAM 30-m telescope. Then, very recently, 20 years after the first water maser line was discovered, a wealth of new lines (all being masers but one, see below) were discovered by Menten, Melnick and collaborators (see Menten for a review).

It was recognized that the most interesting thermal water lines, especially the ground level transitions of both ortho and para forms, could never be observed from the ground or even from the KAO. Thus other paths were tried to determine the water vapor abundance. D- or ^{18}O- substituted water have been used as early as 1973 for HDO (Phillips et al. 1973) but could generally not help to derive a reliable water abundance: concerning HDO, the chemical models are still very uncertain because of the Deuterium fractionation. The only $H_2^{18}O$ line observed presently (the $J_{KK'} = 3_{13}-2_{20}$ transition at 203 GHz) by Jacq et al. (1988, 1990) which corresponds to the 183 GHz line of para $H_2^{16}O$ which is a maser line, and is strongly suspected to be masing also. Thus, all the water vapor lines which have been observed from the ground whether isotopically substituted or not, have not been able to indicate the abundance of water vapor in the interstellar medium. However, some remarkable results have been obtained, among which are the detection of the ($J_{KK'} = 1_{01} - 0_{00}$) ground HDO line at 465 GHz leading to a good determination of its abundance (Schulz et al. 1991), the strong upper limit on redshifted O_2 towards the galaxy NGC 6240 with $O_2/CO < 0.06$ at the 3 σ-level (Combes et al. 1991) and the first detection of a thermal water vapor line (the $J_{KK'} = 2_{11} - 2_{02}$ transition with a rest frequency of 752 GHz) at z = 2.286 towards the very distant galaxy IRAS 10214+4724 (Fig. 5). In this last case, no abundance can be derived because of large opacity (Encrenaz et al. 1993). Two OI fine-structure lines exist in the far infrared, that is, the $^3P_0 - ^3P_1$

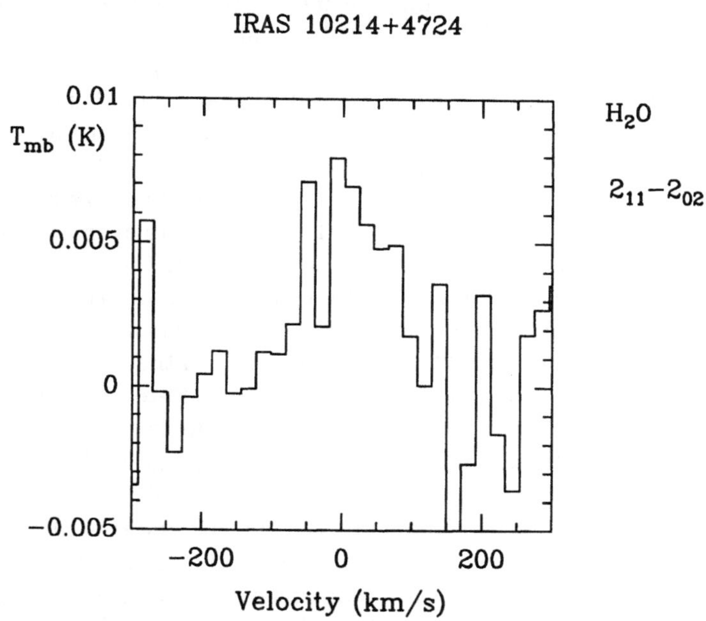

Fig. 5. H_2O thermal transition at 752 GHz detected at z= 2.286 in the very distant galaxy IRAS 10214+4724. This is the first detection of a thermal line of water (Encrenaz et al. 1993). The galaxy has been reobserved since and the line confirmed. We show here the sum of all the observations.

line at 145 μm and the $^3P_1 - ^3P_2$ line at 63 μm; these are observable from the KAO and have been detected in Orion (see Tielens & Hollenbach 1985a for references therein), in the Galactic Center (e.g. Genzel et al. 1984), and many other sources. They often trace a dense ($10^3 - 10^6$ cm^{-3}) and hot (> 100 K) atomic medium, sometimes shocked, which represents one of the intermediate layers of the Photodissociation Region (PDR) models of Tielens & Hollenbach (1985b) between the ionized HII region and the cold molecular cloud. Apperantly, no reliable column density estimates have been obtained from these observations; in any case theses data would not indicate how much neutral atomic oxygen there is in the deep interior of clouds.

4.2 Abundance Estimates

Four oxygen reservoirs have been investgated over the last several years, that is CO, in both in gas phase and in solid phase, OH and water ice. Although CO is easy to detect, its abundance is not so easy to derive because it was found from the ^{13}CO observations that the ^{12}CO (J=1 – 0) 115 GHz line was optically thick. ^{13}CO itself could be optically thick in many places and at least in Orion A (Castets et al. 1990) and only the rarer C^{18}O isotopomer should be optically thin in most cases. However, the latter does not seem to be able to trace all the gas, especially at the borders of the clouds. With great care concerning CO excitation conditions and modeling, we can say that a consensus puts its abundance in the range 5 $\times 10^{-5}$ to 1.5 $\times 10^{-4}$ of that of H_2. In the dense interstellar medium, abundances are often given with respect to H_2. Hence the oxygen cosmic abundance will be expressed as 1.7 $\times 10^{-3}$ that of H_2. Thus, it was known quite quickly that CO could not contain more than 10 % of the total oxygen, with a probable value around 5 %. The reader should note that in the dense interstellar medium, all the hydrogen is converted into molecular hydrogen which is almost impossible to observe directly. Thus the way to estimate the hydrogen column density is very often linked to the estimate of the CO column density itself or to the CO observations in some way and thus doubt is cast on the actual abundance ratio of CO in dark clouds in the most opaque regions. Only in the cloud borders where the extinction is not too low can the H_2 column density be estimated by independent means, such as, UV Lyman-α absorption in front of bright stars or by means of star counts if the dust to gas ratio is known (Dickman 1978; Frerking et al. 1982). In some cases, near infrared measurements of embedded stars can give an amount of extinction towards opaque regions which can be further related to H_2 column density and with the apparition of sensitive millimeter and submillimeter bolometers, dust continuum measurements start to be used. The results can be quite controversial. In the Orion KL case, the H_2 column density has been estimated to be between 2.4 $\times 10^{22}$ cm^{-2} (Scoville et al. 1983) and 3 $\times 10^{23}$ cm^{-2} (Blake et al. 1987). On this basis, the abundance of gaseous CO has been estimated to lie between 2.5 $\times 10^{-5}$ and 3.1 $\times 10^{-4}$ (from Scoville, et al. 1983; see also the discussion in Knacke & Larson 1991). This illustrates the problem of estimating the abundance of CO in particular (see review by Combes 1991 for details) and all the species in general. This is why observed abundances are often referred to that of CO instead of H_2. There seems to be a general agreement that the CO abundance range should be narrower than the range given above. For example, in cold clouds, like TMC-1 or L134N, the usual range given is 6 to 8 $\times 10^{-5}$. Frozen CO on grain mantles have been found to account for about one third of the total CO (gas+grains) in abundance towards cold clouds while no or very little trace could be found towards cloud hot cores (Whittet & Duley 1991). OH excitation conditions in the Λ doubling lines at 18 cm are very difficult to evaluate and thus only in cold clouds, with low and approximately uniform kinetic temperatures, can the OH column density be determined with some confidence. In TMC-1, X(OH) = 3 $\times 10^{-7}$ while in L134N or L183 X(OH) = 7.5 $\times 10^{-8}$ (Irvine et al. 1987). This shows that OH is a minor species, which is not unexpected because it reacts easily with OI or NI to form O_2 or NO (itself reacting again with OH to form O_2). Thus, it does not seem to be important to be able to derive a precise column density here, as we can certainly neglect OH. As we indicated in the introduction, near infrared studies quickly showed that a water ice absorption band at 3.07 μm was clearly visible in Orion BN-KL, and other sources (Gillett & Forrest 1973; Knacke et al. 1982), for which a column density estimate could be derived. Water ice abundance is about half that of gaseous CO in Orion KL, which would represent 2 to 5% of total oxygen (from Knacke & Larson 1991; Knacke et al. 1982). Hasegawa et al. (1992) indicated that in the Taurus cloud the water ice abundance is even higher than that of the gaseous CO; the H_2O-ice/CO-gas ratio would reach 1.25. It is also found that gaseous CO abundance is 1.2 10^{-4} that of H_2, which is higher than usual (see above). For X(CO) = 6 to 8 10^{-5}, water ice would represent 4-6 % of total oxygen in dense and dark cold clouds. Recent technological progress has allowed us to discover new tracers of oxygen-bearing molecules among which H_3O^+, $H_2^{18}O$ ground level transition and $^{16}O^{18}O$. Another way to search for water is to try to measure the H_3O^+ abundance, since water is a direct product of the recombination of this ion and this is probably the main pathway although reactions with neutrals such as ammonia also lead to water. Detecting H_3O^+ is not very easy because only submillimeter lines exist and these are emitted only in warm and dense regions. The most difficult point is to know the branching ratio of H_3O^+ towards O, OH and H_2O when it recombines with an electron. Only the branching ratio to form OH has been measured by Herd et al. (1990) and Adams et al. (1992); this is 0.65. Relating water gas phase abundance to the observations of H_3O^+ is thus difficult (Phillips et al. 1992). In 1986, this species was detected towards dense, warm sources at 307 GHz ($J_K = 1_1^- - 2_1^+$, Hollis et al. 1986; Wootten et al. 1986). With two other detected transitions at 364 GHz ($J_K = 3_2^+ - 2_2^-$, Wootten et al. 1991) and at 396 GHz ($J_K = 3_0^+ - 2_0^-$, Phillips et al. 1992) the latter were able to derive relatively secure column densities leading to H_3O^+ abundances in the range 1 $\times 10^{-10}$ to 3×10^{-9}, close to the theoretical predictions. This confirms that H_3O^+ and HCO^+ are the most abundant oxygen-bearing ions. Since OH and H_2O are directly produced from H_3O^+, it is tempting to predict their abundances from these observations. Phillips et al. (1992) obtained X(O_2) = 10^{-6} - 10^{-5} which is possibly pessimistic (see below) and X(H_2O) = 10^{-7} - 10^{-6} which they find to be low, which seems consistent with the most direct measurements of water vapor abundance towards Orion

(Herd et al. 1990; Knacke & Larson 1991; Wannier et al. 1991; Zmuidzinas 1993, see below). The next reservoir to be investigated was water vapor. H_2O near infrared bands have been marginally detected from the KAO using the maximum velocity shift allowed by Earth's orbital rotation. Knacke & Larson (1991) find a gas-phase abundance ratio of $H_2O/CO = 0.03 \pm 0.02$ and an H_2O gas/solid ratio of < 0.05 towards BN-KL in Orion. Another way to measure the H_2O abundance is to look at the $H_2^{18}O$ ground transition in the ortho state ($J_{KK'} = 1_{10} - 1_{01}$) at 548 GHz (this transition for the main isotope is at 557 GHz). This transition is undoubtedly thermal, in contrast to the one observed at 203 GHz by Jacq et al. (1988, 1990). A first search (Wannier et al. 1991) gave an upper limit of $H_2O/CO < 0.003$ towards Orion, NGC 2024 and NGC 2264. This very low upper limit has been recently confirmed by the actual detection of the line with a SIS receiver (Fig. 6, Zmuidzinas 1993). The detected line indicates that of $H_2O/CO = 0.001$.

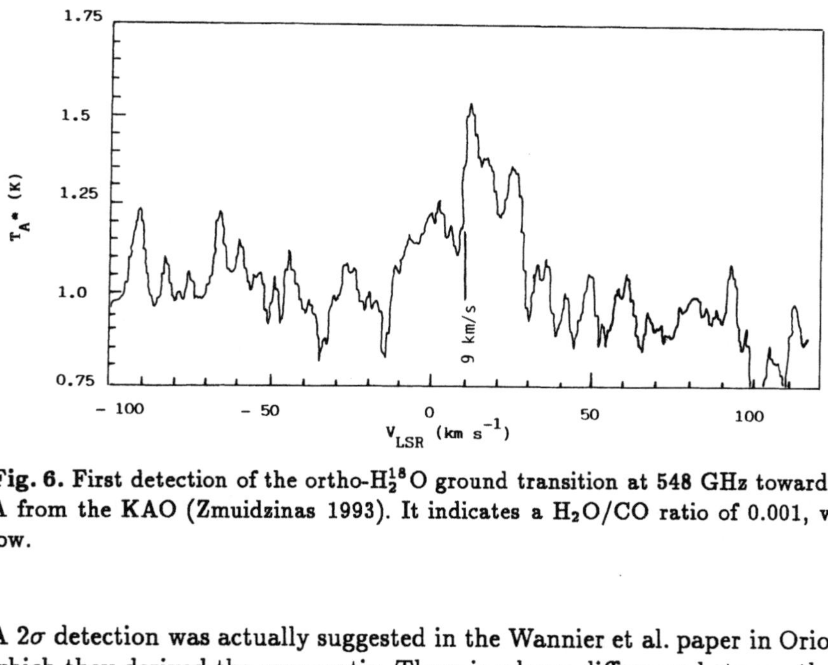

Fig. 6. First detection of the ortho-$H_2^{18}O$ ground transition at 548 GHz towards Orion A from the KAO (Zmuidzinas 1993). It indicates a H_2O/CO ratio of 0.001, which is low.

A 2σ detection was actually suggested in the Wannier et al. paper in Orion from which they derived the same ratio. There is a large difference between these two results (a factor of 30 between the IR band experiment and the $H_2^{18}O$ experiment) which is likely due to a very large beam size difference, weakness of both signals and difficulty of finding the correct interpretation. However both point toward a water vapor abundance in the range 2.5×10^{-8} to 9.3×10^{-6} (compared to H_2), depending on which CO abundance and which H_2O/CO ratio we take. But most of the H_2O must be frozen onto grains. Water vapor can thus be safely ignored in the final balance.

The last reservoir which has been recently explored is molecular oxygen. The search of $^{16}O^{18}O$ started by Goldsmith et al. (1985) and Liszt & Vanden Bout

(1985). This concentrated on the ($N_J = 2_1-0_1$) transition at 234 GHz; this year a tentative detection was reported by Pagani et al. (1993); for L134N and possibly NGC 7538 the O_2 abundance is a few 10^{-5} (Fig. 7). The L134 result gives a 'normal' abundance. This detection must be confirmed but if real, it would be

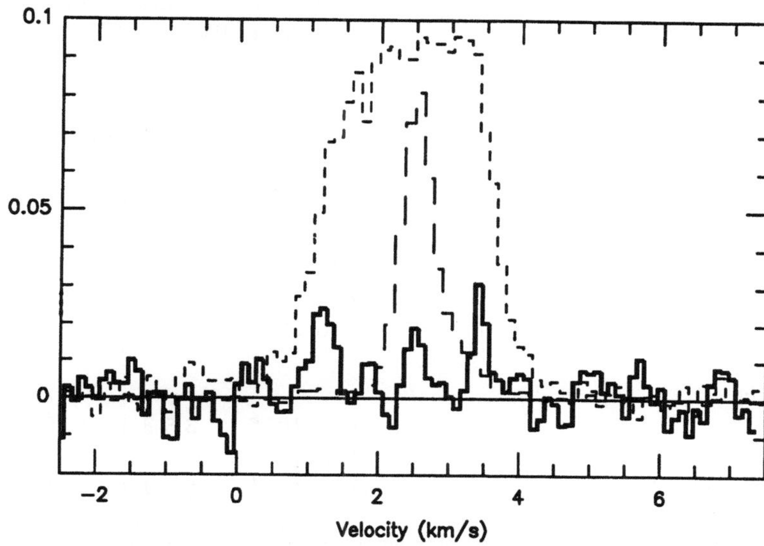

Fig. 7. First tentative detection of $^{16}O^{18}O$ towards L134N (Pagani et al. 1993). Only the central peak is exactly at the $C^{18}O$ velocity but all three lines are within the ^{12}CO line width and could be oxygen lines. This should be checked by observing other transitions of O_2.

indicative of an O_2/CO ratio of 0.2-0.4. This is typical for most of the models we considered here. As we have noted, the Pineau des Forets et al. (1992) model predicts an abundance which is too low by several orders of magnitude. There is no conflict here with the previous (observational) O_2 limits. The interpretation was that the observed ratio was much below the cosmic C/O ratio. Most used 1 σ upper limits (in one case overestimated $C^{18}O$ column densities were reported); after correcting all upper limits in a consistent way it was found that the best limit on galactic $^{16}O^{18}O$ works was only $O_2/CO < 0.8$, compatible with $C/O = 0.4$. In summary, O_2 would represent about 2 to 4 % of the total oxygen in cold clouds and perhaps up to 8% in Orion if the same O_2/CO ratio holds.

Table 3 summarizes the measured or estimated abundances of the main oxygen-bearing species. In this table, we have neglected all minor oxygen-bearing molecules detected, such as HCO^+ and H_3O^+, since together these represent less than 1% of the total oxygen. These are mean values along the line-of-sight; we know that the cloud structure is clumpy and that shielding and chemical effects tend to put species at different depths in the clouds (as partly illustrated by the Viala model shown in Fig. 2) thus all the figures of Table 3 may not be rep-

Table 3. Observed and estimated abundances as a fraction of H_2 of the most important oxygen-bearing species towards Orion A (BN-KL) and typical dark clouds TMC-1, L134 N (= L183) and total observed and observed+estimated abundances as a fraction of cosmic oxygen abundance.

Species	Fractional Abundance Orion A (BN-KL)	Dark Clouds (TMC1, L134 N)
Observed:		
O_2[a]	$< 5\times10^{-5}$	$1.2\text{-}3.2\times10^{-5}$
OH	–	$1.5\text{-}6\times10^{-8}$
CO-gas	$5\text{-}15\times10^{-5}$	$6\text{-}8\times10^{-5}$
CO-ice	$< 1\text{-}3\times10^{-6}$	$1.8\text{-}2.4\times10^{-5}$
H_2O-gas	$1\text{-}90\times10^{-7}$	$< 10^{-5}$
H_2O-ice	$2.5\text{-}7.5\times10^{-5}$	$7.5\text{-}10\times10^{-5}$
Total Observed Abundance/X(O)	4.5 to 14 %	11 to 16%
Estimated:		
OI	$3.5\text{-}8.5\times10^{-5}$	$3.5\text{-}8.5\times10^{-5}$
O_2[a,c]	$1\text{-}6\times10^{-5}$	–
Silicates+Oxydes [b]	$1.2\text{-}4.8\times10^{-4}$	$1.2\text{-}4.8\times10^{-4}$
Total Obsd.+Estd. Abundance/X(O)	14 to 50%	20 to 50%

[a]In the comparison with x(O), O_2 abundance is multiplied by 2.
[b]The abundance here is the sum of Fe, S, Si and Mg cosmic abundances and to estimate the abundance of oxygen locked in these grains, we consider that there are 0.5 to 2 oxygen atoms per atom of anything else to compare with x(O).
[c]We suppose the same O_2/CO ratio in Orion than in dark clouds

resentative of the actual local abundances in the clouds. For the species found so far, adding the most optimistic estimates leads to the identification of <20% of the O atoms available in the interstellar medium (14% towards Orion and slightly more, <16%, in cold clouds where 30% of the total CO is frozen onto grains). The most pessimistic case allows us to identify only 5% to 10% of the O atoms. In both cases, the most O is missing. Of course two reservoirs have not yet been evaluated in the dense medium that is OI in the gas phase and the group of silicates, carbonates and metal oxides which are expected to form the grain cores. However by observing the atomic medium in front of hot stars, one can estimate the minimum depletion on grains (see e.g. Jenkins 1987 for a review or Savage et al. 1992 for more recent work). This depletion seems to increase along line of sights with higher mean density suggesting that in the dense interstellar medium, its effect should be even greater. In the diffuse medium, the oxygen depletion factor is of the order of 50%. Because of the high depletion factors the chemical models suppose, they do not predict very high abundances of atomic oxygen in the gas phase though it is usually the most abundant species or second to CO only by number. In the models it represents typically 2 to 5% of the total oxygen. This is insufficient to fill the gap and direct measurements are needed to estimate its actual column density. Next, apart from H, C and O there are only 5 reactive elements, that is Fe, Mg, N, S and Si which are abundant enough to play a significant role in the chemistry of the interstellar medium (mostly in the solid phase for most of them). Summing their cosmic abundances (Anders & Grevesse 1989) we reach a total of 2.4×10^{-4} (4.8×10^{-4} with respect to H_2) that is 28 % of the cosmic oxygen abundance, N representing half of it. NO itself being not very important; $X(NO) = 2.7 \times 10^{-8}$ in TMC-1 and 2.0×10^{-7} in L134N (Gerin et al. 1993). Only the large family of silicates, carbonates and some important metallic oxydes like FeO, Fe_2O_3 and Fe_3O_4 which should form the refractory core of grains could hold important quantities of oxygen. If we ignore N and the carbonates and suppose that there is typically one oxygen atom per atom of anything else in the grain core, we will account for more than 14 % of the oxygen. A more conservative range would then be from half an oxygen atom to two oxygen atoms per atom, that is 7 to 28 %. Thus up to 33% of oxygen atoms could be added up to what has been identified and we would still be left with more than half of the oxygen atoms unidentified. It must be noted that even the 28% we suggest are already below the oxygen depletion factor which has been observed in the diffuse medium probably indicating that oxydes and silicates are not the only components of grain cores. Are carbonates ($X^{2+} - CO_3^{2-}$, $X^{+} - HCO_3^{-}$) the ultimate refuge of oxygen on grains? Are the grain cores a conglomerate of ice and refractory material which would hide away large quantities of water ice (comets which should trace the original solar nebula are large conglomerates of dirty water ice with a loose structure. They are suggestive of a similar structure for the interstellar grains instead of a hard, compact silicate core with an icy mantle around)? Or does atomic oxygen make up most of O in the ISM? We may learn more about this question in the next decade.

5 Plans for the Future

From the ground, there are two more lines of $^{16}O^{18}O$ which are observable that is at 298 GHz ($N_J = 2_2 - 0_1$) and 401 GHz ($N_J = 3_2 - 1_2$). The problem is that these lines should be even weaker than the 234 GHz line and thus much more difficult to observe as the telescope, receiver and sky are all somewhat worse at higher frequencies. It is planned to replace the KAO airplane by the SOFIA project, which is a 747 Boeing with a 3 m telescope. The SOFIA airplane should fly higher, getting largely above the tropopause, and thus above most of the water vapor. The strongest main isotope lines will still not be observable, but the $H_2^{18}O$ line at 548 GHz should be easier to observe and to map. In the near future, several balloon borne experiments will be launched, to observe O_2 and H_2O. A U. C. Santa Barbara-JPL group has prepared a balloon-borne 1 m telescope with a SIS receiver to observe the ($N_J = 1_1 - 1_0$) line of O_2 at

119 GHz. The first flight is due during summer 1993. Another balloon-borne experiment, PRONAOS (a collaboration between several French laboratories mainly involving CESR (Toulouse), IAS (Orsay) and DEMIRM (Meudon) and CNES) should carry a 0.8 millimeter SIS receiver on a 2 m telescope to observe the ($N_J = 3_2 - 1_1$) line of O_2 at 368 GHz and the 380 GHz line of H_2O. The first flight is due in 1995. The next flight should carry another receiver tuned to 557 GHz to observe the ground rotational line of ortho H_2O. ESTEC (Nordwijk) is also preparing a 60 cm telescope with an uncooled Schottky receiver at 557 GHz to observe the ground transition of the main isotope of ortho water. This is due to fly in the fall of 1993. Finally a series of satellites are being planned to observe water and molecular oxygen. The SWAS satellite (a 0.6 m dish) is the first one to be planned. It will have passively cooled Schottky mixers to observe the ($N_J = 3_2 - 1_2$) O_2 line at 424 GHz and the 547/557 GHz ground rotational lines of ortho H_2O and $H_2^{18}O$ plus the ^{13}CO and $C^{18}O$ (J=5 − 4) lines at 550 GHz. Other small satellites being proposed are ODIN (Sweden), a 1.1 m dish telescope for Earth's limb sounding and sky mapping of several lines, among which the 557 GHz line of H_2O and the 119 GHz line of O_2 (all Schottky receivers), TOUTATIS (France), a 1.2 m dish with SIS receivers at 424 GHz (O_2), 5_{40}-5_{80} GHz (H_2O $H_2^{18}O$) and 752 GHz (H_2O). All these satellites could be launched before 2000. More ambitious are the SMIMM and FIRST projects, both with a 3 m telescope and SIS receivers covering the 500-1200 GHz band. They are planned for the early 2000's.

Acknowledgements Thanks to E. Roueff, G. Pineau des Forets, Y. Viala and M. Gerin for fruitful discussions and critical reading of the manuscript.

References

Adams N.G., Herd C.R., Geoghegan M., et al., 1992, J.Chem.Phys. 94, 4852
Anders E., Grevesse N., 1989, Geochim. et Cosmochim. Acta 53, 197
Blake G.A., Sutton E.C., Masson C.R., Phillips T.G., 1987, ApJ 315, 621
Brown P.D., Millar T.J., 1989, MNRAS 237, 661
Castets A., Duvert G., Dutrey A., et al., 1990, A&A 234, 469
Cernicharo J., Thum C., Hein H., et al., 1990, A&A 231, L15
Charnley S.B., Tielens A.G.G.M., Millar T.J., 1992, ApJ 399, L71
Cheung A.C., Rank D.M., Townes C.H., Thornton D.D., Welch W.J., 1968, Phys. Rev. Letters 21, 1701
Cheung A.C., Rank D.M., Townes C.H., Thornton D.D., Welch W.J., 1969, Nat. 221, 626
Combes F., 1991, ARA&A 29, 195
Combes F., Casoli F., Encrenaz P., Gerin M., Laurent C., 1991, A&A 248, 607
Dickman R.L., 1978, ApJS 37, 407
Draine B.T., 1991. In : Falgarone E., Boulanger F., Duvert G. (eds) Fragmentation of Molecular Clouds and Star Formation, Kluwer Academic Publishers, Dordrecht, p. 185
Draine B.T., Roberge W.G., Dalgarno A., 1983, ApJ 264, 485
Encrenaz P.J., Combes F., Casoli F., et al., 1993, A&A 273, L19
Frerking M.A., Langer W.D., Wilson R.W., 1982, ApJ 262, 590

Genzel R., Watson D.M., Townes C.H., et al., 1984, ApJ 276, 551
Gerin M., Viala Y., Casoli F., 1993, A&A 268, 212
Gillett F.C., Forrest W.J., 1973, ApJ 179, 483
Goldsmith P.F., Snell R.L., Erickson R., et al., 1985, ApJ 289, 613
Graff M.M., Dalgarno A., 1987, ApJ 317, 432
Hasegawa T.I., Herbst E., Leung C.M., 1992, ApJS 82, 167
Heiles C., 1968, ApJ 151, 919
Herbst E., Klemperer W., 1973, ApJ 185, 505
Herd C.R., Adams N.G., Smith D., 1990, ApJ 349, 388
Hollis J.M., Churchwell E.B., Herbst E., Lucia F.C.D., 1986, Nat 322, 524
Irvine W.M., Goldsmith P.F., Hjalmarson A., 1987. In : Hollenbach D.J., Thronson Jr., H.A. (eds) Interstellar Processes, Reidel, Dordrecht, p. 561
Jacq T., Jewell P.R., Henkel C., Walmsley C.M., Baudry A., 1988, A&A 199, L5
Jacq T., Walmsley C.M., Henkel C., et al., 1990, A&A 228, 447
Jenkins E.B., 1987. In : Hollenbach D.J., Thronson Jr., H.A. (eds) Interstellar Processes, Reidel, Dordrecht, p. 533
Knacke R.F., Larson H.P., 1991, ApJ 367, 162
Knacke R.F., McCorkle S.M., Puetter R.C., Erickson E.F., Kratschmer W., 1982, ApJ 260, 141
Langer W.D., Graedel T.E., 1989, ApJS 69, 241
Liszt H.S., 1985, ApJ 298, 281
Liszt H.S., 1992, ApJ 386, 139
Liszt H.S., Vanden Bout P.A., 1985, ApJ 291, 178
Millar T.J., Rawlings J.M.C., Bennett A., Brown P.D., Charnley S.B., 1991, A&AS 87, 585
Pagani L., Langer W.D., Castets A., 1993, A&A, in press
Palmer, P., Zuckerman, B., Buhl, D., Snyder, L.E., 1969, ApJ 156, L147
Phillips T.G., Dishoeck E.F.v., Keene J., 1992, ApJ 399, 533
Phillips T.G., Jefferts K.B., Wannier P.G., 1973, ApLetters 15, 17
Phillips T.G., Kwan J., Huggins P.J., 1979. In : Andrew B.H. (ed.) Interstellar Molecules, Reidel, Dordrecht, p. 21
Pineau des Forets G., Flower D.R., Chieze J.P., 1992, MNRAS 256, 247
Prasad S.S., Tarafdar S.P., Villere K.R., Huntress Jr W.T., 1987. In : Hollenbach D.J., Thronson Jr., H.A. (eds) Interstellar Processes, Reidel, Dordrecht, p. 631
Roueff E., Pineau des Forets G., 1993. In : Rowe B.R., Mitchell J.B.A. (eds) Dissociative Recombination, Theory, Experiment and Applications, Plenum Press, in press, p.
Savage B.D., Cardelli J.A., Sofia U.J., 1992, ApJ 401, 706
Schulz A., Gusten R., Serabyn E., Walmsley C.M., 1991, A&A 246, L55
Scoville N., Kleinmann S.G., Hall D.N., Ridgway S.T., 1983, ApJ 275, 201
Tielens A.G.G.M., Hollenbach D., 1985a, ApJ 291, 747
Tielens A.G.G.M., Hollenbach D., 1985b, ApJ 291, 722
Viala Y.P., 1986, A&AS 64, 391
Wannier P.G., Pagani L., Kuiper T.B.H., et al., 1991, ApJ 377, 171
Waters J.W., Gustincic J.J., Kakar R.K., et al., 1980, ApJ 235, 57
Whittet D.C.B., Duley W.W., 1991, A&AR 2, 167
Wilson R.W., Jefferts K.B., Penzias A.A., 1970, ApJ 161, L43
Wootten A., Boulanger F., Bogey M., et al., 1986, A&A 166, L15
Wootten A., Mangum J.G., Turner B., et al., 1991, ApJ 380, L79
Zmuidzinas J., 1993, Personal Communication

Interstellar CNO Isotope Ratios

C. Henkel[1], *T.L. Wilson*[1], *N. Langer*[2], *Y.-N. Chin*[3], *R. Mauersberger*[1]

[1] Max-Planck-Institut für Radioastronomie, Auf dem Hügel 69, D-53121 Bonn, Germany
[2] Max-Planck-Institut für Astrophysik, Karl-Schwarzschild-Str. 1, D-85740 Garching, Germany
[3] Radioastronomisches Institut der Universität Bonn, Auf dem Hügel 71, D-53121 Bonn, Germany

Abstract:
In an interpretation of interstellar, circumstellar, and solar system CNO isotope ratios, we find two scenarios which are free of internal inconsistencies. The first requires that the early solar system was enriched by material from massive stars, leading to enhanced $^{12}C/^{13}C$ and $^{18}O/^{17}O$ ratios and to a reduced $^{14}N/^{15}N$ ratio. The second involves infall of gas onto the galactic disk after the formation of the solar system. Both scenarios require that the bulk of the interstellar ^{16}O, ^{18}O, and ^{15}N originates from massive stars ($>8\,M_{\odot}$), with ^{18}O and perhaps ^{15}N being destroyed in lower mass stars. ^{17}O is mainly synthesized in stars of intermediate mass while ^{12}C, ^{13}C, and ^{14}N are produced in stars of high and intermediate masses.

With respect to nucleosynthesis, evolutionary models of massive stars applying the Ledoux criterion for convection and semiconvective mixing should be preferred. These yield high ^{18}O abundances above the convective He-burning core. For ^{17}O, theoretical yields and observational data are not consistent, since models predict too much ^{17}O from massive stars. The absence of an $^{18}O/^{17}O$ gradient in the galactic plane remains unexplained. Since ^{18}O and ^{17}O are products of quite different processes, this absence strongly constrains numerical models of the 'chemical' evolution of the Galaxy.

1 Introduction

Spectroscopy of stellar or interstellar gas in its atomic phase does not allow to discriminate between different isotopic species. If one has no direct access to the material (through cosmic rays or solar system samples), isotopic abundances can only be determined from spectroscopy of molecular isotopomers. This is why molecular spectroscopy is of fundamental importance to constrain stellar nucleosynthesis and the 'chemical' evolution of the Galaxy.

The isotopes of carbon, nitrogen, and oxygen are of particular interest, given their large abundances and the presence of seven stable nuclei, each of them being formed under specific physical conditions. Most known interstellar and circumstellar molecules contain at least one of these elements. $^{12}C/^{13}C$, $^{14}N/^{15}N$, $^{16}O/^{18}O$, and $^{18}O/^{17}O$ ratios are thus known not only for the solar system, but also for the galactic nuclear region and disk out to $R_{GC} \sim 10\,kpc$ ($R_{\odot}$=8.5 kpc;

see Table 1). Data from cosmic rays, extragalactic starburst nuclei, and interstellar dust grains have also been collected (e.g., Gibner et al. 1992; Amari et al. 1993; Henkel & Mauersberger 1993). While Table 1 displays the best presently available interstellar and solar system ratios, Fig. 1 illustrates the radial isotope gradients in the galactic disk.

2 Searching for internally consistent concepts

Over the last few decades, several approaches have been developed to interprete the data summarized in Table 1 and Fig. 1. These attempts are the following: (1) The solar system ratios are typical for the local interstellar medium $4.5\,10^9$ yrs ago; effects caused by radial inflow or infall of gas from outside the galactic plane can be neglected; (2) a large amount of radially inflowing gas or, equivalently, a substantial increase in the solar galactocentric distance has decoupled solar and interstellar isotope ratios; (3) ejecta from massive stars have enriched the presolar nebula, and (4) infall of gas onto the galactic disk has altered the ISM abundances after the formation of the solar system.

While a number of numerical studies exist, *a qualitative self-consistent picture is still missing.* The basic equations to describe the 'chemical' evolution, including infall and delayed mass return from evolving stars (see, e.g., Güsten & Mezger 1983), are

$$\frac{\mathrm{d}}{\mathrm{d}t} M_{\mathrm{g}} = A(t) - \Psi(t) + \int_{M_{\mathrm{l}}}^{M_{\mathrm{u}}} \Psi(t-\tau(m))\,\phi(m)\,b(m)\,\mathrm{d}m \qquad (27)$$

and

$$\frac{\mathrm{d}}{\mathrm{d}t}[Z_{\mathrm{i}}(t)M_{\mathrm{g}}(t)] = Z_{\mathrm{A,i}}(t)\,A(t) - Z_{\mathrm{i}}(t)\Psi(t) + \int_{M_{\mathrm{l}}}^{M_{\mathrm{u}}} \Psi(t-\tau(m))\,\phi(m)\,b(m) \times [Z_{\mathrm{i}}(t-\tau(m)) + \Delta m(Z_{\mathrm{i}})]\,\mathrm{d}m. \qquad (28)$$

M_{g} denotes the total gas mass. $A(t)$ is the time dependent infall rate, $\Psi(t)$ the star formation rate, $\phi(m)$ the initial mass function (IMF) normalized for the mass range $M_{\mathrm{l}}...M_{\mathrm{u}}$, $b(m)$ the mass returned back to the interstellar medium after a stellar lifetime $\tau(m)$, Z_{i} the total gas mass fraction of isotope i, and $Z_{\mathrm{A,i}}$ the corresponding fraction for the infalling gas. For a 'primary' nucleus i which is not a seed element for a 'secondary' isotope, the material returned to the ISM is enriched by $\int \delta_{\mathrm{i}}\,\mathrm{d}m = \int \Delta m(Z_{\mathrm{i}})\,b(m)\phi(m)\Psi(t-\tau(m))\,\mathrm{d}m$. For a secondary isotope, this quantity becomes $\int \delta_{\mathrm{i}}\,Z_{\mathrm{j}}(t-\tau(m))\,\mathrm{d}m$, with j being the seed nucleus.

Qualitatively, these equations allow us to differentiate between 'primary', 'secondary', and 'tertiary' nuclei, depending on the number of stellar generations required to synthesize a given nucleus. High mass stars need little time to eject their processed material into the ISM, while lower mass stars need more time.

Table 1. Isotope ratios[a)]

Isotope ratio	Region	Observed ratios
$^{12}C/^{13}C$	Galactic disk	$(7.5\pm1.9)\,R_{GC} + (7.6\pm12.9)$
	Galactic center	25±5
	Inner disk	40±10
	Local ISM	70±10
	Solar system	89
	Starburst nuclei	~50
$^{14}N/^{15}N$	Galactic disk	$(19.0\pm8.5)\,R_{GC} + (288.6\pm65.1)$
	Galactic center	900±200
	Inner disk	375±60
	Local ISM	450±60
	Solar system	270
	Starburst nuclei	–
$^{16}O/^{18}O$	Galactic disk	$(58.8\pm11.8)\,R_{GC} + (37.1\pm82.6)$
	Galactic center	250±30
	Inner disk	300±30
	Local ISM	540±30
	Solar system	490
	Starburst nuclei	~150–200
$^{16}O/^{17}O$	Galactic disk	$(206\pm41)\,R_{CG} + (130\pm289)$
	Galactic center	900±150
	Inner Disk	1050±200
	Local ISM	1900±200
	Solar system	2700
	Starburst nuclei	>1200
$^{18}O/^{17}O$	Galactic disk	~3.5, no gradient
	Galactic center	3.5±0.2
	Inner disk	3.6±0.3
	Local ISM	3.6±0.3
	Solar system	5.5
	Starburst nuclei	≥8

a) The data are taken from Wilson & Matteucci (1992), Henkel & Mauersberger (1993), and Wilson & Rood (1994); for the $^{16}O/^{17}O$ ratios, the non-solar $^{16}O/^{18}O$ values were multiplied by 3.5. Errors for the solar system ratios are negligible when compared with those for the interstellar medium (ISM). R_{GC} is the galactocentric distance in kpc. In those cases where there is a gradient, the inner disk values refer to R_{GC}=4.5 kpc, the local ISM values to R_{GC}=8.5 kpc.

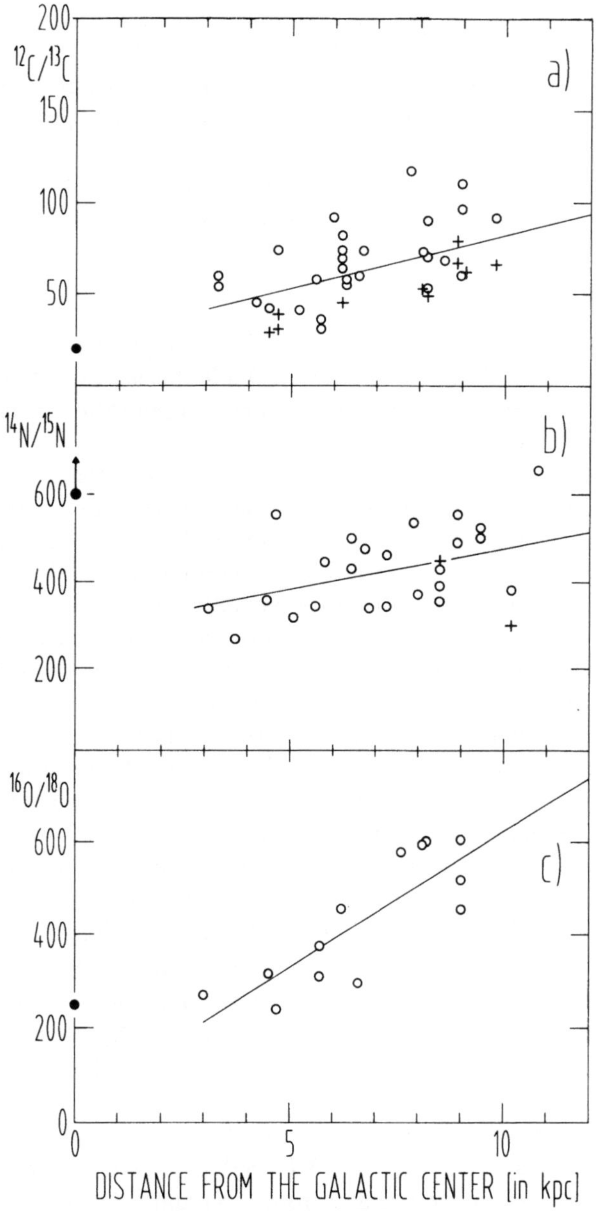

Fig. 1. $^{12}C/^{13}C$ (upper panel), $^{14}N/^{15}N$ (middle panel), and $^{16}O/^{18}O$ (lower panel) disk gradients as a function of radial distance from the galactic center (R_{GC}=8.5 kpc). The figure was taken from Wilson & Rood (1994).

Depending on the number of stellar generations and on the mass and lifetime of the stars involved, the timescale of 'chemical' enrichment differs from nucleus to nucleus. This leads to characteristic nuclear abundance ratios which vary as a function of time and metallicity. To give an example: ^{13}C is a product of incomplete CNO burning and requires pre-existing ^{12}C. Thus ^{12}C/^{13}C isotope ratios tend to decrease with time and metallicity. Such considerations based on models of stellar evolution and a comparison with measured ratios provides fundamental insights into the progress of 'chemical' evolution in the galactic plane. Since stellar yields are highly uncertain, we will limit ourselves to the basic qualitative concepts, leaving numerical aspects to future articles.

2.1 The standard approach

In most studies, the solar CNO abundances were assumed to be representative for the local ISM 4.5 10^9 yrs ago (cf. Wilson & Rood 1994). While this is qualitatively consistent with the carbon isotope ratios, the nitrogen and oxygen ratios remain unexplained. The following problems arise:

(1) ^{14}N is in part a primary and in part a secondary product (e.g. Pagel 1985; Sneden 1985; Diaz & Tosi 1986; Matteucci 1986a,b). The relative amounts synthesized in stars with high ($>8\,M_{\odot}$) and intermediate ($1-8\,M_{\odot}$) masses are not well known. ^{15}N is believed to be produced in novae and perhaps also in Type II supernovae (e.g. Matteucci & d'Antona 1991). It is partially destroyed in stars of intermediate mass (for ^{14}N/^{15}N ratios in the circumstellar shells of late type stars, see e.g. Wannier et al. 1991). Since the local ISM value is larger than the solar system ratio, we can assume that interstellar ^{14}N/^{15}N ratios increase with time and metallicity, indicating that ^{14}N must be more secondary than ^{15}N. This is consistent with the galactic center ^{14}N/^{15}N ratio which is larger than the solar and galactic disk values. This difference is then explained as a consequence of more extensive nuclear processing in the central region of the Galaxy. However, the ^{14}N/^{15}N ratio systematically decreases with decreasing galactocentric radius along the galactic plane toward the inner disk. This is opposite to the expected behavior. An alternative view, taking a decreasing ^{14}N/^{15}N ratio with increasing degree of nuclear processing (cf. Matteucci & d'Antona 1991) leads to two major inconsistencies: First, the local interstellar ratio exceeds the solar value, in contradiction with this assumption of slowly decreasing interstellar ^{14}N/^{15}N ratios. Second, the large ^{14}N/^{15}N ratio in a highly processed region like the galactic center is also not compatible with our assumption.

(2) When analysing oxygen, similar problems arise. ^{16}O is a primary product of nucleosynthesis, mainly synthesized via ^{12}C$(\alpha,\gamma)^{16}$O by He-burning in massive stars. If ^{18}O (synthesized by He-burning from ^{14}N) and ^{17}O (produced by H-burning) are secondary, the galactic disk ^{16}O/^{17}O and ^{18}O/^{17}O gradients are readily explained. Also, the galactic center ratios fit well into this scheme. There are, however, two basic problems: The solar system ^{18}O/^{17}O ratio is 5.5, significantly larger than the local interstellar value, $\sim$3.5 (Table 1). Explaining this by defining ^{18}O as a secondary and ^{17}O as a tertiary nucleus is in conflict with the absence of an ^{18}O/^{17}O gradient as a function of galactocentric radius.

In addition, the $^{16}O/^{18}O$ ratio of the local ISM is slightly larger than the solar value. This is in contradiction with the $^{16}O/^{18}O$ gradient, which requires decreasing $^{16}O/^{18}O$ ratios in time.

In summary, *the assumption of 'normal' solar isotope ratios*, excluding radial inflow or infall of gas onto the galactic plane, *fails to explain the nitrogen and oxygen data.*

2.2 Radial inflow

There are several mechanisms which may lead to a loss of orbital angular momentum in the gas phase, driving a radial inflow. Such a process (neglecting infall) may be caused by the viscosity of the gas layer, a gravitational interaction between the gas and a bar or by scattering of stellar objects in the gravitational field of giant molecular clouds (e.g. Mayor & Vigroux 1981; Lacey & Fall 1985; Russell & Dopita 1992). With a negative galactic disk O/H gradient (i.e. decreasing O/H with increasing galactocentric radius) and a solar oxygen abundance eventually surpassing local interstellar values (see Sect. 3.1.5) this appears to be an attractive scenario: the oxygen-rich gas out of which the solar system formed would now be located well inside the solar circle, while fresh oxygen-poor interstellar gas from outside the solar circle has reached the solar neighborhood. A less likely scenario, involving a gradual increase of the solar galactocentric distance, leads to an equivalent effect.

If the ISM or the solar system have been substantially affected by such radial motions, the galactic disk gradients imply that the interstellar $^{12}C/^{13}C$, $^{14}N/^{15}N$, and $^{16}O/^{18}O$ ratios of the local ISM should be larger than the solar values. For $^{12}C/^{13}C$, the contrary is observed. While the interstellar $^{16}O/^{18}O$ ratios may be consistent with this model, the $^{18}O/^{17}O$ ratios are also not explained: the interstellar ratio, 3.5, is independent of galactic radius, metallicity, and degree of nuclear processing and is inconsistent with the solar system ratio of 5.5.

We conclude that *radial motions cannot be the main cause for the differences between solar system and interstellar CNO isotope ratios.*

2.3 Peculiar solar system CNO isotope ratios

The problems with the standard approach and radial flows as well as the possibility that the solar system O/H abundance may be almost twice as large as that of the local ISM (see Sect. 3.1.5) provide motivation to relax the assumption of 'normal' solar isotope ratios. In order to avoid too many unknown parameters, we must estimate what anomalies are to be expected. A high oxygen abundance implies a significant enrichment by material from massive stars. Thus the assumption of a presolar nebula enriched not only in ^{16}O but also in other massive star constituents is a good first approximation.

Following Henkel & Mauersberger (1993), the interstellar CNO isotope ratios can be interpreted as follows: For the $^{12}C/^{13}C$ gradient, ^{13}C is synthesized from ^{12}C. ^{12}C may be more efficiently produced in high mass stars than ^{13}C (e.g. Weaver & Woosley 1993). A part of the solar $^{12}C/^{13}C$ enhancement over the local interstellar value may thus not be due to the fact that there has been ongoing ^{13}C enrichment of the ISM in the 4.5 10^9 years since the formation of the solar system, but may have been caused by an enrichment of the presolar nebula by massive stars. For first attempts to unravel the relative efficiencies of these effects, see Audouze (1985) and Henkel & Mauersberger (1993; their Sect. 6.4).

As mentioned in Sect. 2.1, ^{15}N is synthesized in novae and destroyed in late type stars of intermediate mass. Adding a significant primary contribution (cf. Audouze 1985; Güsten & Ungerechts 1985), the nitrogen ratios can also be explained: Compared to the ISM, the smaller solar system ratio is due to the peculiar environment of the protosolar nebula. If the initial mass function (IMF) shifts toward massive stars at smaller galactic radii and the equilibrium between ^{15}N destruction and production shifts more in favor of production, then the gradient with smallest ^{14}N/^{15}N ratios in the inner galactic disk can be reproduced. The large galactic center value indicates a lack of massive stars, consistent with the observations.

If ^{18}O is mainly produced by massive stars (cf. Audouze 1985) as a secondary or tertiary product from ^{14}N (which is itself primary or secondary; see Sect. 2.1) and if ^{17}O arises from stars of intermediate mass (e.g., Kahane et al. 1992), the peculiar solar system ^{16}O/^{18}O, ^{16}O/^{17}O, and ^{18}O/^{17}O ratios are immediately understood. Since ejecta from massive stars contribute extra amounts of ^{16}O and ^{18}O to the presolar nebula, the solar system ^{16}O/^{17}O and ^{18}O/^{17}O ratios are significantly enhanced, while ^{16}O/^{18}O remains close to the interstellar value. The ^{16}O/^{18}O and ^{16}O/^{17}O gradients can be understood in terms of the primary and non-primary nature of the nuclei. However, the almost constant ^{18}O/^{17}O ratio must be addressed separately. Here secondary and tertiary products from massive stars, being partially destroyed by lower mass stars, are compared with secondary products mainly synthesized by stars of intermediate mass. The apparently constant interstellar ^{18}O/^{17}O ratio over the entire galactic plane is remarkable and cannot be explained by the model.

Overall, *the assumption of peculiar solar system CNO ratios is consistent with the data.*

2.4 Infall scenarios

As an alternative to presolar nebula enrichment, it is also possible that the nucleosynthesis products in the ISM have been diluted by infalling gas from the halo or a captured dwarf galaxy with little ^{12}C, ^{15}N, ^{16}O, and ^{18}O (cf. de Freitas Pacheco 1993).

If the halo experienced massive star formation only very early in its history, it may contain a large fraction of gas processed in stars of intermediate mass. It is thus justified to assume that its isotopic properties are characterized by small ^{12}C/^{13}C, ^{18}O/^{17}O, and large ^{14}N/^{15}N ratios. These are the properties of the local ISM when compared with the solar system isotope ratios. With a total ISM of $\sim$5 10^9 $M_\odot$ and a star formation rate of a few $M_\odot$ yr^{-1}, the gas consumption time of the present ISM is only 1$-$2 10^9 yrs. While infall rates are not well known (e.g.

Bregman 1980; Mirabel & Morras 1984), a gradual change in the composition of the local ISM due to halo gas is thus possible within a timescale of several 10^9 yrs. This implies that *a change of interstellar abundances caused by infall of gas onto the galactic plane is consistent with the data.* As before, the lack of an $^{18}O/^{17}O$ gradient as a function of galactocentric radius remains unexplained.

2.5 Starbursts in the nuclei of other galaxies

Starbursts are well known for their large numbers of massive luminous stars. A comparison of galactic isotope data with results from extragalactic starburst nuclei thus allows us to discriminate between nuclei synthesized in massive and less massive stars. Here we confine ourselves to oxygen.

Sage et al. (1991) and Henkel et al. (1994) find $^{18}O/^{17}O \geq 8$ in the nuclear starbursts of NGC 253, M 82, and NGC 4945. These results are based on three completely independent data sets. For the 'sub-burst' nucleus IC 342, Sage et al. (1991) tentatively reported $^{18}O/^{17}O \geq 5.5$. This was confirmed ($\sim$6.5) by later measurements (Sage, priv. comm.). These ratios might imply that also in the Galaxy, 10 billion years ago when the disk was forming, massive star and ^{18}O formation was high, yielding large $^{18}O/^{17}O$ ratios ($\gg$5.5). Later, almost five billion years ago and now, the ratios dropped to 5.5 and 3.5, respectively. There are, however, two problems: (1) ^{18}O is a product of He-burning from ^{14}N. Since ^{14}N must have had a low abundance at early epochs (e.g. Wheeler et al. 1989), ^{18}O should only become abundant in stars with metallicities similar to or higher than those of the local ISM. (2) If $^{18}O/^{17}O$ decreases with time, we should observe a galactic gradient with smallest values in the inner disk or near the galactic center, where 'chemical' processing is most advanced. (cf. Sect. 2.1). This is incompatible with the data (see Table 1). Assuming instead a peculiar solar system ^{18}O abundance from the ejecta of massive stars or from infall of ^{18}O-poor gas into the local ISM after the formation of the solar system, the data can be better explained. In the starbursts, the ISM is enriched by ejecta from massive stars, yielding high $^{18}O/^{17}O$ ratios. The $^{16}O/^{18}O$ ratios appear to be low (see Table 1 and Henkel et al. 1993, 1994), thus favoring *highly* enhanced ^{18}O production. This is consistent with a large ^{14}N abundance at the onset of the starburst. The $^{16}O/^{17}O$ limit for starbursts (see Table 1) requires ratios surpassing the galactic center ratio and supports ^{16}O synthesis in massive and ^{17}O production in less massive stars.

2.6 Two possible concepts

Testing several concepts of galactic chemical evolution, we find only two which are consistent with the measured CNO isotope ratios. These infer that either the solar system was substantially enriched by ejecta from massive stars at the time of its formation or that infall of large amounts of gas onto the galactic plane has later altered the composition of the ISM. Both scenarios do not explain the absence of an interstellar $^{18}O/^{17}O$ gradient in the galactic plane.

3 Discussion

In the following, we discuss observational and theoretical aspects which are crucial for a critical evaluation of the two successful scenarios. These include measured abundance inhomogeneities in the ISM, the influence of dust grains on gas phase abundances, the solar 'oxygen anomaly', isotope ratios from cosmic rays, and new results from models of massive stellar evolution.

3.1 Observational aspects

3.1.1 Abundance inhomogeneities in the ISM

The assumption of peculiar solar system abundances requires that small pockets of interstellar gas can be enriched by stellar ejecta of high metallicity. Whether this is a realistic approach can best be studied in a low metallicity environment, where the abundance contrast between normal and contaminated material becomes most pronounced.

O and N abundances in giant HII regions of metal poor dwarf irregular galaxies were modeled by Pilyugin (1992, 1993), who finds that a substantial fraction of the HII regions must be 'self-enriched' due to winds of the ionizing stars and supernovae. A more direct confirmation for this situation has been obtained by Rosa & Mathis (1985) for the LMC: Within 30 Dor, there is a region of only a few pc in size which exhibits a normal nitrogen abundance but shows overabundances in O, Ne, S, Cl, and Ar by a factor of ~5. All these elements with enhanced abundances are believed to be synthesized by massive stars. In the Galaxy abundance contrasts tend to be smaller, but more accurate observational data can be obtained. Studying oxygen abundances in the Orion OB association, Cunha & Lambert (1992) find that the younger sub-associations have higher average O/H ratios. ¿From these examples it is clear that observational evidence favors the concept of metal enrichment within a limited volume of gas.

3.1.2 The role of the dust

Are there specific isotopes preferentially incorporated into dust grains, thus altering gas phase abundances in a way that true interstellar ratios cannot be determined? The influence of dust grains on oxygen abundances appears to be small when comparing results from HII regions and young stars. However, CNO isotope ratios are determined in dense cool molecular clouds where conditions might be different.

Dust grains arising from the winds of late type stars (e.g. Woitke et al. 1993) should be ^{13}C and ^{17}O-rich and ^{15}N and ^{18}O-poor (see, e.g., Kahane et al. 1988). If these stars are the main contributors to dust grain masses, we expect gas phase ^{13}C and ^{17}O abundances to be depleted and ^{15}N and ^{18}O abundances to be enhanced relative to the main species. Comparing interstellar with solar system $^{12}C/^{13}C$, $^{14}N/^{15}N$, and $^{16}O/^{18}O$ ratios, we find, however, the opposite behavior. Dust grains may also form out of the metal enriched ejecta of supernovae (e.g. Seab 1987) and may accrete dust grain mantles in dark clouds (Seab 1987; but see also Chokshi et al. 1993). If such grains are astrated in dense molecular clumps and if sticking probabilities do not strongly depend on isotopic substitution (see Williams 1993 for the case of deuterium), their CNO isotope ratios would be similar to that of the ambient gas. This is consistent with the lack of a significant difference between $^{12}C/^{13}C$ ratios from the dense molecular gas ($H_2^{12}CO/H_2^{13}CO$) and the diffuse ISM ($^{12}CH^+/^{13}CH^+$). Another argument in favor of gas phase data providing reliable abundance estimates for the entire ISM is related to data from 'processed' meteoritic inclusions. The observed variations in oxygen abundances (e.g. Clayton 1993) appear to be decoupled from those of other elements and can be interpreted in terms of a two-phase model, with a protosolar ^{16}O-enriched dust component and a ^{16}O- depleted gas component ($^{18}O/^{17}O$ ratios are almost the same). In this case differences between oxygen abundances of the gas and dust components are only a few percent, well below the sensitivity limits of present interstellar data.

The analysis of small 'unprocessed' interstellar graphite, diamond, and silicon carbide dust grains is also possible (e.g., Amari et al. 1992, 1993; Hoppe et al. 1992; Ott 1993). Deviations from 'normal' isotope ratios can be quite large. Notable are the rare 'grains X' from SiC particles of the Murchison CM2 chondrite with isotopically very light carbon ($^{12}C/^{13}C$ up to ~2000) and very heavy nitrogen ($^{14}N/^{15}N$ down to ~10). Most interesting within this context are the graphite particles (Hoppe et al. 1992): About 25% of the analysed grains have large ^{18}O excesses. In the most extreme case, $^{16}O/^{18}O$ is reduced to ~50. Even more interesting, the grains with reduced $^{16}O/^{18}O$ ratios have also anomalously low $^{14}N/^{15}N$ and high $^{12}C/^{13}C$ ratios, thus yielding a good agreement with the peculiarities postulated in Sect. 2.3 for massive star ejecta. For the grains with $^{16}O/^{18}O{\sim}50$, it is found that $^{12}C/^{13}C=722$ and $^{14}N/^{15}N=134$. According to Hoppe et al. (1992), these grains contain material which must have been ejected from Wolf-Rayet stars.

3.1.3 Meteoritic inclusions and cosmic rays

In Sect. 3.1.2, we used already oxygen data from 'processed' meteoritic inclusions in primitive meteorites. Here we only add that contamination of the protosolar material by short-lived radioactive nuclei (e.g. ^{26}Al, ^{107}Pd, and ^{129}I with half lives of $\sim 10^6 - 10^7$ yrs) is commonly explained by explosions of nearby supernovae or by accretion of stellar wind material from giant stars of intermediate mass (e.g. Cameron 1988). Clayton et al. (1988) favor a supernova scenario to explain ^{48}Ca, ^{49}Ti, ^{50}Ti, ^{54}Cr, and ^{64}Ni anomalies.

Cosmic rays may arise from massive stars. Hence it is interesting to compare their isotope ratios with those in starbursts. To date, only a $^{16}O/^{18}O$ ratio has been determined from cosmic rays: $^{16}O/^{18}O{\sim}100$ (Gibner et al. 1992). This result is close to that obtained from starbursts (Table 1). The solar system ratio (490) is much larger. This and the relatively small solar system $^{18}O/^{17}O$ value when compared with that from starbursts (5.5 versus $\geq$8) indicates that the solar material is diluted if it was enriched by material from massive stars. With solar system $^{16}O/^{17}O{\sim}2700$ and $^{16}O/^{18}O{\sim}100$ from cosmic rays (i.e. high mass stars) we then find that $^{18}O/^{17}O{\geq}30$ should be typical for massive star ejecta (see also Sects. 3.1.2 and 3.2).

3.1.4 The 'oxygen anomaly'

As already mentioned in Sects. 2.2 and 2.3, the solar system oxygen abundance *might* be almost twice as large as that of the local ISM. This potential difference is fundamental for any realistic description of galactic 'chemical' evolution and thus requires a more detailed discussion. There are an impressive number of independent studies which favor a lower oxygen abundance in the ISM and young stellar objects. These are based on indirect evidence from radio recombination lines from HII regions (Shaver et al. 1983; Wink et al. 1983), more direct OI absorption from the diffuse ISM (de Boer 1981; York et al. 1983), early B stars, intermediate mass F-K supergiants, young planetary nebulae, and perhaps young F dwarfs (Luck & Lambert 1985; Peimbert 1987; Dufton et al. 1990; Fitzsimmons et al. 1990; Baldwin et al. 1991; Cunha & Lambert 1992; Gies & Lambert 1992; de Freitas Pacheco 1993; López et al. 1993). Sub-solar oxygen abundances are most thoroughly studied in Orion, where the influence of dust grains affecting the gas phase oxygen abundances is known to be small (e.g. Peimbert 1987; Gies & Cunha 1992). The difference between oxygen abundances from the solar photosphere, the local ISM, and young stellar objects in the solar neighborhood thus seems to be well established. However, metal rich ([Fe/H]> -0.3) F and G dwarfs not belonging to young open clusters have O/H ratios within 0.05 dex of the solar value (e.g. Nissen & Edvardsson 1992; Edvardsson et al. 1993). Compared to most local F and G dwarfs, the solar oxygen abundance thus appears to be normal.

The discrepancy between oxygen abundances from F and G dwarfs and 'chemically' younger interstellar and stellar sources can be reconciled with the infall scenario introduced in Sect. 2.4 (cf. de Freitas Pacheco 1993). An increase in the number of Type I relative to Type II supernovae (also suggested by de Freitas Pacheco 1993) cannot account for this effect: Type Ia supernovae are rare (e.g., Evans et al. 1989; van den Bergh 1990) and do not appear to destroy oxygen (e.g., Nomoto et al. 1984), while massive stars ending their life as Type Ib supernovae should synthesize large amounts of oxygen. A more exotic explanation of the oxygen data is the assumption that the photospheric O/H values of nearby F and G stars (including the sun) are not representative for the bulk of their mass. Following Peimbert (1987), deviations may be caused by two processes: (a) the solar surface might be enriched by capture of planetesimals or cometary material and (b) the solar surface oxygen may be enhanced by an oxygen poor solar wind.

The accretion of cometary material would increase oxygen and decrease helium abundances. This agrees with the photospheric anomalies as observed by Peimbert (1987). Note, however, that Baldwin et al. (1991) favor a high solar helium abundance. If the latter result is correct, the cometary accretion scenario has to be rejected.

The heavy element compositions of the solar corona and the solar wind differ from photospheric values according to their first ionization potential (FIP). H, C, N, and O all belong to the group of elements with 'high' (>11 eV) FIP, implying that CNO abundances relative to hydrogen are almost normal (e.g. Meyer 1985; McKenzie & Feldman 1992). This does not favor evolutionary modifications of CNO abundances in the solar photosphere due to the solar wind (since Si belongs to the elements with 'low' first ionization potential, the peculiar O/Si ratios discussed by Luck & Lambert (1985) are not relevant here). Since Fe has, unlike O, a low first ionization potential, its abundance may be enhanced in the solar wind (cf. Meyer 1985). If the wind can significantly alter photospheric abundances, one can expect that differences in the wind properties from star to star will introduce significant variations in [Fe/O]. However, [Fe/O] ratios in nearby F and G dwarfs vary by only ± 0.04 dex for [Fe/H]> -0.3 (Nissen & Edvardsson 1992).

To summarize: Alternative views exist, but it is likely that the solar oxygen abundance (and that of other F and G dwarfs) does not require significant systematic corrections.

3.2 Theoretical aspects

The three oxygen isotopes ^{16}O, ^{18}O, and ^{17}O are produced in quite different environments and provide a sensitive measure of stellar mass and stellar convection. As noted by Henkel & Mauersberger (1993), models of massive stars using the Schwarzschild criterion for convection, with or without overshooting (e.g. Maeder 1985; Schaller et al. 1992), do not provide ^{18}O enrichment of the ISM by massive stars; instead, enrichment by ^{17}O tends to be favored. In contrast, massive star models computed with the Ledoux criterion for convection and semiconvective mixing, including the effect of 'molecular' weight gradients on convection, allow for a significant production of ^{18}O.

Weaver & Woosley (1993) calculated the evolution of a variety of massive star models, varying the $^{12}C(\alpha,\gamma)^{16}O$ reaction rate, the stellar mass, and the efficiency of semiconvective mixing. Deriving the integrated yields, there appears an interesting pattern for those intermediate mass isotopes ($^{16}O \ldots {}^{40}Ca$) which are expected to be produced in massive stars. For a specific range of $^{12}C(\alpha,\gamma)^{16}O$ rates, the 'production factors' (see Weaver & Woosley 1993 for a definition) are $\sim$10. With $\sim$10% of the material cycled through a massive star, solar abundances for almost all these nuclei are thus reproduced. The exception is ^{18}O: The models with a fast semiconvective mixing rate do not show any peculiar behavior. However, the models with strongly reduced semiconvection predict an ^{18}O overproduction by roughly an order of magnitude, almost independent of the $^{12}C(\alpha,\gamma)^{16}O$ rate. Production factors $\sim$50 as predicted in these models could lead to a much higher ^{18}O abundance than what is typical for the ISM. Note, however, that ^{18}O is produced from ^{14}N which was underabundant relative to ^{16}O when metallicities were low. Also, much of the ^{18}O may be destroyed in late type stars of intermediate mass, where commonlly $^{16}O/^{18}O \gg 500$ (e.g., Kahane et al. 1992). While the grid of parameters in the Weaver & Woosley (1993) models may be somewhat coarse, the results obtained with a low efficiency of semiconvection fit nicely into our concept of ^{18}O production in massive stars.

Models of Wolf-Rayet stars including the effect of 'molecular' weight gradients on convection (Langer 1991; Woosley et al. 1993) predict the presence of an ^{18}O-rich transition layer between the He-burning core and the outer mantle. This leads to temporary surface abundance ratios of order $^{16}O/^{18}O \sim 10$, while $^{18}O/^{17}O \sim 200$, which is consistent with the cosmic ray and graphite data (see also Biermann & Cassinelli 1993).

Such a good agreement between theory and observations cannot be obtained for ^{17}O. The Weaver & Woosley (1993) ^{16}O production factors tend to be slightly smaller than those for ^{17}O. However, starburst $^{16}O/^{17}O$ ratios are larger than the corresponding value for the galactic center region (see Table 1). Furthermore, ^{17}O is known to be strongly enhanced in evolved stars of intermediate mass (e.g. Kahane et al. 1992). Finally, the solar system $^{16}O/^{17}O$ ratio is larger than the local ISM value. Since the observational data are clear, in this case it is the theory of stellar nucleosynthesis which has to be modified in order to account for the observational results. Within this context it is to be mentioned that the CNO-equilibrium abundance of ^{17}O is strongly affected by the highly uncertain nuclear reaction rates of $^{17}O(p,\alpha)^{14}N$ and $^{17}O(p,\gamma)^{18}F$ (cf. Landré et al. 1990).

Almost independent of stellar mass, $^{12}C(\alpha,\gamma)^{16}O$ rate, and the treatment of convection, about half of the ^{15}N is found to be destroyed in the ejecta of massive stars prior to the iron core collapse (Weaver & Woosley 1993). For explosive nuclear processing during the supernova event, Weaver & Woosley obtain ^{15}N enhancements by factors of 2–6. This is not sufficient to compensate for the ^{15}N losses in the outer shell prior to the supernova. Note, however, that calculations of explosive nucleosynthesis may give larger uncertainties. Even for hydrostatic burning, which should be easier to handle, computed massive stellar ^{17}O yields appeared to be too large to be consistent with the observations.

3.3 Summary

As we have seen, there is compelling evidence for small interstellar regions with peculiar abundances. Dust grains appear not to distort gas phase CNO isotope ratios. There is theoretical evidence for enhanced ^{18}O production in massive stars. Graphite grains have been found which show all the CNO isotope anomalies required for the idea of peculiar solar abundances (Sect. 2.3). While all this is consistent with the view that the presolar nebula has been enriched by ejecta from massive stars, stellar and interstellar oxygen data seem to favor the alternative scenario: a variation of ISM abundances due to infall of gas from outside the galactic plane. To discriminate between both concepts, it is thus mandatory to study the 'oxygen anomaly' in even greater detail and to establish the isotopic composition of the solar wind.

4 Improving the molecular database

4.1 Extragalactic sources

The amount and precision of the extragalactic CNO isotope data does not match those from the galactic ISM and the solar system. However, extragalactic data are of high importance, because they allow to discriminate between products of massive and less massive stars. We have some approximate estimates for the $^{12}C/^{13}C$ ratios from two galactic nuclei, the molecular 'goldmines' NGC 253 and NGC 4945 (Henkel et al. 1993, 1994). Two or three more nearby ($D \leq 5$ Mpc) sources are still within reach of the most sensitive cm- and mm-wave instruments. For $^{12}C/^{13}C$ studies of nearby galaxies, perhaps the most important molecular species is CN. ^{12}CN emission is relatively strong and almost always optically thin (thus providing a linear correlation between abundance and line intensity). The hyperfine structure of the rotational lines allows a direct estimate of the main isotope's opacity. However, $^{12}C/^{13}C$ ratios from CN have never been determined for the central region of the Galaxy. While fractionation or selective photoionization may not be significant, an observational test is clearly required before taking extragalactic CN based $^{12}C/^{13}C$ ratios seriously.

A very important experiment could be made with nitrogen. If $^{14}N/^{15}N$ disk gradients are a common phenomenon in spirals (elemental abundance gradients in spirals appear to be similar to those observed in the galactic disk; e.g. Shields 1990) and if starbursts are triggered by infall of disk gas toward the nuclear regions (due to bars; e.g. Henkel et al. 1991), the pre-burst $^{14}N/^{15}N$ ratio should be of order 350–400 (the inner galactic disk value; see Table 1). If ^{15}N is synthesized in significant quantities by massive stars, the $^{14}N/^{15}N$ ratio should thus become <350. With a J=1–0 HCN line opacity of order 2.5 (Henkel et al. 1993) and $^{14}N/^{15}N \sim 100$, the HCN/HC^{15}N intensity ratio is ~40 and HC^{15}N should be detectable.

The determination of $^{16}O/^{18}O$ and $^{16}O/^{17}O$ ratios from extragalactic systems will remain very difficult in the foreseeible future. However, there is a chance to further confirm that $^{18}O/^{17}O$ ratios are substantially enhanced in starbursts: If ratios of 3–4 are found in quiescent extragalactic nuclear regions, this would provide the needed evidence.

4.2 The galactic ISM

Interstellar $^{12}C/^{13}C$ ratios from the galactic disk and center are approximately known. Radio data from at least two molecular species, C^{18}O and H_2CO, are available which appear to bracket the actual ratios (e.g. Langer et al. 1984; Wilson & Rood 1994). Perhaps most urgent are good optical CH^+ data from the local ISM, since $^{12}CH^+/^{13}CH^+$ abundance ratios are believed to be free of fractionation effects. There appears to be a convergence to ratios ~70 (e.g. Stahl et al. 1989; Stahl & Wilson 1992), but recent results by Vladilo et al. (1993) underline the need for more and better spectroscopic data.

Most $^{14}N/^{15}N$ measurements have been collected from HCN. Including more results from other molecules (NH_3, HNC or CN) would be desirable, although the few $^{14}N/^{15}N$ ratios obtained from NH_3 seem to be consistent with those from HCN (Wilson & Rood 1994)

Almost all $^{16}O/^{18}O$ and $^{18}O/^{17}O$ ratios are based on only H_2CO and CO, respectively. Only toward the galactic center sources Sgr A and Sgr B2, $^{16}O/^{18}O$ was also estimated from OH (e.g. Whiteoak & Gardner 1981). For Sgr B2, $^{18}O/^{17}O$ was also obtained from HCO^+ (Guélin et al. 1982). These ratios are consistent with those derived from H_2CO and CO. While this is encouraging and while fractionation is not expected to affect the results (e.g., Langer et al. 1984), more data from OH, HCO^+, and other species are needed.

5 Conclusions

The determination of CNO isotope gradients as a function of galactocentric radius (e.g. Wilson et al. 1976; Gardner & Whiteoak 1979; Henkel et al. 1982; Tosi 1982) has deeply altered our view of chemical processing and evolution in the Galaxy. The recently obtained data from extragalactic starbursts have triggered a similarly fundamental change of perspective: Somehow, *the solar system CNO ratios must have decoupled from those of the ISM.*

The two concepts which can explain the measured data in a self-consistent way are related: They associate a specific CNO isotope with the same stellar cooking site. Even without the use of a numerical code, many features of these models are clear. The following points are of particular interest:

(1) To explain the measured $^{16}O/^{18}O$ and $^{18}O/^{17}O$ ratios, massive star models computed with the Ledoux criterion for convection and semiconvective mixing should be preferred. Low efficiencies of semiconvection seem to provide the best agreement with observational data. Consistent with these models, evidence from cosmic rays and interstellar graphite grains indicates that $^{18}O/^{17}O$ should temporarily reach values $\geq$30 in the ejecta of massive stars.

(2) Measured $^{16}O/^{17}O$ and $^{18}O/^{17}O$ ratios indicate that ^{17}O abundances are inconsistent with the values predicted from models of stellar evolution. The calculated ^{17}O yields from massive stars are too large.

(3) A significant part of the ^{15}N nuclei must arise from massive stars.

(4) The lack of a $^{18}O/^{17}O$ gradient in the galactic plane is not explained by any model. ^{18}O appears to be a secondary and tertiary product from massive stars. Some ^{18}O must be destroyed in stars of intermediate mass. ^{17}O is predominantly a secondary product from stars of intermediate mass. The lack of a gradient will thus provide an important input for the fine-tuning of numerical models simulating the 'chemical' evolution of the Galaxy.

Acknowledgements. N. Langer acknowledges partial support through DFG grant La 587/8–1; Y.-N. Chin was supported by the DAAD.

References

Amari, S., Hoppe, P., Zinner, E., Lewis, R.S., 1992, ApJ 394, L43
Amari, S., Hoppe, P., Zinner, E., Lewis, R.S., 1993, Nat 365, 806
Audouze, J., 1985, Production and Distribution of C, N, O Elements, ESO Conference and Workshop Proceedings No. 21, eds. I.J. Danziger, F. Matteucci, K. Kjär, Garching, p373
Baldwin, J.A., Ferland, G.J., Martin, P.G., Corbin, M.R., Cota, S.A., Peterson, B.M., Slettebak, A., 1991, ApJ 374, 580
Biermann, P.L., Cassinelli, J.P., 1993, A&A 277, 691
Bregman, J.N., 1980, ApJ 236, 577
Cameron, A.G.W., 1988, ARA&A 26, 441
Chokshi, A., Tielens, A.G.G.M., Hollenbach, D., 1993, ApJ 407, 806
Clayton, R.N., 1993, Ann. Rev. Earth & Plan. Sci. 21, 115
Clayton, R.N., Hinton, R.W., Davis, A.M., 1988, Phil. Trans. Roy. Soc. Lond. 325, 483
Cunha, K., Lambert, D.L., 1992, ApJ 399, 586
de Boer, K.S., 1981, ApJ 244, 848
de Freitas Pacheco, J.A., 1993, ApJ 403, 673
Diaz, A.I., Tosi, M., 1986, A&A 158, 60
Dufton, P.L., Brown, P.J.F., Fitzsimmons, A., Lennon, D.J.: 1990, A&A 232, 431
Edvardsson, B., Andersen, J., Gustafsson, B., Lambert, D.L., Nissen, P.E., Tomkin, J., 1993, A&A 275, 101
Evans, R., van den Bergh, S., McClure, R.D., 1989, ApJ 345, 752
Fitzsimmons, A., Brown, P.J.F., Dufton, P.L., Lennon, D.J., 1990, A&A 232, 437
Gardner, F.F., Whiteoak, J.B., 1979, MNRAS 188, 331
Gibner, P.S., Mewaldt, R.A., Schindler, S.M., Stone, E.C., Webber, W.R. 1992, ApJ 391, L89
Gies, D.R., Lambert, D.L., 1992, ApJ 387, 673
Guélin, M., Cernicharo, J., Linke, R.A., 1982, ApJ 263, L89
Güsten, R., Mezger, P.G., 1983, Vistas Astr. 26, 159
Güsten, R., Ungerechts, H., 1985, A&A 145, 241
Henkel, C., Baan, W.A., Mauersberger, R., 1991, A&AR 3, 47
Henkel, C., Mauersberger, R., 1993, A&A 274, 730
Henkel, C., Mauersberger, R., Wiklind, T., Hüttemeister, S., Lemme, C, Millar, T., 1993, A&A 268, L17
Henkel, C., Whiteoak, J.B., Mauersberger, R., 1994, A&A, in press
Henkel, C., Wilson, T.L., Bieging, J., 1982, A&A 109, 334
Hoppe, P., Amari, S., Zinner, E., Lewis, R.S., 1992, Meteoritics, 27, 235
Kahane, C., Cernicharo, J., Gómez-Gonzáles, J., Guélin, M., 1992, A&A 256, 235
Kahane, C., Gómez-Gonzalez, J., Cernicharo, J., Guélin, M., 1988, A&A 190, 167
Lacey, C.G., Fall, S.M., 1985, ApJ 290, 154
Landré, V., Prantzos, N., Aguer, P., Bogaert, G., Lefebvre, A., Thibaud, J.P., 1990, A&A 240, 85
Langer, N., 1991, A&A 248, 531
Langer, W.D., Graedel, T.E., Frerking, M.A., Armentrout, P.B., 1984, ApJ 277, 581
López, R.J., Rebolo, R., Herrero, A., Beckman, J.E., 1993, ApJ 412, 173
Luck, R.E., Lambert, D.L., 1985, ApJ 298, 782
Maeder, A., 1985, Production and Distribution of C, N, O Elements, ESO Conference and Workshop Proceedings No. 21, eds. I.J. Danziger, F. Matteucci, K. Kjär, Garching, p187

Mayor, M., Vigroux, L., 1981, A&A 98, 1
Matteucci, F., 1986a, PASP 98, 973
Matteucci, F., 1986b, MNRAS 221, 911
Matteucci, F., D'Antona, F., 1991, A&A 247, L37
McKenzie, D.L., Feldman, U., 1992, ApJ 389, 764
Meyer, J.-P., 1985, ApJS 57, 151
Mirabel, I., Morras, R., 1984, ApJ 279, 86
Nissen, P.E., Edvardsson, B., 1992, A&A 261, 255
Nomoto, K., Thielemann, F.K., Yokai, K.: 1984, ApJ 286, 644
Ott, U., 1993, Nat 364, 25
Pagel, B.E.J., 1985, Production and Distribution of C, N, O Elements, ESO Conference and Workshop Proceedings No. 21, eds. I.J. Danziger, F. Matteucci, K. Kjär, Garching,p155
Peimbert, M., 1987, IAU Symp. 115, Star Forming Regions, eds. M. Peimbert, J. Jugaku, Reidel, Dordrecht, p111
Pilyugin, L.S., 1992, A&A 260, 58
Pilyugin, L.S., 1993, A&A 277, 42
Rosa, M.R., Mathis, J.S., 1985, Production and Distribution of C, N, O Elements, ESO Conference and Workshop Proceedings No.21, eds. I.J. Danziger, F. Matteucci, K. Kjär, Garching, p207
Russell, S.C., Dopita, M.A., 1992, ApJ 384, 508
Sage, L.J., Mauersberger, R., Henkel, C., 1991, A&A 249, 31
Schaller, G., Schaerer, D., Meynet, G., Maeder, A., 1992, A&AS 96, 269
Seab, C.G.,1987, Interstellar Processes, eds. D.J. Hollenbach, H.A. Thronson, Reidel, Dordrecht, p491
Shaver, P.A., McGee, R.X., Newton, L.M., Danks, A.C., Pottasch, S.R. 1983, MNRAS 204, 53
Shields, G.A., 1990, ARA&A 28, 525
Sneden, C., 1985, Production and Distribution of C, N, O Elements,ESO Conf. and Workshop Proceedgs. No.21, eds. I.J.Danziger, F.Matteucci, K.Kjär, Garching, p1
Stahl, O., Wilson, T.L., Henkel, C., Appenzeller, I., 1989, A&A 221, 321
Stahl, O., Wilson, T.L., 1992, ApJ 254, 327
Tosi, M., 1982, ApJ 254, 699
van den Bergh, S., McClure, R.D., 1990, ApJ 359, 277
Vladilo, G., Centurión, M., Càssola, C., 1993, A&A, 273, 239
Wannier, P.G., Andersson, B.G., Olofsson, H., Ukita, N., Young, K., 1991, ApJ 380, 593
Weaver, T.A., Woosley, S.E., 1993, Phys. Rep. 227, 65
Wheeler, J.C., Sneden, C., Truran, J.W., 1989, ARA&A 27, 279
Whiteoak, J.B., Gardner, F.F., 1981, MNRAS, 197, 39p
Williams, D.A., 1993, Dust and Chemistry in Astronomy, eds. T.J. Millar, D.A. Williams, IOP Publishing LTD, Bristol, p143
Wilson, T.L., Bieging, J., Downes, D., Gardner, F.F., 1976, A&A 51, 303
Wilson, T.L., Matteucci, F., 1992, A&AR 4, 1
Wilson, T.L., Rood, R.T., 1994, ARA&A 32
Wink, J.E., Wilson, T.L., Bieging, J.H., 1983, A&A 127, 211
Woitke, P., Dominik, C., Sedlmayr, E., 1993, A&A 274, 451
Woosley, S.E., Langer, N., Weaver, T.A., 1993, ApJ 411, 823
York, D.G., Spitzer, L., Bohlin, R.C., Hill, J., Jenkins, E.B., Savage, B.D., Snow, T.P., 1983, ApJ 266, L55

Angular Structure and Motions of Interstellar Masers

James M. Moran

Harvard–Smithsonian Center for Astrophysics, 60 Garden Street, Cambridge, MA 02138

Abstract: A brief historical account is given of the discovery of OH emission and the experiments that led to the identification of the radiation mechanism as maser amplification. An outline is presented of the progress by the MIT, CfA and Jodrell Bank groups in mapping the prototypical maser, W3(OH). The measurements of the structure and proper motions of masers are described and their limitations explained. Studies of the structure of extragalactic water vapor masers are summarized.

1 Introduction

Masers have been studied intensively for almost 30 years and the literature is very extensive. In this review I concentrate on some of the work on the measurement of the angular distribution of masers, the proper motions of masers and the characteristics of extragalactic water vapor masers. Because this meeting is partly an historical celebration of molecular cloud radio astronomy, I will also take the opportunity to review the work that led to the discovery of masers and the interferometric observations that produced the first images of them. More comprehensive reviews of masers can be found in Clegg and Nedoluha (1993) and Elitzur (1992a,b). A discussion of current topics of importance in the physics of masers can be found in the following article in this volume by Watson.

2 The Discovery of OH Maser Emission

Shklovskii (1953) and Townes (1954, 1957) first drew attention to the possibility of detecting radio lines of molecules. The ground state lambda doublet transitions of OH and CH were prime candidates, but early searches for them were unsuccessful (e.g., Barrett and Lilley, 1957), primarily because the transition frequencies were not well known. Ehrenstein, Townes and Stevenson (1959) measured the frequencies of OH in the laboratory to a quoted accuracy of 30 kHz (they were off by 27 kHz); not an easy feat because of the difficulty of making OH in sufficient quantities in laboratory environments. Weinreb *et al.* (1963) made the first detection of OH in absorption against Casseopeia A with the Millstone Hill radar antenna and Weinreb's novel autocorrelation spectrometer, which he built for his Ph.D. thesis.

I was in Alan Barrett's course in radio astronomy as a first-year graduate student at MIT in the fall of 1963 when the discovery was made. The excitement surrounding the new findings made a great impression on me. The ratio of the absorption strength between the $F = 2 \rightarrow 2$ line at 1667 MHz and the $F = 1 \rightarrow 1$ line at 1665 MHz was approximately 9:5, as expected. Several groups began studying OH and unusual absorption ratios were found in other sources, as well as unexpected velocity distributions compared to those of HI. Groups at MIT, Harvard, CSIRO, and Bell Labs began searching for OH in emission. They usually concentrated on the 1667 MHz line in regions of low thermal emission and high HI abundance. Nothing turned up, and limits on the excitation temperature were pushed below 10 K, based on comparisons with nearby absorption profiles.

In the pursuit of absorption measurements of OH, the thermal sources from the Westerhout catalog were obvious candidates for background sources. Emission wasn't expected from them because OH should not exist in the hot ionized environment of HII regions. Ellen Gundermann, a student of Ed Lilley and Sam Goldstein was using the 60-ft. telescope at the Agassiz station of Harvard College Observatory for her Ph.D. research, which involved studying OH in absorption and searching for emission. In April 1964 she found emission towards W49, at a level of about 3 K at 1665 MHz and 2 K at 1667 MHz. She noticed that with the narrowest filters available to her, 25 kHz, corresponding to a velocity resolution of about ~ 5 km s^{-1}, the spectra had at least two components, which were unresolved in frequency. This was not the main topic of her thesis and she placed little emphasis on it when she presented her thesis in June 1965 (Gundermann 1965).

Meanwhile, Weaver and his colleagues at the University of California at Berkeley began observing the Westerhout objects in January 1965 with a new 100 channel spectrometer at the Hat Creek Observatory that had 2 and 10 kHz filters (see Dieter *et al.* 1966). Weaver recounts their discovery in his oral memoirs (Shields 1991). He said that at about 3 a.m. one morning the observer on duty at the telescope noted that W49 showed a familiar absorption but the spectrum also had a few positive spikes; he wanted to know what he should do about this apparent spectrometer malfunction. Weaver and his colleagues followed this up aggressively and found that the spectra of W49, W3, Orion A and NGC6334 had complex emission-line structure with antenna temperatures as high as ~ 30K (Weaver *et al.* 1965). To draw attention to their discovery they attributed the lines to an unknown emitter, "mysterium", (a name inspired by the phantom element "nebulium," identified earlier in the century as doubly ionized oxygen). They said in their paper that "the rest frequency of the mysterium line is higher than the 1665 MHz line of OH," so they were probably serious about the concept of an unknown molecule. There was general agreement almost immediately that the lines were due to OH (e.g., Weinreb *et al.* 1965). In Australia, McGee and his collaborators had noticed narrow features in the spectrum of Sgr A, which were probably the emission from the maser in Sgr B2 entering the sidelobes of the Parkes antenna. They belatedly reported this result in the December 1965 issue of *Nature* (McGee *et al.* 1965). So, of the three groups that stumbled onto OH masers at about the same time, only the Berkeley group followed it up promptly.

The unusual characteristics of the OH emission soon became apparent: complex spectra with little similarity among the four hyperfine transitions. The lines were narrow, corresponding to thermal temperatures as low as 5K, but the brightness temperatures were at least a few hundred Kelvins. Moreover, the lines were highly polarized with both linear (Weinreb *et al.* 1965) and circular (Davies *et al.* 1966 and Barrett and Rogers 1966) polarization (see Figure 1).

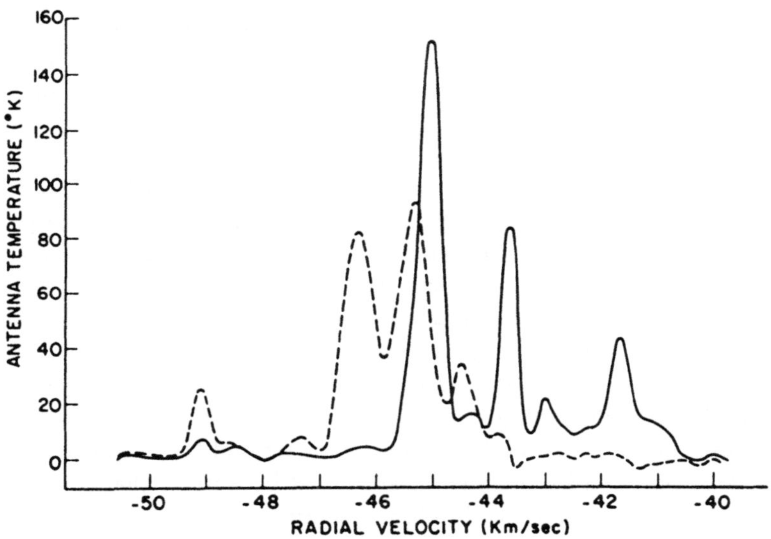

Fig. 1. The spectrum of the 1665 transition of OH towards W3(OH). The solid and dashed lines denote right and left circular polarization, respectively (Barrett and Rogers 1966).

Maser action was proposed by Perkins, Gold and Salpeter (1966) and a detailed model was developed by Litvak *et al.* (1966). Both models envisioned ultraviolet pumping, which has since been superceded by more viable schemes. The first interferometric experiments at MIT (Rogers *et al.* 1966) and Caltech (Cudaback *et al.* 1966) were reported in back-to-back articles in the *Physical Review Letters.* They showed that the brightness temperatures were at least 10^6K, which convincingly confirmed the maser hypothesis. These experiments and the follow-up work (Rogers *et al.* 1967; Raimond and Eliasson 1969) also pinned down the location of the masers to an accuracy of about 10″. Ellen Hardebeck (nee Gundermann) came to Cal Tech about 1970 after a postdoctoral position at Cornell and began using the OVRO interferometer to measure systematically the positions of OH masers. Her results are reported in a series of papers (e.g., Hardebeck 1972).

The first known maser in W3, now called W3(OH), was found to be 14′ from the center of the HII region, at a location where nothing was visible on the Palomar plates. Several years later a small radio source was found at this position and Mezger *et al.* (1967) postulated the connection between compact HII regions and masers. Shklovskii proposed that masers were associated with protostars or newly formed stars, a suggestion that found immediate support when the maser in Orion was found to be near the BN object (see Shklovskii 1969).

3 The Structure of the OH Maser in W3(OH)

At MIT we began to plan interferometry with longer baselines, and we conceived the idea of connecting the Millstone antenna and the Agassiz antenna which were separated by about 14 km (see Moran 1987). Alan Rogers rigged up a microwave link to bring the signal from Agassiz back to the correlator at the Haystack Observatory. The local oscillators ran more-or-less independently, which was good psychological preparation for later VLBI work. Our goal was to measure a limit on the brightness temperature; we never thought we could separate or resolve the components. The phase stability of the interferometer was very poor. However, it was immediately obvious that the *relative* phases of the various frequency components could be measured and that they were not zero. It slowly dawned on us that the maser spots were in fact separated in angle. The first maser map was published in the *Astrophysical Journal* in May 1967 (Moran *et al.* 1967a). The Jodrell Bank group was also working on W3(OH) at the same time with the Jodrell Bank–Malvern interferometer and they could tell that the features were separated by roughly an arcsecond because of the fringe beating. They also set a lower limit on the brightness temperature of about 10^{10}K (Davies *et al.* 1967).

The next step in resolution clearly required "very" long baselines. In the spring of 1967 the NRAO–Cornell group had tried unsuccessfully to get fringes with their new VLBI system at 600 MHz. Burke suggested that OH masers provided the ideal test source for VLBI because the lines were only about 5 kHz wide. The temporal correlation length was therefore 0.2 milliseconds and only coarse clock synchronization would be required. The NRAO–Cornell group obtained their first results on the NRL–NRAO baseline in May 1967. In June 1967 we observed with the NRAO–Cornell group on the Haystack–NRAO baseline at a resolution of 0.″04 at 1665 MHz. We had built a VLBI terminal for Haystack largely from the available equipment at the station. We processed the data at Haystack using tapes that had been flown back from Green Bank while the experiment was still in progress. The entire processing was done in a general-purpose computer, a CDC 3200. The program, which I wrote, had the model geometric delay entered with the wrong sign. However at the exact instant when the geometric delay was zero, fringes appeared. The first set of fringes are reproduced in Figure 2. We also had a wider band of 120 kHz from which we constructed a map of the maser, which is shown in Figure 3.

During the summer of 1968 we also attempted a series of impromptu experiments between NRAO and Jodrell Bank using a Ferrograph audio tape recording system developed at Jodrell Bank, but these did not produce any fringes. The first VLBI experiments with multiple baselines were conducted in early 1968. In an experiment involving Haystack, Hat Creek, NRAO and Onsala, we showed that an apparently isolated feature in W3(OH) at –43.7 $\mathrm{km\,s^{-1}}$ was resolved and had an angular size of about 0.0045 arcseconds (see Figure 4). W3(OH) is now known to have an unusually low amount of interstellar scattering. Most other masers in the galactic plane suffer substantial broadening from interstellar scattering and masers can be used as a tool to study its properties (Gwinn *et al.* 1988; van Langevelde *et al.* 1992).

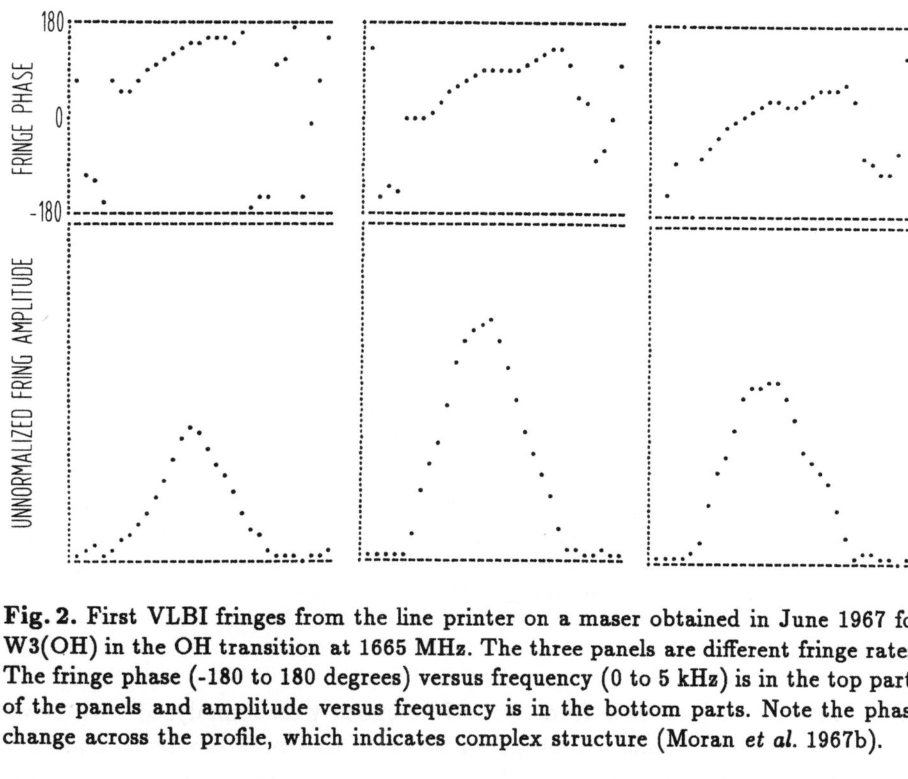

Fig. 2. First VLBI fringes from the line printer on a maser obtained in June 1967 for W3(OH) in the OH transition at 1665 MHz. The three panels are different fringe rates. The fringe phase (-180 to 180 degrees) versus frequency (0 to 5 kHz) is in the top parts of the panels and amplitude versus frequency is in the bottom parts. Note the phase change across the profile, which indicates complex structure (Moran *et al.* 1967b).

Since the earliest experiments the groups at CfA and MIT have competed with the one at Jodrell Bank to produce ever more detailed and accurate maps of W3(OH) at 1665 MHz. A list of these maps is given in Table 1. The first synthesis map of W3(OH) was produced by Reid *et al.* (1980). As sensitivity and processing techniques improved the number of maser spots detected has increased. It is important to note that although W3(OH) varies in time the overall character of the spectrum has remained the same since 1965. For example the relative position of the -43.7 and -45.1 $\mathrm{km\,s^{-1}}$ have remained nearly constant for 20 years, as shown in Table 2 and Figure 5. A recent study by Bloemhof

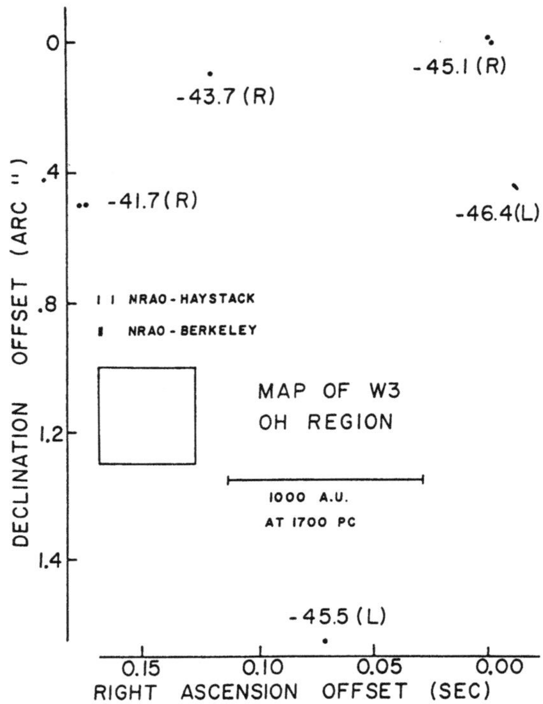

Fig. 3. First VLBI map of a maser, W3(OH). The box indicates the 2σ error region. The fringe spacings of the two interferometers used in the experiment are indicated. The spot positions are plotted with some indication of structure and are labelled by their velocities and sense of polarization (Moran*et al.* 1968).

et al. (1992), in which data from the 1978.8 epoch were reanalyzed with newer data from 1986.3, reveals the proper motions of the maser spots, as shown in Figure 6. These motions are on the level of about 0.5 mas yr^{-1} corresponding to a linear velocity of 5 $km\,s^{-1}$(see Eq. 1). The dominant motion is clearly expansion. It is also clear that the shapes of the maser spot remain the same between the two epochs. There can be little doubt that the motions are ballistic. These motions are also evident in the VLBI maps of CH_3OH (Menten *et al.* 1992). There are two difficulties in interpreting the proper motion results. First, the systemic velocity of the HII region cannot be determined easily with high precision. The recombination line velocities shift with frequency and position across the HII region (Berulis and Ershov 1983; Wilson *et al.* 1991). Since the masers are predominantly on the front side of the HII region, which is optically thick at 18 cm wavelength, the distance and systemic parameters are highly correlated.

This makes it difficult to derive a distance from the proper motions without independent knowledge of the systemic velocity (see section 4). Second, because the positions of the maser spots are all determined relative to a reference feature, one constant velocity vector can be added to the velocity field. Hence the flow could be radially outward or it could be a cometary flow with respect to the underlying star.

Proper motions are much more difficult to measure in OH masers than in H_2O masers (which are discussed in the following section) because of the lower frequency, larger spot size, and lower velocities of the OH masers. The only other sources besides W3(OH) with detected proper motions are Cepheus A (Migenes *et al.* 1992) and U Orionis (Chapman *et al.* 1991).

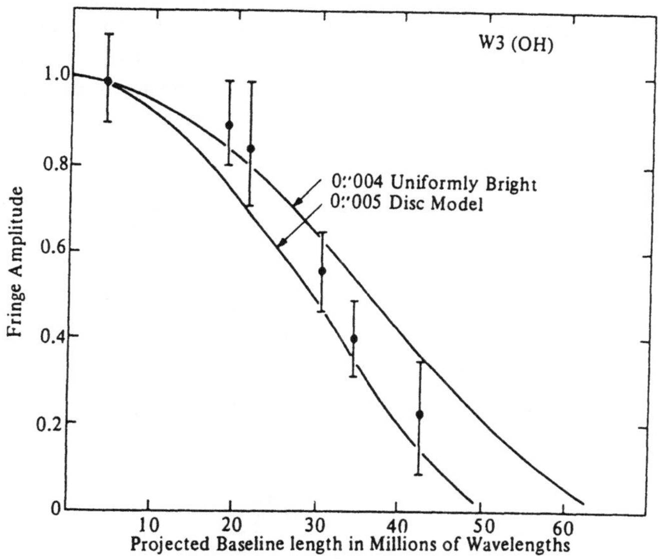

Fig. 4. Fringe amplitude of the -43.7 km s^{-1} feature in W3(OH) from a VLBI experiment formed with the telescopes at Haystack Observatory, Hat Creek Observatory, Onsala Space Observatory and NRAO (Moran 1968).

4 Proper Motions and Distances of H_2O Masers

In this section the techniques, limitations and results of the measurements of proper motions of H_2O masers are discussed, along with methods for estimating distances to the masers. H_2O maser emission arises in the outflows surrounding newly formed stars. The angular velocity of a clump of masing gas at a distance D from the earth, moving with transverse velocity V_T is

$$\theta = 206\left(\frac{V_T}{\text{km s}^{-1}}\right)\left(\frac{D}{\text{kpc}}\right)^{-1} \mu\text{as yr}^{-1} \ . \quad (1)$$

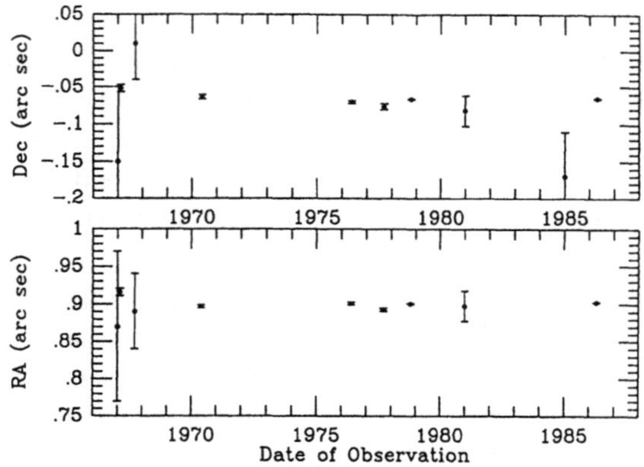

Fig. 5. The relative position [top, declination; bottom, right ascension x cos(declination)] of the -43.7 and -45.1 km s^{-1} as a function of time (see Table 2). The proper motion is barely perceptible on this plot (compare the 1976.4, 1978.8 and 1986.3 epochs). An earlier suggestion of proper motion in declination (Norris and Booth 1981) based on the 1967 – 1978 epochs is probably not correct because of insufficient angular resolution to resolve the structure in the -45.1 km s^{-1} feature.

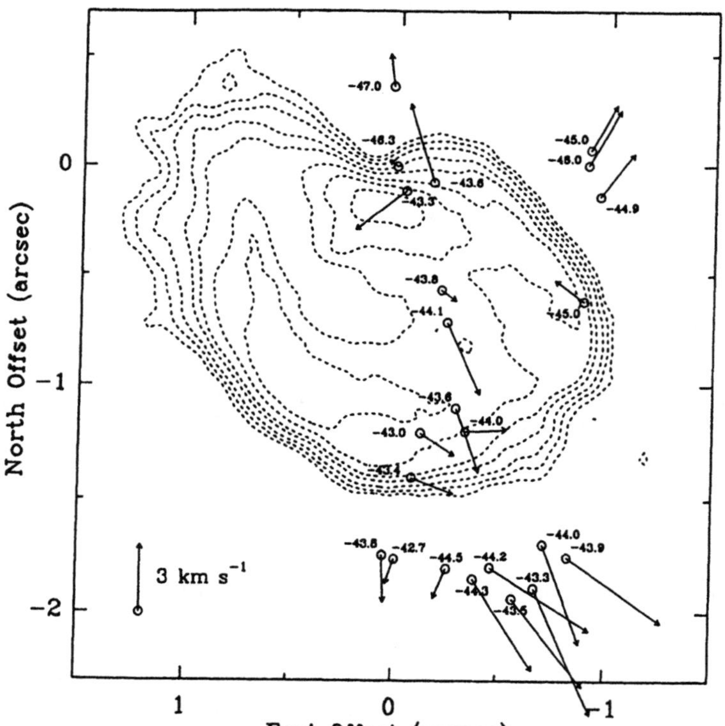

Fig. 6. Map of the OH maser in W3(OH) at 1665 MHz with the spots labelled by radial velocity and transverse velocities (for a distance of 2.2 kpc) indicated by arrows. The dotted contours denote thermal continuum emission (Bloemhof*et al.* 1992).

Table 1. Maps of OH Maser distribution in W3(OH) at 1665 MHz

instrument	stations	pol[1]	res[2] ($''$)	ΔS[3] (Jy)	map[4]	spots	ref[5]
Agassiz-Millstone	2	L	2.8	150	P	3	1
VLBI	3	RC/LC	0.01	60	F	7	2
Jodrell-Defford	2	RC/LC	0.3	25	P	16	3
Jodrell(MkII-III)	2	RC/LC	1.6	13	P	22	4
VLBI	8	L	0.01	5	S	70	5
Jodrell (MKIA-III)	2	RC/LC	1.6	5	P	20	6
Merlin	6	RC/LC	0.3	0.6	S	88	7
VLA[6]	27	RC/LC	1.0	0.6	S	28	8
VLBI	5	RC/LC	0.01	2	S	81	9
VLBI[7]	7	RC/LC	0.005	3	S	45	10

[1]polarization: L, one sense of linear polarization, RC, right circular polarization, LC, left circular polarization;
[2]minimum fringe spacing;
[3]weakest feature mapped;
[4]mapping technique: P = phase fitting, F = fringe rate fitting S = synthesis imaging;
[5]References: 1. Moran *et al.*(1967a); 2. Moran *et al.*(1968); 3. Cooper *et al.*(1971); 4. Harvey *et al.*(1974); 5. Reid *et al.*(1980); 6. Norris and Booth (1981); 7. Norris *et al.*(1982); 8. Gaume and Mutel (1987); 9. Garcia-Barreto *et al.*(1988); 10. Bloemhof *et al.*(1992)
[6]poor spectral resolution;
[7]only those features with measurable proper motions reported.

Hence a velocity of 40 $\mathrm{km\,s^{-1}}$ results in an angular velocity of 1 mas yr^{-1} and 10 μas yr^{-1} at distance of 8 kpc (~Galactic Center) and 800 kpc (nearby galaxy), respectively. The synthesized beam of an intercontinental VLBI array has a width θ_r that is limited to about 300 μas for the 22 GHz transition of H_2O. However, in the absence of systematic errors, the relative positions of masers can be estimated to an accuracy of about $\delta\theta \sim 0.5\theta_r/\mathrm{SNR}$, where SNR is the signal-to-noise ratio and θ_r is the angular resolution. Only masers with reference features strong enough to be detectable within the interferometer's coherence time can be readily studied. With current instrumentation the coherence

Table 2. Relative positions of -43.7 and -45.1 km s^{-1} OH maser components in W3(OH) at 1665 MHz as a function of time [(1)]

Date	Velocities Km s^{-1}	Resolution km s^{-1}	Δx ($''$)	Δy ($''$)	Reference[(2)]
1967.0	-43.7/-45.1	0.2	0.87 ± 0.1	-0.15 ± 0.1	1
1967.1	-43.7/-45.1	0.5	0.916 ± 0.005	-0.052 ± 0.005	3
1967.7	-43.7/-45.1	0.2	0.89 ± 0.05	0.01 ± 0.05	2
1970.4	-43.7/-45.1	0.1	0.897 ± 0.002	-0.064 ± 0.003	4
1976.4	-43.9/-45.1	0.15	0.901 ± 0.002	-0.071 ± 0.002	5
1977.7	-43.8/-45.0	0.05	0.893 ± 0.002	-0.077 ± 0.004	6
1978.8	-43.78/-44.9	0.15	0.901 ± 0.0005	-0.067 ± 0.0005	9
1981.0	-43.8/-45.0	0.1	0.898 ± 0.02	-0.082 ± 0.02	7
1985.0	-43.9/-45.0	1.1	0.60 ± 0.06	-0.17 ± 0.06	8
1986.3	-43.8/-44.9	0.15	0.903 ± 0.0001	-0.066 ± 0.0001	10

[(1)]Proper motion can be seen by comparing the 1978.8 and 1986.3 measurements, which were processed in identical fashion;
[(2)]same as in Table 1.

time is about 2 minutes and limiting sensitivity is about 1 Jy. There are about 300 known interstellar H_2O masers with flux densities exceeding this level. After a reference feature is detected, the SNR improves as the square root of the integration time, and features as weak as $\sim$ 20 mJy can be detected.

A prime source of systematic error is due to structural changes in the maser as a function of time. That is, two features with velocities separated by less than the linewidth and positions separated by less than the interferometric beamwidth cannot be readily distinguished, and any change in their relative amplitudes will give rise to an apparent shift in position. In addition, although phase referencing removes geometric errors to first orders, second order systematic errors remain, given by

$$\delta\theta \sim \frac{\Delta B}{B}\Delta\theta + \frac{\Delta\nu}{\nu}\frac{c\Delta\tau}{B} \ , \tag{2}$$

where $\Delta\theta$ and $\Delta\nu$ are the offsets from the reference feature in angle and frequency, respectively, B is the baseline length, ν is the frequency, and ΔB and $\Delta\tau$ are the errors in the baseline length and instrumental delay. For ΔB and $c\Delta\tau \sim 10$ cm, $\Delta\theta = 2''$ and $\Delta\nu = 2$ MHz, $\delta\theta$ is below 1 μas. Hence geometric errors are significant primarily for the study of extragalactic masers and for space VLBI (e.g., Reid 1984).

If the motions of the maser cloudlets are random and isotropic, then the distance to the source can be estimated by the method of statistical parallax. If σ_μ is the dispersion in transverse angular velocities in one coordinate and σ_z is the dispersion in line-of-sight velocity, then the distance estimate is given by $D = \sigma_z/\sigma_\mu$. Note that the observed dispersion in transverse angular velocity, $\sigma_{\mu_{obs}}$, will be larger than σ_μ because of the effects of error in the position measurements, such that $\sigma_\mu^2 = \sigma_{\mu_{obs}}^2 - \sigma_m^2$, where σ_m is the measurement error in angular velocity. Hence, an underestimate of σ_m leads to an underestimate in the distance. This dispersion correction is important primarily for extragalactic work. The statistical error in the distance estimate is given by the relation

$$\frac{\sigma_D}{D} \sim \frac{1}{\sqrt{2N}}\left[\frac{1}{1-(\sigma_m/\sigma_{\mu_{obs}})^2}\right]^{1/2} \tag{3}$$

where N is the number of maser spots measured in each coordinate. Note that if the measurement errors are small, then $\sigma_D/D \sim 1/\sqrt{2N}$. This technique was first applied by Genzel *et al.* (1981) to W51M. Greenhill *et al.* (1993) give the generalized expression for the distance error when the measurement errors vary among the spots. If the flow is not isotropic and random, then the statistical parallax method can be biased. For example, if the maser cloudlets were flowing from a central point with constant flow velocity, V_e, but were confined to the hemisphere facing us, then the spread in angular velocity would be $2V_e/D$ and the spread in radial velocity would be V_e. Hence, the distance estimate would be one-half the true distance. Comparison of the ambient velocity of the molecular cloud and the mean velocity of the maser features is an important diagnostic test for assessing the isotropy and randomness of the maser emission.

When the masers show organized motion such as outflow, their velocities can be fit to a parameterized model where a distance parameter relates angular and linear velocities. This analysis was first applied to the masers in Orion by Genzel *et al.* (1981a). Such an analysis is not entirely straightforward, because it is nonlinear and the line-of-sight positions for the features are unknown parameters. Hence an observation of 25 masers features yields 75 velocity-component measurements (and 50 coordinate measurements). A model of simple outflow at constant velocity has 7 global parameters and 25 line-of-sight position parameters. The least-mean-square analysis involves both radial and transverse velocities, which are experimentally quite distinct quantities, and the solution depends on the relative weights assigned to them. The quantity to be minimized can be written as (e.g., Gwinn *et al.* 1992)

$$\chi^2 = \frac{1}{3N_m - N_p}\sum\left[\frac{(\dot{\theta}_i - \frac{V_{m_i}}{D})^2}{\sigma_i^2} + \frac{(V_{z_i} - V_{zm_i})^2}{\sigma_{z_i}^2}\right] , \tag{4}$$

where $\dot{\theta}_i$ and σ_i are the measured angular velocities and their errors; V_{z_i} and σ_{z_i} are the measured line-of-sight velocities and their errors; V_{m_i} and V_{zm_i} are the model velocities in the transverse and radial directions, respectively; N_m is the number of maser spots; and N_p is the number of model parameters. The weights are given by $\sigma_i^2 = \sigma_V^2/D^2 + \sigma_{n_i}^2$ and $\sigma_{z_i}^2 = \sigma_V^2$, where σ_V is the rms turbulent velocity of the flow and σ_{n_i} is the measurement error. Normally, σ_V is greater than measurement errors in radial velocities. Note the unusual situation wherein the weighting depends on the distance parameter. If the values for σ_{n_i} can be calculated from the residuals in the position-versus-time analyses for individual features, then σ_V can be found by requiring that $\chi^2 = 1$. An example of the proper motions in W49N is shown in Figure 7. It was also possible to solve for the velocity as a function of radius in this source, as shown in Figure 8. A similar case of apparent acceleration was also found in the maser in Orion (Genzel *et al.* 1981). MacLow *et al.* (1992) have investigated possible causes of this acceleration.

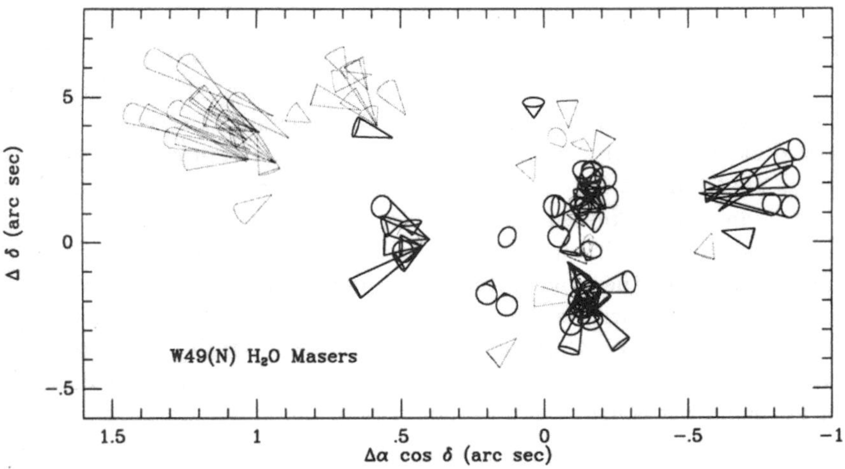

Fig. 7. Proper motions of the water vapor masers in W49N (Gwinn *et al.* 1992).

The proper motions of eight Galactic masers have been analyzed (see Table 3). Note that the flow velocities fall in the narrow range of 18–45 $\mathrm{km\,s^{-1}}$. The flow in W3(OH) is highly collimated and its distance cannot be estimated reliably. The distance estimates for the other masers can be used to derive a value for R_o based on a χ^2 minimization of the form appropriate when there are errors in the independent and dependent variables (e.g., Press *et al.* 1992),

$$\chi^2 = \frac{1}{(N-1)} \sum_{i=1}^{N} \frac{(D_{PM_i} - bD_{K_i})^2}{\sigma_{PM_i}^2 + b^2\sigma_{K_i}^2} , \tag{5}$$

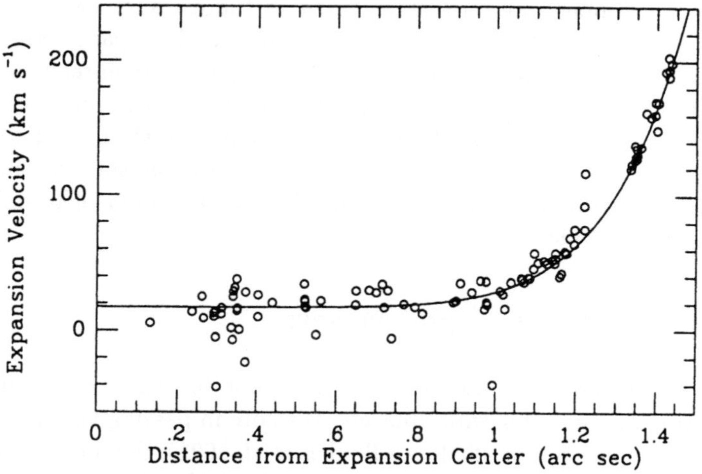

Fig. 8. Expansion velocity as a function of radius for W49N (Gwinn *et al.* (1992)).

Table 3. Distances of Galactic H_2O Masers Derived from Proper Motions.

Source	Fit[(1)]	N[(2)]	$V_e^{(3)}$ km s^{-1}	$V^{(4)}$ km s^{-1}	$D_{PM}^{(5)}$ kpc	$D_K^{(6)}$ kpc	$\chi^{(7)}$	Ref.[(8)]
Orion	M	26	18	8	0.5±0.1	0.7±1.0	−0.5	1
W51M	SP	27	-	63	7.0±1.5	5.0±2.0	0.8	2
W51N	SP	10	-	63	8.3±2.5	5.0±2.0	1.1	3
Sgr B2N	M	24	45	50	7.1±1.5	7.7±0.2	−0.4	4
Sgr B2M	M	27	35	26	6.5±1.5[(9)]	7.7±0.2	−0.9	5
W49N	M	105	18	7	11.4±1.2	10.8±0.7	0.6	6
W3(OH)	M	42	18	−48	–	–	–	7
Cep A	SP	11	–	−11	0.5± 0.2[(9)]	1.1±0.9	−1.4	8

[(1)]SP = statistical parallax; M = model fit.
[(2)]Number of maser features.
[(3)]Outflow velocity in km s^{-1}. In Orion and W49N there are additional higher velocity flows.
[(4)]Systemic velocity.
[(5)]Proper motion distance.
[(6)]Distance from kinematic model: v = 220 km s^{-1}, R_0 = 7.7 kpc and random velocity = 4 km s^{-1}.
[(7)]Weighted residual.
[(8)]References: 1. Genzel *et al.*(1981a); 2. Genzel *et al.*(1981b); 3. Schneps *et al.*(1981); 4. Reid *et al.*(1988b); 5. Reid *et al.*(1988a); 6. Gwinn *et al.*(1992); 7. Alcolea *et al.*(1993); 8. Cawthorne *et al.*(1993).
[(9)] Preliminary estimate.

where D_{PM} is the proper motion distance, D_K is the distance based on the kinematic model of the Galaxy, and b is the desired scaling parameter. The result of the minimization based on the data in Table 3, which requires a random velocity error of 4 km s^{-1} for the estimation of the values of σ_K in order to make $\chi^2 = 1$, is $R_0 = 7.7 \pm 0.6$ kpc. Note that this result depends almost entirely on the data for SgrB2 and W49N and only weakly on the other sources and the kinematic model. A summary of measurement of R_0 by a variety of techniques can be found in Reid (1993).

5 Extragalactic Water Vapor Masers

In this section we review the work that has been done to measure the angular structure of extragalactic water vapor masers. We note in passing that about 15 OH megamasers have been identified (see Henkel *et al.* 1991), the most luminous ones being IRAS 20100-4156 (Staveley-Smith *et al.* 1989), which has isotropic luminosity of about 10^4 $L_\odot$ and a distance of about 500 Mpc ($z = 0.13$), and IRAS 14070+0525 (Baan *et al.* 1992), which has a luminosity of 10^4 $L_\odot$ and a distance of 1 Gpc ($z = 0.26$). Baan *et al.* suggest that the luminosity of the OH megamaser is proportional to the infrared luminosity of the host galaxy. For a general discussion of extragalactic masers and their environments see Henkel *et al.* (1991).

About 15 extragalactic water vapor masers have been found (see Greenhill *et al.* (1990) for a list). These masers seem to be of two types: (1) those in spiral arms of nearby galaxies and (2) those in the centers of galaxies with active nuclei. The latter type have very high luminosity and are often called megamasers. The properties of extragalactic water vapor masers whose angular structure has been measured are listed in Table 4. We summarize the results on the sources.

5.1 M33

There are several sites of water vapor masers in M33; only the one in IC133 is strong enough to have been studied with VLBI. The proper motions of 6 features were identified from observations at two epochs and the distance was determined to be between 300 and 900 Mpc (Greenhill *et al.*1993). Data have been acquired by the CfA group for five more epochs and a more accurate distance should be available soon.

5.2 IC10

The stronger of the two masers in IC10, known as IC10 SE, was observed with VLBI on 3 occasions (Argon *et al.* 1994). The maser lies close to a larger complex of HII regions near the galactic nucleus and coincident with a radio source that could be either a thermal or nonthermal source. This maser is probably the most highly variable extragalactic maser known. During one of the VLBI experiments,

Table 4. Extragalactic H_2O Masers with Measured Angular Structure

galaxy(1)	type(2)	BG(3)	v(4) km s^{-1}	D(5) Mpc	S(6) Jy	Δ v(7) km s^{-1}	Δr(8) pc	N(9)	ref (10)
M33/IC133	S	T		0.7	2	-260/-200	0.13	32	1
IC10/SE	IBm	T	-350	1.2	20	-350/-320	0.1	6	2
NGC4258	SAB	S	460	6.7	4	400/550 (11)	0.01	37	3,4
NGC3079	Sc	S	1200	16	7	940/1040	0.8	15	5,6

(1) Galaxy name with maser designation;
(2) morphological type of galaxy;
(3) continuum source associated with maser: T = thermal HII region, S = nuclear synchrotron source;
(4) systemic velocity of galaxy;
(5) distance to galaxy;
(6) peak flux density;
(7) velocity range of maser;
(8) linear spread of maser spots;
(9) number of spots identified;
(10) references: 1. Greenhill *et al.* (1993), 2. Argon *et al.* (1994), 3. Claussen *et al.* (1988), 4. Greenhill *et al.* (1994a), 5. Haschick *et al.* (1990), 6. Greenhill *et al.* (1994b);
(11) does not include high velocity features at -500 and 1200 km s^{-1}(Nakai *et al.* 1993).

a feature at -338.3 km s^{-1} was observed to decrease in flux density by 60% over a period of 14 hours and then rebound slightly. (see also Becker *et al.* 1993)

5.3 NGC3079

The maser in the nucleus of NGC3079 is the archtypical example of a water vapor megamaser. The image of the maser shows about 15 maser spots spread over region of diameter 0.8 pc with the principal concentration clustered within 0.08 pc, similar to the strongest Galactic masers such as W49N (Greenhill *et al.* 1994b). However, the apparent isotropic luminosity of NGC3079 is ~500 $L_\odot$, compared to 1 $L_\odot$ for W49N. The generation of so much power in a nuclear megamaser would seem to be a problem. One solution is that the masers might amplify the nuclear continuum source at 1.35 cm and may consequently be highly beamed (Haschick *et al.* 1990). To understand this beaming consider a shell of masing regions surrounding the nucleus at a radius D_m. If the nuclear continuum source has diameter d_n then the maser will be beamed into solid angle $\Omega_b \sim (d_n/D_m)^2$, and the observed size of the maser will be $\Omega_s \sim (d_n/D)^2$ where D is the distance to the galaxy. The maser is unsaturated if $T_b < \frac{h\nu}{2k}\frac{\Gamma}{A}\frac{4\pi}{\Omega_b}$, where Γ is the maser decay rate and A is the transition coefficient. The observed flux density is $S = 2kT_b\Omega_s/\lambda^2$. Using these relations one can show that the maser will be unsaturated as long as

$$D_m > \left[\frac{\lambda^3 ASD^2}{4\pi hc\Gamma}\right]^{1/2}, \tag{6}$$

which does not depend on the size of the continuum source, as long as the maser cloud covers it. For NGC3079, with D=16 Mpc, $\lambda = 1.35$ cm, A=1.9×10^{-9}s^{-1}, $\Gamma = 1$ s^{-1} and S = 7 Jy, the condition is $D_m > 1$ pc. Suppose D_m= 50 pc and d_n=0.1 pc (such that the continuum source has a brightness temperature close to the Compton limit of 10^{12}K). The beam angle and luminosity of the maser we observe would be 1.5×10^{-9} steradians and 10^{-7} $L_\odot$, corresponding to a photon rate of only 3×10^{42} s^{-1}, which is modest by galactic standards. In order to have a reasonable chance of seeing such a maser, there would have to be about 10^9 of them in the shell, which means the total luminosity would be ~ 100 $L_\odot$, close to the apparent isotropic luminosity. The 10^9 masers could be associated with individual stars or a shell of shocked gas. In either case the pump power can be generated over a much larger volume (i.e., 50 pc in radius) than the observed maser extent. The concept of unsaturated maser amplification has been unattractive because it is inefficient compared to a saturated amplifier which can produce one maser photon per pump photon. However, the small beaming angle makes this model tractable. (see also Deguchi and Watson, 1989)

5.4 NGC4258

The VLBI maps of the megamaser in NGC4258 show about 37 spots spread over an area of 0.01 pc (Greenhill *et al.* 1994a). The masers exhibit a smooth shift in velocity with position of 80 km s^{-1} over 0.01 pc, which would imply a virial mass of at least 10^3 $M_\odot$. The Nobeyama group, with their new 16000 channel AOS, discovered that the maser has two sets of satellite lines removed from systemic velocity by $\pm$ 1000 km s^{-1} (Nakai *et al.* 1993; see also Moran 1993). Preliminary VLBI measurements show that these lines are coincident with the systemic lines to within an angular separation corresponding to 2 pc. If the masers were embedded in a rotating disk, the enclosed mass within this radius required to produce Keplerian motion would be about $10^8 M_\odot$, which might be associated with a black hole. The intriguing aspects of the spectrum, shown in Figure 9, are the symmetry in spacing of the high velocity lines and the absence of intermediate velocity lines. Optical observations of Hα emission reveal intertwined filamentary features with velocities on the order of 600 km s^{-1} on size scales of 1000 pc. (Cecil *et al.* 1992). There are several nonkinematic effects that might explain the high velocity features. The following effects seem unlikely: (1) hyperfine components of H_2O transition – their spread is only 6 km s^{-1}; (2) other molecular transitions – there are HDO and $HO^{37}Cl$ lines at frequencies corresponding to -970 and +912 km s^{-1} respectively, but they are expected to be

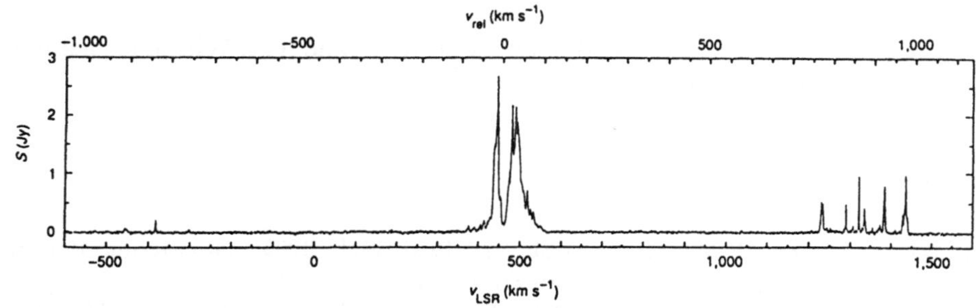

Fig. 9. Spectrum of H_2O megamaser in NGC4258. Note the high velocity lines that are symmetrically offset from the systemic velocity of 460 km s^{-1}(Nakai *et al.* 1993).

very weak; (3) Zeeman effect – since H_2O is a non-paramagnetic molecule a field strength of 10^5G would be required to produce the shift; (4) resonant Stark effect – a brightness temperature of 10^{23} K would be required; (5) stimulated Raman scattering from intermediate molecules – possible with NH_3, but this would require 10th ordering scattering. An intriguing mechanism suggestioned by Nakai *et al.* (1993) is stimulated Raman scattering in an intervening plasma cloud. The scattering frequency ($\pm$ offset from the maser frequency) in the absence of magnetic field is

$$\nu_s = \nu_p \left[1 + 3 \left(\frac{4\pi D_e}{\lambda} \right)^2 \right]^{1/2} , \tag{7}$$

where ν_p is the plasma frequency and D_e is the Debye length. A related phenemenon has been seen in ionospheric back-scattering experiments (Perkins *et al.* 1965). The required electron density would be about 10^7 cm^{-3}. Unusual conditions would probably be required to generate sufficient turbulent energy at the plasma frequency to give the observed line strength (Deguchi (1994)).

6 Conclusions

Although masers are generally very strong, high sensitivity and large interferometric arrays and excellent calibration are essential for obtaining high quality images and accurate determinations of proper motions. The measurement of proper motions of water vapor masers with the VLBI Network has been an extremely laborious, frustrating and protracted task. This situation is changing with the advent of the VLBA and introduction of new generations of faster computers. It should be possible to measure the proper motions of many masers of various molecular species throughout the Galaxy with relative ease. However, it will remain a challenging task for ground-based arrays to make sufficiently accurate measurements to determine the dynamics and distances to extragalactic masers. Understanding the role of masers in the nuclei of active galaxies will be a rewarding endeavor in the coming years.

I thank R. Booth, G. Garay, W.M. Goss, M.J. Reid, and T.L. Wilson for helpful comments on the manuscript.

References

Alcolea, J., Menten, K.M., Moran, J.M., and Reid, M.J., 1993, *Astrophysical Masers*, ed. A. Clegg and G. Nedoluha, Springer-Verlag, Berlin p. 225.

Argon, A.L., Greenhill, L.J., Moran, J.M., Reid, M.J., Menten, K.M., Henkel, C., and Inoue, M., 1994, *Ap.J.*, in press.

Baan, W.A., Rhodes, J., Fisher, K., Altschuler, D.R., 1992, *Ap.J.(Lett.)*, **396**, L99.

Barrett, A.H. and Lilley, A.E., 1957, *A.J.*, **62**, 5.

Barrett, A.H. and Rogers, A.E.E., 1966, *Nature*, **210**, 188.

Barrett, A.H., 1967, *Science*, **157**, 881.

Becker, R., Henkel, C., Wilson, T.L. and Wouterloot, J.G.A., 1993, *Astron. & Astrophys.*, **268**, 483.

Berulis, I.I. and Ershov, A.A., 1983, *Sov. Astron. Lett.*, **9**, 341.

Bloemhof, E.E., Reid, M.J., Moran, J.M., 1992, *Ap.J.*, **397**, 500.

Cawthorne, T.V., Moran, J.M., Reid, M.J., 1993, *Astrophysical Masers*, ed. A. Clegg and G. Nedoluha, Springer-Verlag, Berlin p. 237.

Cecil, G., Wilson, A.S., and Tully, R.B., 1992, *Ap.J.*, **390**, 365.

Chapman, J.M., Cohen, R.J., and Saikia, D.J., 1991, *M.N.R.A.S.*, **249**, 227.

Claussen, M.J., Reid, M.J., Schneps, M.H., Moran, J.M., Gusten, R., 1988 in "The Impact of VLBI on Astrophysics and Geophysics" proceedings of IAU symposium 129 ed. M.J. Reid and J.M. Moran, Kluwer, Dordrecht.

Clegg, A.W., and Nedoluha, G.E., 1993, *Astrophysical Masers*, Lecture Notes in Physics, Vol. 412, Springer-Verlag, Berlin.

Cooper, A.J., Davies, R.D., Booth, R.S., 1971, *M.N.R.A.S.*, **152**, 383.

Cudaback, D.D., Read, R.B., and Rougoor, G.W., 1966, *Phys.Rev.Lett.*, **17**, 452.

Davies, R.D., de Jager, G., Verschur, G.L., 1966, *Nature*, **209**, 974.

Davies, R.D., Rowson, B., Booth, R.S., Cooper, A.J., 1967, *Nature*, **214**, 1109.

Deguchi, S., 1994, *Ap.J.*, **420**, 551.

Deguchi, S., and Watson, W.D., 1989, *Ap.J.*, **340**, L17.

Dieter, N.H., Weaver, H., Williams, D.R.W., 1966, *Sky and Telescope*, **31**, 132.

Ehrenstein, G., Townes, C.H., and Stevenson, M.J., 1959, *Phys.Rev.Lett.*, **3**, 40.

Elitzur, M., *Astronomical Masers*, 1992a, Kluwer, Dordrecht.

Elitzur, M., 1992b, *Ann. Rev. Astronomy and Astrophysics*, **30**, 75.

Garcia-Barreto, J.A., Burke, B.F., Reid, M.J., Moran, J.M., Haschick, A.D., and Schlizzi, 1988, *Ap.J.*, **326**, 954.

Gaume, R.A. and Mutel, R.L., 1987, *Ap.J. (Suppl.)*, **65**, 193.

Genzel, R., Reid, M.J., Moran, J.M., and Downes, D.,1981a, *Ap.J.*, **247**, 884.

Genzel, R., Downes, D., Schneps, M.H., Reid, M.J., Moran, J.M., Kogan, L.R., Kostenko, V.I., Matveyenko, L.I., and Ronnang, B., "Proper Motions and Distances of H_2O Maser Sources. II: W51 Main," 1981b, *Ap.J.*, **247**, 1039-1051.

Greenhill, L.J., Jiang, R.D., Moran, J.M., Reid, M.J., Claussen, M.J., Lo, K.Y., 1994a in preparation.

Greenhill, L.J., Moran, J.M., Reid, M.J., Gwinn, C.R., Menten, K.M., Eckart, A., and Hirabayashi, H., 1990, *Ap.J.*, **364**, 513.

Greenhill, L.J., Moran, J.M., Reid, M.J., Haschick, A.D., Hirabayashi, H., and Baan, W.A., 1994b, *Ap.J.*, in preparation.

Greenhill, L.J., Moran, J.M., Reid, M.J., Menten, K.M., Hirabayashi, H., 1993, *Ap.J.*, **406**, 482.

Gundermann, E.J., 1965 Ph.D. thesis Harvard University, p. 50.

Gwinn, C.R., Moran, J.M., and Reid, M.J., 1992, *Ap.J.*, **393**, 149.

Gwinn, C.R., Moran, J.M., Reid, M.J., and Schneps, M.H., 1988, *Ap.J.*, **330**, 817.

Hardebeck, E., 1972, *Ap.J.*, **172**, 583.

Harvey, P.J., Booth, R.S., Davies, R.D., Whittet, D.C.B., McLaughlin, W., 1974, *M.N.R.A.S.*, **169**, 545.

Hashick, A.D., Baan, W.A., Schneps, M.H., Reid, M.J., Moran, J.M., and Gusten, R., 1990 *Ap.J.*, **356**, 149.

Henkel, C., Baan, W.A., and Mauersberger, R., 1991, *Astron. & Astrophys. Rev*, **3**, 47.

Litvak, M.M., McWhorter, A.L., Meeks, M.L., and Zeiger, H.J., 1966, *Phys.Rev.Lett.*, **17**, 821.

MacLow, M., and Elitzur, M., 1992, *Ap.J.(Lett.)*, **393**, 33.

McGee, R.X., Robinson, B.J., Gardner, F.F., Bolton, J.G., 1965, *Nature*, **208**, 1193.

Menten, K.M., Reid, M.J., Pratap, P., Moran, J.M., and Wilson, T.L., 1992, *Ap.J.(Lett.)*, **401**, 39.

Mezger, P.G., Altenhoff, W., Schraml, J., Burke, B.F., Reifenstein, E.C., and Wilson, T.L., 1967, *Ap.J.(Lett.)*, **150**, L157.

Migenes, V., Cohen, R.J., Brebner, G.C., 1992, *M.N.R.A.S.*, **254**, 501.

Moran, J.M., 1968, PhD thesis, M.I.T.

Moran, J.M., 1987, in Interstellar Matter, Proceedings of the Second Haystack Observatory Meeting, ed. J.M. Moran and P.T.P. Ho, Gordon and Breech, New York, p. 387.

Moran, J.M., 1993, *Nature*, **361**, 17.

Moran, J.M., Barrett, A.H., Rogers, A.E.E., Burke, B.F., Zuckerman, B., Penfield, H., Meeks, M.L., 1967a, *Ap.J.(Lett.)*, **148**, L69.

Moran, J.M., Burke, B.F., Barrett, A.H., Rogers, A.E.E., Ball, J.A., Carter, J.C., Cudaback, D.D., 1968, *Ap.J.*, **152**, L97.

Moran, J.M., Crowther, P.P., Burke, B.B., Barrett, A.H., Rogers, A.E.E., Ball, J.A., Carter, J.C., Bare, C.C., 1967b, *Science*, **157**, 676.

Nakai, N., Inone, M., and Miyoshi, M., 1993, *Nature*, **361**, 45.

Norris, R.P., Booth, R.S., and Diamond, P.J., 1982, *M.N.R.A.S.*, **201**, 209.

Norris, R.P., and Booth, R.S., 1981, *M.N.R.A.S.*, **195**, 213.

Perkins, F., Gold, T., and Salpeter, E.E., 1966, *Ap.J.*, **145**, 361.

Perkins, F., Salpeter, E.E., and Yngvesson, K.O., 1965, *Phys.Rev.Lett.*, **14**, 579.

Press, W.H., Teukolsky, S.A., Vetterling, W.T., and Flannery, B.P., *Numerical Recipes*, second edition, Cambridge University P., Cambridge, p. 660.

Raimond, E., and Eliasson, B., 1969, *Ap.J.*, **157**, 881.

Reid, M.J., 1984, in QUASAT-A VLBI Observatory in Space Proceeding of a Workshop at Gross Enzersdorf, Austria, European Space Agency, p. 181.

Reid, M.J., Gwinn, C.R., Moran, J.M., and Matthews, A.H., 1988, *Bull. AAS*, **20**, 1017

Reid, M.J., Haschick, A.D., Burke, B.F., Moran, J.M., Johnston, K.J., Swenson, G.W., 1980, *Ap.J.*, **239**, 89.

Reid, M.J., Schneps, M.H., Moran, J.M., Gwinn, C.R., Genzel, R., Downes, D., and Ronnang, B., 1988, *Ap.J.*, **330**, 809.

Reid, M.J., 1993, *Ann. Rev. Astronomy Astrophysics*, **31**, 345.

Rogers, A.E.E., Moran, J.M., Crowther, P.P., Burke, B.F., Meeks, M.L., Ball, J.A., and Hyde, G.M., 1966, *Phys.Rev.Lett.*, **17**, 450.

Schneps, M., Lane, A.P., Downes, D., Moran, J.M., Genzel, R., and Reid, M.J., 1981, *Ap.J.*, **249**, 124.

Sheilds, J.C., 1991, "Harold F. Weaver: California Astronomer, An Oral History" p. 67.

Shklovskii, I.S., 1953, *Dok. Akad. Nauk. USSR*, **92**, 25.

Shklovskii, I.S., 1969, *Soviet AJ*, **13**, 1.

Staveley-Smith, L., Allen, D.A., Chapman, J.M., Norris, R.P., and Whiteoak, J.B., 1989, *Nature*, **337**, 625.

Townes, C.H., 1954, *J.G.R.* **59**, 191. (Note: this is a reference to a title for an oral conference paper).

Townes, C.H., 1957, in Radio Astronomy, Proceedings of IAU Symp. 4, ed. H.S. Van de Hulst, Cambridge U. Press, Cambridge p. 92.

van Langevelde, H.J., Frail, D.A., Cordes, J.M., and Diamond, P.J., 1992, *Ap.J.*, **396**, 686.

Weaver, H., Williams, D.R.W., Dieter, N.H., and Lum, W.T., 1965, *Nature*, **208**, 29.

Weinreb, S., Barrett, A.H., Meeks, M.L., and Henry, J.C., *Nature*, **200**, 829.

Weinreb, S., Meeks, M.L., Carter, J.C., Barrett, A.H., Rogers, A.E.E., 1965, *Nature*, **208**, 440.

Wilson, T.L., Johnston, K.J., and Mauersberger, R., 1991, *Astron. & Astrophys.*, **251**, 220.

Interpretations for Observations of Astronomical Masers

W. D. Watson

Department of Physics, University of Illinois at Urbana-Champaign, 1110 West Green Street, Urbana, IL 61801-3080

Abstract: Studies of astronomical masers have progressed in parallel with research on complex molecules, beginning with the original detections of both which occurred some twenty-five or so years ago. Basic aspects of the interpretation of the observational data are discussed with particular application to the 22 GHz water masers in regions of star formation. Specific issues include the nature of the gas dynamical environment based on pumping and other considerations, the structure and energetics from information in the spectral line profiles, and inferences about the strengths of magnetic fields at the highest gas densities from the polarization of maser radiation.

1 Introduction

The study of astronomical masers has developed in parallel with that of complex molecules during the past 25 or so years. The original detections of the strong masers occurred near the beginning of this era—OH at 18 cm, H_2O at 1.3 cm and SiO at 3.5 mm. Numerous additional strong masing transitions of these and other (especially CH_3OH) species have subsequently been detected and studied at wavelengths into the submillimeter regime. Because of their small sizes and high surface brightnesses, masers are ideal probes for astronomical phenomena at the highest spatial resolution using the techniques of long baseline, radio interferometry. Astronomical masers are associated with active regions of the molecular gas, mainly related to star formation and to the return of matter to the gas from stars. They also occur near the centers of activity in the nuclei of other galaxies. Some of the earliest indications of the bipolar flows of gas in regions of star formation came from the observation of the high velocity masers. The structure of circumstellar matter near late-type giants has been delineated and the sizes of the circumstellar shells have been determined from the time delays in the response of the maser emission to the time variations in the infrared emission of the central star. I will focus upon issues centered around the interpretation of observations of a particluar type of astronomical maser—the interstellar (and now extragalactic) 22 GHz water masers. The relatively recent detection of a number of other masing transitions of the water molecule confirms that water masers are a basic component of the astronomical environment. The underlying issues for the 22 GHz masers are generally, and in some cases quite specifically, relevant to the interpretation of other types of astronomical masers, as well. Of the types of masers, the 22 GHz water masers seem to provide the greatest challenge to the interpretations. Their study does, however, offer the opportunity for significant rewards in astronomical information. The 22 GHz masers tend to be nearest to the centers of star formation and are also located near the nuclei of other galaxies. They possess the highest surface brightness, greatest luminosities and smallest angular sizes—qualities which make them especially valuable probes of astronomical phenomena at the highest spatial resolution. Information is obtained on high velocity outflows and, possibly, on disks associated with protostellar environments and with the nuclei of other galaxies. Astronomical distances on the galactic (and potentially extragalactic) scale are obtained from a statistical analysis of the motions of individual masing spots. Strengths for the magnetic fields at the highest densities in the star forming gas have been inferred from observations of the circular polarization of the radiation. Understanding the very high brightness temperatures and the tight beaming of the maser radiation places the most severe restrictions upon the theory for the pumping and the radiative transfer, as well as upon the nature of the astronomical environments. If limitations due to time, space or the author's ability to provide a coherent presentation require that the focus be limited to only one type of maser, the 22 GHz water masers are a reasonable choice.

The presentation here is organized into three parts which are centered around the gas dynamical nature and the pumping of the water masers (Section 2), the structure and energetics based on information in spectral features (Section 3) and the inferences about the strengths of magnetic fields in the star forming gas from the polarization characteristics of the radiation (Section 4).

2 The Nature and Environment of the Astronomical Water Masers

The 22 GHz masing transition of water occurs between the 6_{16} and 5_{23} rotational energy levels which lie at an energy of about $E/k \simeq 600$ K above the ground state. Each rotational level consists of three hyperfine states. There are then six radiative transitions. Calculations indicate that the "22 GHz maser line" probably is a merged profile due to the strongest three of these hyperfine transitions (Nedoluha and Watson 1991). Other, newly detected masing transitions (e.g., Cernicharo et al. 1990; Menten et al. 1990) occur between energy levels that are even higher as well as lower in energy. Infrared transitions also connect these masing states to other rotational (and vibrational) states and tend to have wavelengths of about 100 microns or shorter. Gas temperatures greater than 300 - 400 K, which are warmer than the 50 K or so that are typical of molecular clouds, are immediately suggested in order to elevate the water molecules into these high lying states.

To maintain an inverted population for masing, it is necessary that the populations of the masing molecules be kept away from thermodynamic equilibrium.

This might be done by a difference between the kinetic temperature of the gas and the effective temperature of the infrared radiation which is emitted and absorbed by dust grains (hence, the "hot grains/cold gas" and "hot gas/cold grains" scenarios). It might also be achieved by a temperature difference between the particles that collisionally excite the water molecules—hydrogen molecules and the electrons, or the positive ions. At this time, none of the specific scenarios that utilize these somewhat subtle causes for the pumping of the 22 GHz masers seem to be adequate (e.g., Anderson and Watson 1990). As a result, attention has returned to the old idea that non-equilibrium populations are maintained by the escape of infrared, spectral line radiation associated with transitions between the masing states and other states of the water molecule. Collisions with hydrogen elevate the water molecules into excited states from which infrared, radiative cascades occur. The molecular physics of the cascade, in which the optical depths of the infrared transitions typically are large, determines the inversion of the radio frequency masing transitions. This, too, can be viewed as a "two-temperature" scenario. The effective temperature of the radiation is lowered from the kinetic temperature by the escape of radiation, and can vary from one infrared transition to another as a result of differences in their optical depths. This "line leaking", or IR escape scenario for the pumping of masers was recognized long ago (Jefferies 1971) and calculations were performed at that time which provide the essential results for the predicted pumping for H_2O masers (deJong 1973, Strelnitskij 1973). Subsequent calculations of this type in recent years differ by utilizing new collision cross sections that have been provided by chemical physicists, by utilizing escape probabilities for the infrared radiation based upon the large-velocity-gradient approximation instead of those for a static gas, and by including more molecular states to provide predictions about the new masing transitions of water that have been discovered in recent years. The accuracy in the application of the "new" collision cross sections also is unclear since they are actually calculated for "helium-like" molecular hydrogen (i.e., the internal rotational and vibrational structure of the H_2 is ignored). Differences between the type of escape probability treatment that is employed for the infrared radiation are almost certainly less important (they both tend to be proportional to [optical depth]$^{-1}$ at high optical depths which are relevant here) than the uncertainty in applying a simple escape probability treatment at all. In any case, all of the calculations of this IR escape scenario yield results that are not meaningfully different with regard to the main issue—the high apparent luminosities and extreme brightness temperatures of the 22 GHz water masers. This is not at all surprising since this type of maser scenario is constrained by the "thermodynamic limit" that was recognized in early investigations (Goldreich and Kwan 1974b; Rank, Townes and Welch 1971).

The idea of the "thermodynamic limit" is that, for every maser photon that is produced, at least one (approximately) infrared photon must escape from the masing region to produce the inversion . The relevant infrared photons would be expected to arise from one or a few transitions involving the masing states or nearby states. Since the infrared transitions are not masing, the maximum number of infrared photons that can be emitted from the surface of the masing gas is that due to optically thick transitions in thermodynamic equilibrium—the number given by the intensity of the Planck distribution (at the temperature of the gas) for frequencies within the breadth of a spectral line. To see what this means quantitatively, consider a masing cylinder of gas. The "cylinder" is not necessarily a cylindrically shaped gas cloud but is more likely to be a cylindrical region in the distribution of excitation or in which the velocity differences are small enough to create "coherent" paths for the maser radiation. Recognizing that the maser radiation must be significantly beamed, we choose the length ℓ for the cylinder to be much greater than its diameter d. Because of the exponential amplification for maser radiation, most of the maser radiation emerges from the ends of the cylinder whereas most of the infrared radiation escapes through the sidewalls. The number of maser photons that escape per unit time is then approximately

$$\frac{2\nu^2}{c^2} \times \frac{kT_b}{h\nu} \times \Delta\nu \times \Delta\Omega \times \pi d^2/2 \tag{29}$$

where T_b, ν, and $\Delta\nu$ are the brightness temperature, frequency and spectral line breadth of the maser radiation. It is beamed into a solid angle $\Delta\Omega$ as it emerges through the caps (with area $\pi d^2/4$) of the cylinder. The Rayleigh-Jeans approximation is utilized here as well as in equation (2) for the infrared radiation. The analogous expression for the rate of escape of infrared photons in a spectral line is approximately

$$\frac{2\nu_{ir}^2}{c^2} \times \frac{kT_{ir}}{h\nu_{ir}} \times \Delta\nu_{ir} \times \Delta\Omega_{ir} \times \pi\ell d \tag{30}$$

For beaming of maser radiation by a long cylinder $\Delta\Omega \simeq d^2/\ell^2$, whereas for escape of infrared from the sidewalls $\Delta\Omega_{ir} \simeq 2\pi$. The spectral linebreadths $\Delta\nu$ and $\Delta\nu_{ir}$ should be comparable when expressed in terms of Doppler velocities. The requirement that the expression of equation (2) be greater than that of equation (1) is then

$$T_b < 4\pi\left(\frac{\nu_{ir}}{\nu}\right)^2\left(\frac{\ell}{d}\right)^3 T_{ir} \tag{31}$$

The infrared transitions between the masing and the nearby states have wavelengths of about 100μ and the gas kinetic temperature is less than about 10^3K so that $T_b \lesssim 10^8(\ell/d)^3$ for the 22 GHz masers. Thus, aspect ratios $(\ell/d) \gtrsim 100$ are required to understand the observed brightness temperatures $T_b \gtrsim 10^{14}$K. Detailed calculations yield similar results.

It has thus been recognized for some twenty years that the pumping required to produce the high brightness temperatures of the 22 GHz water could be "understood" if one were willing to make the extreme assumption that $(\ell/d) \gtrsim 100$ *for masing cylinders (e.g., Goldreich and Kwan 1974b; Strelnitskij 1973, 1984)!* In the absence of significant independent evidence, the assumption that $(\ell/d) \gtrsim 100$

has seemed to be an implausibly favorable assumption. Until recent years, the effort has thus been directed toward identifying pumping mechanisms that are inherently more powerful than this "infrared escape" scenario. Lack of success has apparently motivated recent investigators to return to the earlier scenario and to advocate $(\ell/d) \gtrsim 100$. Until very recently (see Section 3), there have only been qualitative arguments for such aspect ratios that are not significantly more advanced or compelling than have been available for the past twenty years. The large number of additional masing transitions of H_2O that have been detected in recent years does, however, require a quite general and robust cause for the inversion. This also seems to favor an "infrared escape" scenario.

The cause for the elevated kinetic temperatures of the masing gas above the 50 K or so that is typical of molecular clouds is another key issue in understanding the water masers. Because the 6_{16} and 5_{23} levels lie some 600 K above the ground state, the populations of these states are quite sensitive to the kinetic temperature. Calculations indicate that minimum temperatures of 300-400 K are necessary for the 22 GHz masers (e.g., Kylafis and Norman 1991). Somewhat higher temperatures seem to be required for certain of the newly detected masing transitions of water (Neufeld and Melnick 1991; Melnick et al. 1993). Shock waves are commonly associated with elevated kinetic temperatures in the interstellar gas. Water masers typically exhibit velocities up to 10-20 km s^{-1}, and sometimes 100-200 km s^{-1}, relative to the velocity of the molecular cloud in which they are located. They are frequently known to be in the neighborhood of HII regions which can provide the pressure to drive shock waves into the surrounding neutral gas. It was thus immediately recognized in the 1970's that the warm gas behind a shock is a natural way in which to understand the cause for the elevated gas temperatures. Calculations were performed at that time in which it was concluded that the collisional/infrared escape scenario for excitation in a shock wave would produce brightness temperatures as high as 10^{15} K in a gas with a density of about 10^9 cm^{-3} at about 500 K if the maser radiation is beamed into a small solid angle of $10^{-4} - 10^{-5}$sr (Strelnitskij 1973; Schmeld, Strelnitskij and Muzylev 1976). Current calculations give almost identical results for 22 GHz masers in shocks.

Where current calculations for shock excitation do offer the potential for some significant advances over the earlier work is in relation to the size of the masing region and in delineating more precisely the type of shock. To obtain paths of sufficient optical depth in the warm gas for the required maser amplification, the maser radiation is imagined to propagate parallel to the shock front. The length of the region behind the shock front (i.e., measured parallel to the shock velocity) in which there is strong inversion is then the approximate diameter of the masing spot that is observed. A problem arises for "ordinary" shocks. At the relevant gas densities in the 22 GHz masers ($\simeq 10^9$ cm^{-3}), the gas cools rapidly through the temperature regime ($300 - 1000$ K) that seems to be appropriate for the masing. The length of this regime is clearly less than the diameters of $10^{13} - 10^{14}$ cm that are observed. This has led to the proposal that dissociating J-shocks provide the necessary environment for the water masers (McKee and Hollenbach 1980). In these shocks, the hydrogen molecules that have been dissociated at the shock front are found to recombine when the gas has cooled to about 400 K—an ideal temperature for water masers. If the recombination energy goes into heating the gas, the gas will remain near 400 K for a longer time and hence a longer distance behind the shock front, leading to a larger diameter for the masing region. There is then a "shoulder" in the graph of temperature versus distance behind the shock front. This region may be as long as the 10^{13} cm which is needed for most masers. To obtain a lengthier masing gas behind the shock front that is required for larger masers, it seems to be necessary to invoke magnetic fields in addition to the recombination energy. Pressure due to magnetic fields of an appropriately chosen strength can, in principle, aid in extending the region of the warm gas by halting the increase in gas density in the post-shock region when the optimal physical conditions have been achieved. Dissociating shocks certainly are likely and sometimes may well impinge upon pre-shock gas with a density of about 10^7 cm^{-3} as is necessary to yield the conditions for the 22 GHz masers in the postshock region. On the other hand, for a number of reasons it is unlikely that dissociating J-shocks are the cause for the bulk of the radiation of interstellar and extragalactic water masers as has recently been advocated (Elitzur, Hollenbach and McKee 1989), especially at 22 GHz. i) Shocks are introduced into the interpretation of water masers mainly to provide a cause for the enhancement in the kinetic temperature. This increase is only about 0.03 eV per molecule to reach 400 K whereas dissociation requires an energy input that is at least equal to the binding energy of H_2 (4.5 eV per molecule). Less than one percent of the energy that is injected is actually required for achieving the temperatures for masers. That is, to heat the gas by only a few hundred Kelvins, the gas is first raised to some 10^5 K in this scenario. It seems likely that nature finds a more "energetically efficient" means than dissociating shocks for creating these masing environments which occur commonly. ii) Shock velocities greater than 30 km s^{-1} are required for dissociating J-shocks to produce the masing environment. According to the review of Genzel (1986), most of the 22 GHz maser radiation arises from gas with velocities near the rest velocity of the associated, quiescent molecular cloud (that is, less than 20 km s^{-1}). In a straightforward scenario, the shock would be caused by an enhancement in the pressure due to a stellar outflow (e.g., a wind or jet) or an HII region, most likely associated with a proto- or newly formed star with the velocity of the molecular cloud. The masers would then ordinarily be imagined as in the gas that is swept up by a uniformly expanding shock. In this picture, the observed velocities of the masers relative to the molecular cloud (the expansion velocity) must be essentially the same as the shock velocity. Dissociative shocks would be then excluded as the cause for the bulk of the 22 GHz maser emission. More complicated scenarios can, no doubt, be imagined in which the observed expansion velocity of the masers is not equal to the shock velocity. The necessity for additional complicating assumptions does, however, reduce the likelihood for such as the general interpretation for a widely occurring phenomenon. iii) The key aspect of dissociating shocks—the release of energy to heat the gas from the recombination of H_2—is open to doubt. It requires that a large fraction of the energy that is released when an H_2 molecule forms on the surface of a dust grain (at grain temperatures of about 50 K) remain with the newly formed molecule as translational and vibrational energy. It might instead be absorbed by the dust grain. There seems to be no clear evidence in favor of either possibility. iv) Magnetic pressure is involved to understand the sizes of the larger 22 GHz water masers. The required strengths of these magnetic fields (Elitzur, Hollenbach and McKee 1989, equation 4.6) are larger by about a factor of ten than the fields that have actually been detected in 22 GHz water masers by Fiebig and Güsten (1989) [see also Section 4]. v) A detailed comparison of the observed intensities of several masing transitions of water indicates that the gas temperature in the masing region is significantly higher than the 400 K of the post-shock, recombination region of dissociating shocks in which the water masers are proposed to occur (Melnick et al. 1993). Apparently, this temperature is tightly constrained in dissociating J-shocks and cannot be raised to the necessary temperatures simply by adjusting parameters.

In summary, the outstanding issue for interpreting the 22 GHz water masers has historically been understanding the brightness temperatures that range up to and perhaps somewhat beyond 10^{15} K. On this issue, the recent calculations for the maser emission associated with shocks yield results that are little different from the early work of Strelnitskij and collaborators. All are based upon the infrared escape scenario and are thus constrained by the "thermodynamic limit" (equation 3). Apparent differences arise mostly from the beaming angle (or "aspect ratio") that is employed—a quantity that is obtained simply by assumption and does not follow from such calculations. To obtain the beaming, the masing region ordinarily is considered to be a cylinder—formed either from (i) actual variations in the density and excitation, or (ii) from gradients in the velocity. For the same numerical value of the "aspect ratio" for the cylinder in the two cases, the beaming angle at high brightness temperatures is smaller in case (i) by a factor of about 30 than is the beaming angle obtained when the aspect ratio is expressed as the ratio of the velocity gradients that delineate the cylinder (based on the commonly utilized LVG approximation). Calculations in which the aspect ratio refers to cylinders of case (i) will thus appear to achieve higher brightness temperatures than calculations for the same aspect ratio in case (ii), but this is an artifact due to the difference in the beaming angles that is thus being assumed (see Kylafis and Norman 1991).

3 Luminosity, Geometry and the Degree of Saturation from the Interpretation of Spectral Line Profiles

The spontaneous or incident continuum "seed" radiation, that is amplified to become the maser radiation, has a relatively broad range of frequencies. In the initial, unsaturated regime of the amplification, the optical depth is greatest for frequencies that correspond to molecular velocities near the center of the Maxwellian distribution. Since the amplification varies exponentially with the optical depth, intensities near the center grow more rapidly than those in the wings with the result that the spectral line narrows to a breadth given approximately by $\Delta v \simeq v_{th}\tau_o^{-\frac{1}{2}}$ (FWHM expressed as a velocity) where v_{th} is the thermal breadth (FWHM) of the Maxwellian distribution and τ_o is the optical depth at line center ($\simeq$ 30 for intense 22 GHz water masers). When the maser becomes saturated at line center, the difference between the populations of the upper and lower molecular states is reduced near the center of the velocity distribution below its value for a Maxwellian distribution. The amplification in the wings is then increased relative to that at line center and the spectral profile of the maser radiation rebroadens as indicated in Figure 1 to the thermal breadth of the gas. If the rate for the relaxation of molecular velocities Γ_v is faster than

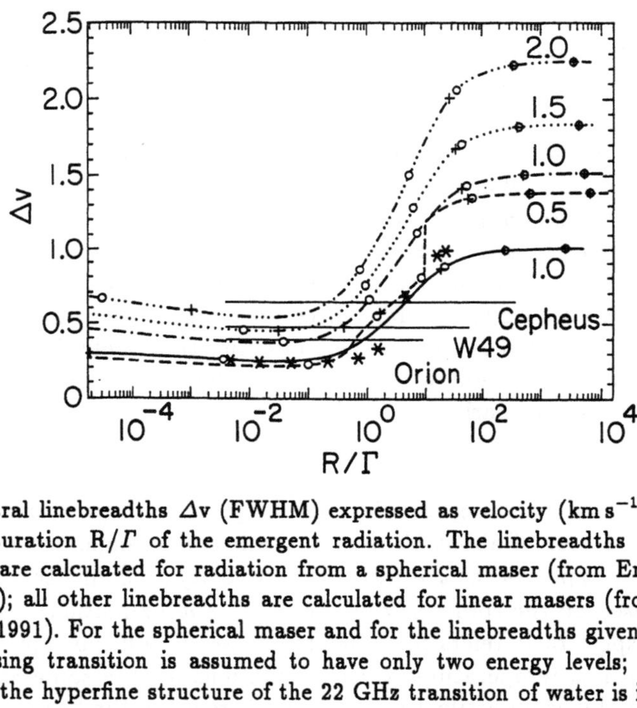

Fig. 1. Spectral linebreadths Δv (FWHM) expressed as velocity (km s^{-1}), versus the degree of saturation R/Γ of the emergent radiation. The linebreadths indicated by asterisks (*) are calculated for radiation from a spherical maser (from Emmering and Watson 1993); all other linebreadths are calculated for linear masers (from Nedoluha and Watson 1991). For the spherical maser and for the linebreadths given by the solid line, the masing transition is assumed to have only two energy levels; in the other calculations, the hyperfine structure of the 22 GHz transition of water is included and causes an increase in the linebreadth. The curves are labeled according to the molecular thermal velocity (FWHM) in km s^{-1}; the spherical maser is calculated for a thermal velocity of 1 km s^{-1} (1 km s^{-1}corresponds to 400 K for the water molecule). Also shown are the observed breadths (horizontal lines) for some of the strongest 22 GHz water masers (from Garay, Moran and Haschick 1989; Moran *et al.*1973; Rowland and Cohen 1986). In the calculations for the two level masers, the linebreadth Δv is proportional to the thermal velocity and can thus be obtained by scaling for the temperatures other than 400 K.

the decay rate Γ for the excitation, the rebroadening is postponed until the rate R for stimulated emission exceeds Γ_v. For the 22 GHz water masers, Γ_v does exceed Γ by a factor of ten or more. Velocity relaxation is mostly due to trapped infrared radiation (Goldreich and Kwan 1974a), though elastic collisions make a minor contribution (Goldreich, Keeley and Kwan 1973). The magnitude for Γ_v can be determined reliably and is about $2s^{-1}$ (Anderson and Watson 1993). Spectral line profiles for even the brightest 22 GHz water masers are clearly less than the 1 km s^{-1} (FWHM) breadth of a Maxwellian at 400 K—the approximate minimum temperature required for the excitation (because of the blending of the hyperfine components, Δv for the fully rebroadened spectral line should actually be about 1.7 km s^{-1} as shown in Figure 1 when the gas temperature is 400 K). This immediately means that $R < \Gamma_v (\simeq 2\ s^{-1})$. The rate R for stimulated emission is given by

$$R \simeq AkT_b \Delta\Omega / 4\pi h\nu \tag{32}$$

where A is the Einstein coefficient for the masing transition, and T_b and $\Delta\Omega$ are defined for equation (1). From the upper limit to R based on the observed linebreadth, together with the observed value for T_b, an upper limit for $\Delta\Omega$ follows immediately from equation (4) as does the luminosity L ($L \simeq \pi d^2 \nu^2 k T_b \Delta\Omega \Delta\nu / 4c^2$ for a maser with an observed diameter d and a linebreadth $\Delta\nu$). Although the observed spectral line profiles might be "contaminated" by various additional effects, all such effects that have been imagined tend to increase the linebreadth (see below). The upper limits obtained by the foregoing reasoning are thus unaffected by such uncertainties. I emphasize that there is minimal uncertainty in the value of R at which rebroadening occurs. Regardless of the importance of velocity relaxation, the spectral line rebroadens when R is approximately equal to the Einstein A coefficients for the infrared transitions that involve the masing state ($\simeq 1s^{-1}$) (Nedoluha and Watson 1991).

Application of the foregoing reasoning to some of the most intense 22 GHz masers (the outbursts in Orion and W49) for which $\Delta v \simeq 0.6$ km/s and $T_b \gtrsim 10^{15} K$ yields $R \lesssim 1s^{-1}$ and $\Delta\Omega \lesssim 10^{-5}$ sr. If the masers are cylinders of length ℓ, $\Delta\Omega \simeq (d/\ell)^2$ so that $\ell \gtrsim 3 \times 10^{15}$ cm when $d \simeq 10^{13}$ cm as is typical. Such long, narrow tubes are difficult to imagine. It is, perhaps, easier to imagine the small beaming angle as due to the interaction of distinct, well separated masing components—a scenario that we have shown to be plausible (Deguchi and Watson 1989). The longstanding puzzle of the high brightness temperatures of the 22 GHz masers is thus largely resolved. It is not due to some highly efficient, unrecognized pumping mechanism. It is probably a result of the simplest possibility—the infrared escape from a warm gas that is most likely heated by a shock, essentially as calculated by Strelnitskij and collaborators in the 1970's (also de Jong 1973). With independent evidence based on the spectral linebreadths, the extremely small beaming angles $\Delta\Omega$ (e.g., equation 4) that are required—which previously seemed implausible—now seem plausible. As noted in Section 2, problems do arise in attempting to construct a specific type of shock scenario that can produce masers with the observed sizes and other properties. It is, as yet, an unresolved issue.

As observers, we are presumably not located in an especially favorable direction in relation to the molecular clouds that contain the masing spots. There must then be about as many masing spots beamed into any other direction. The total number in all directions is thus approximately the number that we see multiplied by $(4\pi/\Delta\Omega)$ for those with a particular value of $\Delta\Omega$. The total number can be quite large since $\Delta\Omega$ is small—evidence in favor of the idea of separated, interacting components as opposed to long, continuous tubes. For the entire statistical collection of masing spots within a molecular cloud, the total luminosity is independent of $\Delta\Omega$. Although the total number of masing spots must increase as $(4\pi/\Delta\Omega)$, the luminosity of each spot is proportional to $T_b \Delta\Omega$ and T_b is already specified by the observations.

The above description of how the spectral line profile first narrows and then rebroadens with increasing saturation is, implicitly, based on the idealization that the rays of maser radiation are parallel (the "linear maser approximation"). It is expected to be a useful guide because the radiation tends to be tightly beamed within a maser. Detailed calculations were presented some time ago (Bettwieser 1976) which seemed to indicate otherwise—in particular, that rebroadening would be suppressed for a spherical maser. To assess this possiblity, we have recently performed computations for the spectral line profile of the radiation that would be observed from a spherical masing region (Emmering and Watson 1994) which are more accurate than those of previous investigations. Based on our calculation, the spectral line profile for a spherical maser does rebroaden in essentially the same manner as that for a linear maser. Its behavior is indicated by the asterisks in Figure 1. Other issues were also clarified in our study of the frequency dependent, radiative transfer in a spherical maser. In contrast to the results from certain recent investigations in the literature, we find i) that the increase in apparent size when measured at frequencies away from line center is small and probably not significant at intensities that are appreciable, and ii) that the "standard approximation" for most of the frequency dependent radiative transfer is adequate for the basic features of spherical masers that have been investigated.

4 Magnetic Fields in Regions of Star Formation

Considerable progress has been made in recent years in delineating the strength of the magnetic field in the relatively diffuse interstellar gas (densities less than about 10^3 cm^{-3}) utilizing the "thermal" emission of H at 21 cm and OH at 18 cm (e.g., Heiles et al. 1993). Masers seem to provide the best opportunity for extending our knowledge of magnetic fields to the high gas densities that are most directly associated with centers of activity related to star formation and other phenomena. Identification of Zeeman pairs of lines in the spectra of 18 cm OH masers in these regions indicates the presence of magnetic fields that typically are a few milligauss. These fields are relevant for gas densities no greater than about 10^7 cm^{-3} because the inversion for the 18 cm masing transitions of OH is quenched at approximately this density. For data at still higher densities and closer to the centers of the active phenomena, the 22 GHz water and the SiO masers which seem to occur at about $10^9 - 10^{10}$ cm^{-3}, must be utilized. Evidence that the information about magnetic fields obtained from masers is generally relevant (and not indicative only of peculiar conditions in masing environments) includes the observations which indicate that the directions of these fields tend to be similar to the directions of the galactic magnetic field (Reid and Silverstein 1990). If masers are in a post-shock gas, the strengths of the fields that are inferred can potentially be related to the structure of the shock, to the gas dynamics in the region, and to the strength of the magnetic field in the ambient, pre-shock gas. The strengths of the magnetic fields at the highest gas densities bear upon the role of magnetic fields in the formation of stars.

Unlike the 18 cm OH masers for which magnetic fields of a few milligauss cause the transition to split into distinct Zeeman components, the Zeeman components of H_2O and SiO masers are separated by much less than a spectral linebreadth. The magnetic moment of OH is larger by a factor of about 1000 due to its unpaired electron. What is then sought in the spectra of the H_2O and SiO masers is net circular polarization within the single spectral feature—an antisymmetric profile for the Stokes V polarization parameter in the ideal case with no velocity gradients or hyperfine structure. The expected fractional circular polarization is small and difficult to measure. It has, however, been detected in recent years at a level of about 10^{-3} (fractionally) in the spectra of the 22 GHz masers in regions of star formation (Fiebig and Güsten 1989; Zhao, Goss and Diamond 1993) and up to nearly 0.1 fractionally in the 43 GHz, circumstellar SiO masers (Barvainis, McIntosh and Predmore 1987).

A straightforward Zeeman interpretation for these data yields magnetic fields up to about 0.1 G and 100 G, respectively. A number of considerations might cause this interpretation to be open to question. These have been examined in several investigations (Nedoluha and Watson 1990; 1992; 1994). I describe below a likely process by which circular polarization can be created in masers that is quite different from the standard Zeeman effect.

When optical starlight passes through an interstellar medium in which the direction of alignment of the dust grains changes, the light becomes weakly circular polarized (Kemp and Wolstencroft 1972). It is an example of the creation of circular polarization in classical optics that occurs when linearly polarized radiation passes through an anisotropic medium (e.g., Rossi 1957) —the "theory of the quarter-wave plate". There are several related effects that might influence the interpretation of the circular polarization that is observed in the H_2O and SiO masers. One seems especially worthy of consideration because it is inherent to the transport of intense maser radiation in the presence of a magnetic field.

When the Zeeman frequency $g\Omega$ (i.e., the Zeeman energy shift divided by $\hbar$) is much greater than the rate R for stimulated emission due to the beam of maser radiation and the rate Γ for all other decay processes of the masing states, the direction of the magnetic field provides the axis of symmetry for the molecular quantum states. In the other extreme ($R >> g\Omega, \Gamma$), the direction of the beam of maser radiation provides an axis of symmetry. In either of these two extremes, the excitation of the gas and the resulting maser amplification can be calculated by considering only ordinary populations of the substates of the angular momentum ("magnetic substates") and rate equations. In the first case, the "z-axis" for the substates is parallel to the magnetic field and in the second, it is parallel to the direction of propagation. When the maser is at least partly saturated ($R \tilde{>} \Gamma$) and neither of these two extremes is appropriate (i.e., when $R \simeq g\Omega$), a convenient axis of symmetry no longer exists. Ordinary populations are no longer adequate for calculating the transfer of polarized radiation. The quantum mechanical density matrix (with "off-diagonal" matrix elements) commonly is used in this regime to describe the excitation. As a beam of maser radiation in a magnetic field propagates and is amplified so that R changes from $R << g\Omega$ to $R \simeq g\Omega$ along its path, the direction changes for the linear polarization of this radiation. By analogy with the example of starlight and the changing direction of alignment of the dust grains, circular polarization might be expected to be created.

Detailed calculations have been performed for this "intensity dependent circular polarization" described above for parameters appropriate for the 22 GHz water masers (Nedoluha and Watson 1990) and for the 43 GHz SiO masers (Nedoluha and Watson 1994). Significant circular polarization can be created. It has a spectral line profile for the Stokes V parameter that is antisymmetric about line center and thus resembles that produced by the ordinary Zeeman effect when there is no hyperfine structure. Such calculations are a challenge because the differential equations of radiative transfer must be integrated as a function of frequency. The solution requires the calculation of the full quantum mechanical density matrix (i.e., including off-diagonal elements) as a function of molecular velocity. Analytic estimates can be made that verify the numerical results (Nedoluha and Watson 1994). For both the SiO and H_2O masing transitions, fractional circular polarizations similar to what is observed can be calculated for magnetic fields that are much smaller than what is inferred based on the standard Zeeman interpretation. Only about 10 mG may be necessary to understand the SiO observations. Whether the observed circular polarizations are caused by this intensity dependent effect depends upon the magnitude of R in comparison with $g\Omega$ when the maser is at least partially saturated ($R > \Gamma$). Various evidence indicates that the 43 GHz SiO masers are at least partly saturated. The observed fractional polarization can then be created by magnetic fields that are weaker by a factor of about 10^3 than what is inferred with the standard Zeeman effect. For the 22 GHz water masers, the evidence for the intensity dependent effect is less compelling. After the original investigation (Nedoluha and Watson 1990), it was recognized—as a result of the analysis of linebreadths described in Section 3—that R for the 22 GHz masers is much smaller than the prevailing estimates so that typically $R < (\Gamma, \Gamma_v)_> \simeq 2s^{-1}$. The magnetic moment for the 22 GHz transition ($F = 7 - 6$) also is larger than that of SiO by a factor of nearly ten. It is thus more likely that $R \ll g\Omega$ for the 22 GHz water masers than for the 43 GHz SiO masers. In addition, the calculations that have been performed to represent the 22 GHz transition actually are performed for an angular momentum $J = 2 - 1$ transition. The difficulty of the calculation increases significantly with increasing angular momentum and with the presence of overlapping hyperfine components in the 22 GHz transition. The likelihood for deviations from the standard Zeeman circular polarization also tends to increase when the fractional linear polarization is large. No circular polarization was detected in the Orion flare even though it has unusually high linear polarization. High linear polarization is common for the SiO masers. In summary, the standard Zeeman interpretation seems to be valid for the 22 GHz observations. When modified slightly for the effect of overlapping hyperfine components (Nedoluha and Watson 1992), magnetic fields are inferred that typically are a few tens of milligauss. The circular polarization of 43 GHz SiO masers probably is not due primarily to the standard Zeeman effect.

5 Optimism for the Future

There is increasing emphasis on probing the small scale structure of astronomical phenomena as indicated by the current commitments to interferometry at higher frequencies and with long baselines, potentially extending into space. Because of their high surface brightnesses and small dimensions, masers will continue to provide ideal sources of radiation with which to probe astronomical phenomena utilizing these instruments. We can reasonably expect to understand better the mechanism by which mass is returned to the gas from late-type stars with more refined imaging of the circumstellar SiO, H_2O and OH masers. Spectral information and images of masers will better delineate the gas dynamics of star formation and, possibly, disk structures associated with stars and galactic nuclei. Space interferometry of extragalactic masers can provide directly distances to nearby galaxies that are free of the usual uncertainties. The role of the magnetic field in the gas dynamics of star formation and in other regions of activity in the molecular gas will be understood more clearly as a result of improved capabilities for the observation of polarized maser radiation.

The author's research is supported in part by NASA Grant NAGW-1104 and NSF Grant AST89-19614.

References

Anderson, N. and Watson, W. D. 1990, ApJ, 348, L69.
Anderson, N. and Watson, W. D. 1993, ApJ, 407, 620.
Barvainis, R. E., McIntosh, G. and Predmore, R. 1987, Nature, 329, 613.
Bettwieser, E. 1976, A&A, 50, 231.
Cernicharo, J., Thum, C., Hein, H., John, D. and Mattioco, F. 1990, A&A, 231, L15.
Deguchi, S. and Watson, W. D. 1989, ApJ, 340, L17.
Elitzur, M., Hollenbach, D. J. and McKee, C. F., 1989, ApJ, 346, 983.
de Jong, T. 1973, A&A, 26, 297.
Emmering, R. T. and Watson, W. D. 1994, ApJ, in press (April 1, 1994 issue).
Fiebig, D. and Güsten, R. 1989, A&A, 214, 333.
Genzel, R. 1986, in Masers, Molecules and Mass Outflows in Star Forming Regions, ed. A. D. Haschick (Westford, MA: Haystack Obs.), p 233.
Goldreich, P., Keeley, D.A. and Kwan, J.Y. 1973, ApJ, 182, 55.
Goldreich, P. and Kwan, J. Y. 1974a, ApJ, 190, 27.
Goldreich, P and Kwan, J. Y. 1974b, ApJ, 191, 93.
Heiles, C., Goodman, A. A., McKee, C. F. and Zweibel, E. G. 1993, Protostars and Planets III, ed. E.H. Levy and J. I. Lunine (Tucson, AZ: Univ. Arizona Press), p 279.
Jefferies, J. T. 1971, A&A, 12, 351.
Kemp, J. C. and Wolstencroft, R. D. 1972, ApJ, 176, L115.
Kylafis, N. D. and Norman, C. A. 1991, ApJ, 373, 525.
McKee, C. F. and Hollenbach, D. J. 1980, Ann. Rev. Astr. Ap., 18, 219.
Melnick, G. J., Menten, K. M., Phillips, T. G. and Hunter, T. 1993, ApJ, 416, L37.
Menten, K. M., Melnick, G. J. and Phillips, T. G. 1990, ApJ, 350 L41.
Nedoluha, G. E. and Watson, W. D. 1988, ApJ, 335, L19.
Nedoluha, G. E. and Watson, W. D. 1990, ApJ, 361, L53.
Nedoluha, G. E. and Watson, W. D. 1991, ApJ, 367, L63.
Nedoluha, G. E. and Watson, W. D. 1992, ApJ, 384, 185.
Nedoluha, G. E. and Watson, W. D. 1994, ApJ, in press (March 1, 1994 issue).
Neufeld, D. A. and Melnick, G. J. 1991, ApJ, 368, 215.
Rank, D. M., Townes, C. H. and Welch, W. J. 1971, Science, 174, 1083.
Reid, M. J. and Silverstein, E. 1990, ApJ, 361, 483.
Rossi, B. 1957, Optics (Reading, MA: Addison-Wesley), p 282.
Schmeld, I. K., Strelnitskij, V. S. and Muzylev, V. V. 1976, Astron. Zh., 53, 728.
Strelnitskij, V. S. 1973, Astron. Zh., 50, 1133.
Strelnitskij, V. S. 1984, MN, 207, 339.
Zhao, J. H., Goss, W. M. and Diamond, P. 1993, in Astrophysical Masers, eds. A. W. Clegg and G. E. Nedoluha (Berlin: Springer-Verlag), p 180.

Molecular Excitation and Galactic Interstellar Clouds

T. L. Wilson

Max-Planck-Institut für Radioastronomie, Auf dem Hügel 69, D-53121 Bonn, Germany

Abstract: A simplified account of the determination of kinetic temperatures, H_2 densities and column densities is given, with emphasis on quasi-thermal excitation. Some examples of the determination of parameters of clouds are presented.

1 Introduction

Since H_2 usually does not emit spectral lines, the properties of molecular clouds must be determined by measurements of gas phase polar species or dust continuum. Four effects may prevent one from completely determining the distribution of H_2 by using polar molecules: (1) Deviations from local thermodynamic equilibrium (LTE), (2) Small scale structure in the clouds, (3) Interstellar chemistry and (4) Depletion. The first effect is spectacular in the case of strong masers (see, e.g., the contributions of Baudry, Moran and Watson in this volume). The second effect was surmised from NH_3 data by Cheung *et al.* (1968); later measurements showed this directly. For example, in the Orion KL source,structure is found on scales of $\leq$0.05 pc, or $\leq 10^{17}$cm (Migenes *et al.* 1989). It is assumed that maps of NH_3 trace structure in H_2. It connection with effect (3), for many sources, linewidths $\Delta v_{1/2}$, and radial velocities, V_{lsr}, differ by much more than the errors, and there are differences in the spatial distributions of different species. One example is given by maps of the cold, nearby quiescent cloud Lynds 183 (Swade 1989), another is the marked separation of the cyanopolyene (HC_nN, n=3,5,7..) and NH_3 maxima in the cold Taurus cloud TMC-1 (see, e.g., Olano *et al.* 1986). Effect (4) is more likely to be found in cold dense regions. In a simple picture, the free-fall time varies as $1/\sqrt{n(H_2)}$, while the molecule-grain collision time varies as $1/n(H_2)$. Because of the low dust temperature, T_d, the molecules will remain on the grain surfaces; see the contribution of Mezger (this volume for a discussion).

For a given direction, spectra provide weighted, line-of-sight averaged, estimates of linewidth, $\Delta V_{1/2}$, radial velocity, v_{LSR}, beam averaged intensity in Kelvins, polarization percentage, and in some cases, line optical depths. For most molecular regions, the $\Delta V_{1/2}$ values indicate supersonic motions; in a few cases, the presence of shock waves can be deduced from $\Delta V_{1/2}$ and v_{LSR} maps. In cold dense cores, motions barely exceed Doppler thermal values (see Benson & Myers 1989). Detailed measurements of $\Delta V_{1/2}$ of lines with moderate to large optical depths show that uniform cloud models in which unsaturated lineshapes have gaussian shapes are not correct; more realistic are models in which shapes are determined by the relative motion of a large number of small condensations, that is, clumps. Spectral line polarization data can provide estimates of magnetic field strengths and directions (see, e.g., Crutcher 1988), but these data may require a detailed analysis (Watson, Baudry, this volume).

2 Molecular Excitation

Detailed treatments of excitation are given by Goldsmith (1988) and Genzel (1992); there is also a brief account of the determination of H_2 densities, $n(H_2)$ in Walmsley & Güsten (this volume). If collisions determine the molecular populations, estimates of local H_2 densities are possible. This requires a knowledge of collision cross sections; the simplest approach is to assume $\sigma=10^{-16}cm^{-2}$ (see Flower 1990 for a discussion and references). Then for a two level system, and a H_2 velocity of 1 km s^{-1}, the critical density is $n_{cr} < \sigma v >=10^{-10}n_{cr}=A$. Here A is the spontaneous decay rate. This applies to an optically thin line; for transitions with optical depth τ, the A coefficient should be replaced by A/τ. A fundamental limitation of density determinations using molecules is that energy levels which are populated at higher densities require higher values of the kinetic temperature, T_{kin} (see, e.g., Wilson 1990; Townes, this volume; Walmsley & Güsten, this volume). There is no method using molecular spectroscopy to detect regions with T_{kin}=10K having densities of $10^9 cm^{-3}$. For this, one needs molecules with dipole moments of 50 Debye, and these do not exist!

Populations can be limited either by densities or by temperatures. In extreme cases, n_{cr} is very low because the A coefficient is small, and then T is the only factor. If collisions completely dominate, the molecular populations are determined by T_{kin}.

Historically, T_{kin} was obtained at first from the peak intensities of rotational transitions of CO, which has a small spontaneous decay rate. The ratio of CO to ^{13}CO line intensities, the CO transitions are optically thick in most regions. For optically thick transitions of CO, the critical density is reduced by the value of the line optical depth to $\sim$50 cm^{-3}. Since the level populations are determined by collisions, the rotational excitation temperature, T_{ex} is T_{kin}. In the Rayleigh-Jeans approximation, the measured line intensity T_L, in Kelvins for a uniform slab is given by

$$T_L = f(T_{ex} - 2.7) \cdot (1 - e^{-\tau}) \qquad (1)$$

where T_{ex} is the excitation temperature, which characterizes the population across the two levels, and τ is the optical depth in the line. Usually, it is assumed that f, the beam filling factor is unity; for distant clouds, this is clearly less. Radiative transitions between different K ladders in symmetric top molecules are forbidden. Then the populations are determined by collisions, so an alternative

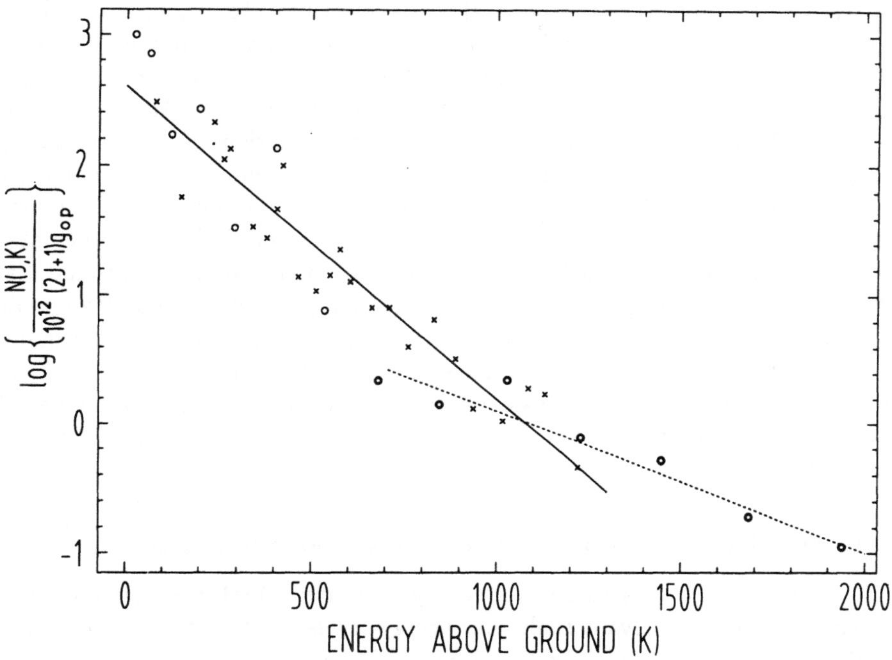

Fig. 1. A plot of the normalized column density of NH_3 for the Hot Core in the Orion KL nebula. These results were obtained from the inversion lines of NH_3. The open circles are metastable ($J = K$) inversion line data, the crosses are non-metastable ($J > K$) line data. That both metastable and non-metastable lie on the same line shows that the NH_3 populations are thermalized in this case.

method of determining T_{kin} is to use the ratio of populations in different K ladders of molecules such as NH_3, CH_3CCH or CH_3CN (see Fig. 2 of Walmsley & Güsten (this volume) for spectra). Since ratios are involved, beam filling factors play no role. For symmetric top molecules, transitions between K ladders are radiatively forbidden. For NH_3, the rotational transitions occur in the far infrared and have A coefficients of order 1 $\sec^{-1}$; for inversion transitions, A values are $\sim 10^{-6} \sec^{-1}$. For NH_3, the decay from a state $(J+1, K)$ to (J, K) occurs quickly and thus non-metastable transitions (those between states with $J > K$) are not suitable for T_{kin} determinations. Rather, one makes use of the metastable inversion transitions between ($J = K$) doublet levels. These populations cannot decay to lower levels via allowed transitions. If the NH_3 transitions are optically thick, one can use the ratios of satellite to main quadrupole hyperfine components to determine optical depths. A large number of T_{kin} determinations have been made using NH_3 in dark dust clouds (see, e.g. Danby *et al.* 1988). Usually these involve the (J, K)=(1,1) and (2,2) inversion transitions. For warmer clouds such as Orion KL Hot Core, more than 30 transitions have been measured (Wilson *et al.* 1993). In a suitable plot of this data (Fig. 1), a straight line indicates a single rotational temperature, T_{rot}, for the populations between K ladders. T_{rot} is given by the inverse of the slope; this is a lower limit to T_{kin} (see, e.g., Danby *et al.* 1988). Given a H_2 density estimate, one can uniquely relate T_{rot} to T_{kin}. The large H_2 densities in the Orion KL Hot Core insure that the NH_3 populations, both metastable and non-metastable, are thermalized. As a consequence of thermalizing the non-metastable populations, for Orion KL, T_{rot} equals T_{kin}. Emission from vibrationally excited NH_3 has been found in this source (see, e.g., Schilke *et al.*1992). It has been surmised that population transfers to vibrationally excited levels may disturb the ground state populations but the data in Fig. 1 are completely consistent with a thermal population in the ground state. The change in the slope is strong evidence for a gradient in T_{kin}. Presumably, the hottest gas is closest to the heating source. For optically thick, thermalized inversion lines, such as those measured in Orion KL, one can set $T_{ex}=T_{rot}$ in Eq.(1) and determine an equivalent size; for Orion KL the same result has also been found directly from maps made with the Very Large Array (Migenes *et al.*1989).

2.1 Density, Column Density and Mass Determinations

The estimates of n_{cr} based on two level systems are at best qualitatively correct. Usually, all of the relevant molecular energy levels must be taken into account, and some of the molecular lines will have a larger optical depth. Generally, the photon transport problem has been treated using the Large Velocity Gradient (LVG) approximation (Goldreich & Kwan 1974, Solomon & Scoville 1974). In the simplest form of the LVG approximation, the cloud is assumed to have a single T_{kin} and $n(H_2)$. The velocity, V, is proportional to cloud radius, R; this gradient must be large compared to thermal broadening; then the photon transport occurs on a local scale and not between widely separated parts of the cloud. For determinations of H_2 densities, the most commonly used molecules are CS (Mundy *et al.* 1986, Stutzki & Güsten 1990, Cesaroni *et al.* 1991), HC_3N (vanden Bout *et al.* 1983) and H_2CO (Mangum & Wootten 1993). These molecules have abundances of $\sim 10^{-9}$ relative to H_2, dipole moments>1 Debye, many transitions in the millimeter and submillimeter region, and reasonably simple energy level schemes.

Although the LVG approximation allows a simple treatment of photon transport, lineshapes cannot be reproduced. Dense clumps probably are surrounded by low density molecular or atomic envelopes. Even within the denser condensations, there may be gradients in T_{kin} and $n(H_2)$. For example, to model protostars realistically, collapse proceeds from the denser inner region to the lower density outer layers (see, e.g. Shu *et al.* 1987), more detailed treatment of photon transport is needed .

Aside from photon transport, there may be additional problems for small regions near embedded sources. For a number of species, emission from vibrationally excited states have been found near intense IR sources, so that a coupling with the ground state is possible. This could mimic large H_2 densities (see Hauschildt *et al.* 1993 for a discussion).

Column densities and masses can be determined by measurements of easily thermalized, abundant polar molecules such as CO, or dust continuum emission, or use of the virial theorem. Using cloud sizes and local H_2 densities leads to poor estimates because of fine scale structure and selection effects.

Genzel (1992) and Mauersberger *et al.* (1992) have noted that for an optically thin transition, and $(C^{18}O/H_2)=1.7\ 10^{-7}$, the relation $N(H_2)=1.2\ 10^{22}\cdot I(C^{18}O, J=2-1)$ holds to within a factor of two, for T_{kin} from 15K to 80K, and $n(H_2)$ from $\sim 10^3$ to $\sim 10^6 cm^{-3}$. This simple result is due to properties of CO and the LVG approximation.

As pointed out by Guélin & Cernicharo (1988), the CO isotopomers may be subthermally excited in the lower density envelopes of clouds. In large surveys of the galaxy, CO has been used, and justifications of this approach has been given on the basis of cloud models which involve extensive clumping (van Dishoeck & Black 1988). Comparisons with γ ray measurements have been made. There are differences of opinion about the factor 'X' used to convert from the integrated line intensity of CO to the column density of H_2 for our galaxy (see Bloemen 1988, Wolfendale 1988). It is possible that 'X' may vary as a function of galactocentric distance, D_{GC}, or from one galaxy to another.

Another method to determine column densities is the use of sub-millimeter dust continuum emission. For this, one requires a conversion from dust continuum intensity to H_2 column density. According to Mezger *et al.* (1988, 1992), for hot gas with low or moderate densities, there is a reasonable agreement between H_2 column densities obtained from dust continuum and CO data. However, this agreement breaks down in very dense cold gas, because the molecules condense onto grains. Walmsley & Schilke (1992) argue that grains coagulate to form fluffy conglomerates, and this changes the relation between emissitiy and mass.

A practical method has been prescribed by Mezger (this volume) to obtain the column density of H_2 from dust continuum emission, using $N(H_2)=1.65\ 10^{25} S/T_{dust}$.

Finally one can apply the virial theorem. If magnetic fields can be neglected, the virial mass is given by $M= 250\ (\Delta V_{1/2})^2\cdot(R(pc))$, where R is the cloud radius. If magnetic fields have a large effect on line widths, the virial mass is an upper limit. Myers & Goodman (1988) have assumed equipartition between magnetic, kinetic and gravitational energies to obtain $B(\mu G)=15.2\cdot(\Delta V_{1/2}{}^2/R(pc))$.

3 Summary of Results

The vast amount of work concerned with molecular clouds over the past 25 years is briefly summarized in, e.g., Wilson & Walmsley (1989) or Genzel (1992). Even though there has been a large increase in the number of clouds mapped, all clouds seem in some sense to be 'peculiar', each with its own special characteristics. This may be caused by the influence of surroundings, local variations in the abundances of C, N, or O, changes in dust characteristics, the presence of embedded stars and outflow sources or the ages of the clouds. As noted by some authors (see Falgarone & Phillips 1991) clouds seem to be fractal. For clouds in the galactic plane, it is thought that virial equilibrium holds; this is *not* the case for high latitude objects (see Blitz 1990). There are some dissidents (Scalo 1990) who argue that all structures found in clouds are transient. Cloud lifetimes are discussed by Larson (this volume). As noted previously, motions in most parts of most clouds are supersonic, but may be sub-Alvénic. One would expect rapid dissipation of such motions on crossing timescales. There must be a replenishment of such motions. There are two extremes: either energy is input on small scales and these in turn serve to power large scale motions, or is input on a large scale and this trickles down to small scales. One possibility is winds from embedded young stellar objects (Norman & Silk 1980, Lada 1985, McKee 1989, Fukui 1991). At the other extreme it is supposed that energy is input on a large scale, and this trickles down to small scales, such as tidal forces or differential galactic rotation. In our galaxy, this last process is more likely to occur where large differential rotation is present, such as in the center of our galaxy (Hüttemeister *et al.* 1993).

Since initially proposed by Larson (1981), the linewidth–size relation has been investigated over a large range of parameters. The basic result now found to be $\Delta V \sim D^2$ (Sanders *et al.* 1985, Dame *et al.* 1986, Solomon *et al.* 1987, Tatematsu *et al.* 1993). When combined with virial equilibrium, this relation implies that all clouds have a constant H_2 column density. This has been attributed to magnetic support (Shu *et al.* 1987) or pressure equilibrium with surroundings (Chièze 1987, Maloney 1988, Elmegreen 1989), or observational selection (Scalo 1990). Distributions of masses give a relationship proportional to $M^{-1.5}$. This has been found for studies with different angular resolutions and in different regions; in-addition to the large scale surveys cited above, see also Stutzki & Güsten (1990), Tatematsu *et al.*(1993). This effect is found in many data sets, which seems to indicate that it is a fundamental part of cloud physics, although it is surprising that the same dependence is found for cold low density dust clouds as for the hot high density Orion cloud.

4 Individual Sources: The Real World of Molecular Clouds

In the following a few sources are discussed as illustrations of the generalities discussed above.

4.1 Molecular Clouds Toward Cassiopeia A

There are 4 groups of molecular clouds along the line of sight to the supernova remnant (SNR) Cassiopeia A, with a diameter $\sim 5'$. Historically, these clouds are interesting since the first molecule found in the radio wavelength region, OH, was detected toward Cas A by Weinreb *et al.* (1963). Clouds with V_{lsr} near -36, -39 and -47 km s^{-1} are in the Perseus arm, ~ 3 kpc from the sun. There is no indication that these molecular clouds are not directly interacting with the material expelled in the SN explosion of Cas A, but are line-of-sight objects. In the centimeter wavelength range, the brightness temperature of Cas A is more than a few hundred Kelvins. This facilitated the detection of OH, and allows high resolution mapping using aperture synthesis techniques as well as allowing the determination of the line-of-sight magnetic field strength using the Zeeman effect for OH and HI. Maps of HI with an angular resolution of 50″ (Schwartz *et al.*1986), OH with a resolution of 8″ (Bieging & Crutcher 1986) and H_2CO with a resolution of 10″ (Goss *et al.*1984) show that molecules are concentrated toward the southern part of the source and are clumped, while the HI is distributed over the whole region. Maps of the $J = 1-0$ line of ^{13}CO for selected velocities are shown in Fig. 2.

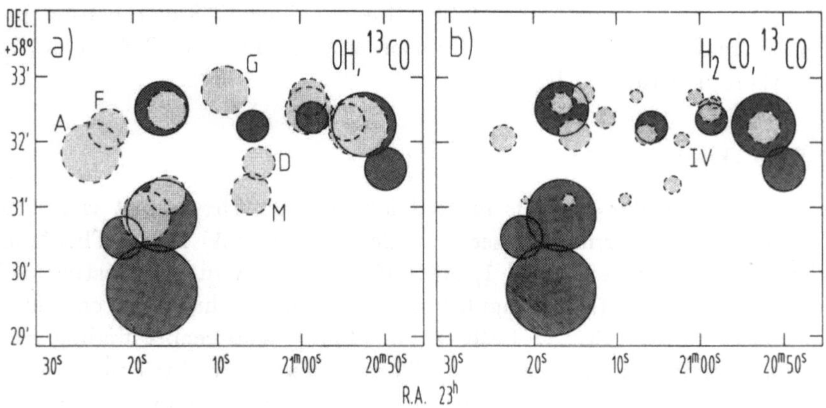

Fig. 2. a) A plot of the peak positions and FWHP sizes of the molecular clouds toward Cassopeia A (shown as darker shaded circles), obtained from maps of the $J = 1-0$ line of ^{13}CO; the lighter shaded circles refer to clouds whose positions and FWHP sizes were obtained from OH absorption line data (Bieging & Crutcher 1986). **b)** Results for cloud sizes and positions obtained from 6cm H_2CO absorption line data, shown with lighter shading (Goss *et al.*1984); the ^{13}CO results as in **a)**. The T_{kin} of these clouds is in the range 15-20K, except for cloud 'F' which has a T_{kin} of 6K. The H_2 densities are in the range 1-4 $10^3 cm^{-3}$.

The T_{kin} of the molecular gas, from the peak temperature of CO, is between 15 and 20K. From ^{13}CO data for the Perseus arm gas, the beam averaged H_2 column density is between 0.2 and $9\,10^{21}$ cm^{-2} and $n(H_2)$ is ≤ 4000 cm^{-3}. The CO masses of the clouds are at most one fifth of that required for virial equilibrium. The line-of-sight magnetic field strengths are too small to affect this conclusion; presumably the total B field is also too small. Thus, even if the B fields caused one-half of the line broadening, the molecular clouds toward Cas A would still be transient phenomena, destined to disperse. These may resemble high latitude molecular clouds (see, e.g., Blitz 1990). Falgarone & Phillips (1991) have proposed a cloud model in which there are a large number of dense small clumps even in the outermost regions; the H_2 densities obtained for the

Cas A clouds are lower than predicted by this model. For the Cas A clouds, the HI/H_2 ratio is $\geq$10 %. This is larger than the typical value found from HI self absorption in cloud cores, but most of the HI may arise from cloud envelopes (see, e.g., Andersson *et al.*1991).

¿From CO absorption line data, Lequeux *et al.* (1993) concluded that there are clouds with $T_{kin}\sim$3K, and in the outer galaxy (and in other galaxies) a large amount of mass is contained in such clouds, which would be missed in CO emission line surveys (cf. Eq(1)). Measurements of CO toward Cas A can be used to test this hypothesis, since these clouds were found on the basis of *absorption* line searches, and so these results are *not* biased toward warm clouds. These clouds are ~10 kpc from the galactic center, so should be considered as parts of the outer galaxy. Only one of the more than 20 clouds found toward Cas A appears to have T_{kin} as low as 6.2K; this cloud has 3% of the total mass in molecular material toward Cas A (Wilson & Mauersberger 1994). Thus, toward Cas A there is little cold H_2.

4.2 The Orion Region

The Orion molecular cloud is located 500 pc from the sun; it is the nearest region of O-B star formation (see, e.g. Genzel & Stutzki 1989). This is located in the Orion Molecular Cloud 1, or OMC1 region. A map consisting of 33000 spectra, measured with an angular resolution of 1.7′ has been carried out in the J=1-0 line of ^{13}CO by Bally *et al.* (1987), who report finding over 100 individual condensations (see Fig. 3). The southern part of the cloud appears to be more diffuse, while structure in the northern part is dominated by a north-south oriented, "*S*" shaped streamer of gas. In this streamer is the HII Region Orion A, the IR nebula Orion KL and the Hot Core; fine scale maps of these sources are discussed by Felli *et al.*1993, and Welch (this volume).

Up to 25′ north and south of Orion KL, the streamer has been investigated in (J,K)=(1,1) and (2,2) inversion lines of NH_3 by Cesaroni & Wilson (1994): Within 5′ of the center of Orion A and Orion KL, the clouds are warmer and denser At larger offsets, the average T_{kin} is ~20K. A comparion of data for the $J=1-0$ line of CS in the streamer (Tatematsu *et al.*1993) with (1,1) and (2,2) inversion lines of NH_3 shows that in the southern and middle regions the CS maxima are more spread out in α than in the region between Orion KL and OMC 2. Since the critical density for NH_3 is ~10% of that for this transition of CS, one might expect the NH_3 to be found in all CS peaks. As suggested by Tatematsu *et al.*(1993), this may be an indication that the CS $J=1-0$ line radiation emitted from the central part of the streamer, where H_2 densities are large, is absorbed and re-emitted in clouds further from the center (see Zhou *et al.*1989 who discuss this effect for dust cloud cores).

Near Orion KL, There have been two models for the cloud dynamics. In one, there is a single rotating cloud (Hasegawa *et al.* 1984, Vogel *et al.* 1985). In the other, there are two clouds, which overlap near the location of KL (see, e.g., Womack *et al.* 1991, Wang *et al.* 1993). In recent years, the weight of evidence has favored the two cloud picture, but it must be remarked that the structure of

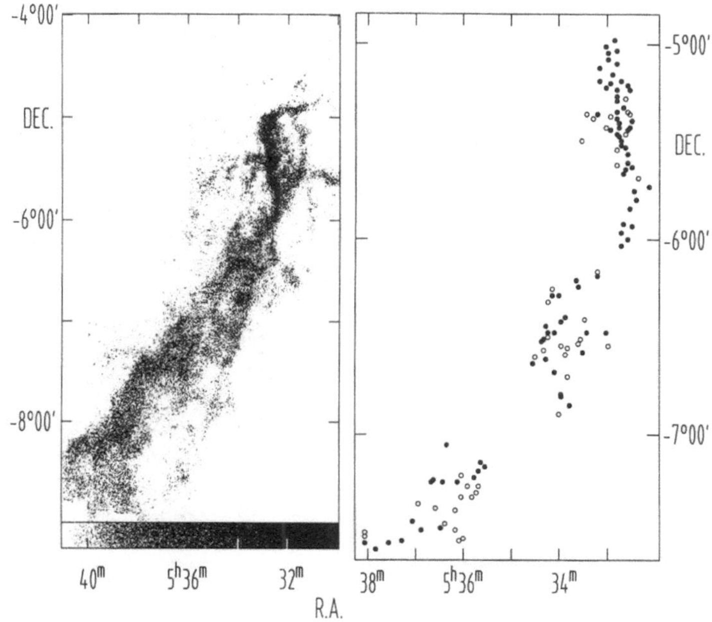

Fig. 3. On the left is a gray-scale map of the ^{13}CO integrated intensity in the Orion molecular cloud (Bally *et al.* 1987). In the middle is a representation of the maxima found in the $J = 1 - 0$ CS line survey (Tatematsu *et al.* 1993); the filled circles are those regions where NH_3 inversion lines were detected with the 100m telescope (Wilson & Testi, unpublished). On the right is a contour map of the integrated intensity of the (1,1) inversion line of NH_3 (Cesaroni & Wilson 1994).

the region close to the Hot Core is different depending on the molecular tracer: the Hot Core is prominent in CH_3CN but not in CH_3CCH. This is caused by chemistry, but not by excitation (see, Wang *et al.* 1993). According to models by T.J. Millar (private comm.), there is a distinct difference between species like CH_3CN and CH_3CCH, with freeze-out on grains being followed by efficient hydrogenation on the surface; CH_3CCH is too unsaturated to survive this process.

Near Orion KL, narrow finger-like structures with long axes oriented in a north-south direction have been found by Hasegawa (1986) in CO and have been mapped by Martín-Pintado *et al.* (1990) in HC_3N. Similar structures appear in the near IR images of Allen & Burton (1993).

3′ north of Orion KL is a region where molecular ions, such as HCO^+ and N_2H^+ are prominent. Away from the dust peaks, the column densities obtained from dust are in good agreement with those obtained from $C^{18}O$. Toward the peaks, dust continuum data gives a much higher H_2 column density. In addition, the dust map shows larger spatial variations. The larger contrast in the dust map is caused by the effect of beam chopping, which tends to eliminate slowly varying emission; this chopping would tend to eliminate a smoothly varying component, which is large. If this smoothly varying component were eliminated, the column

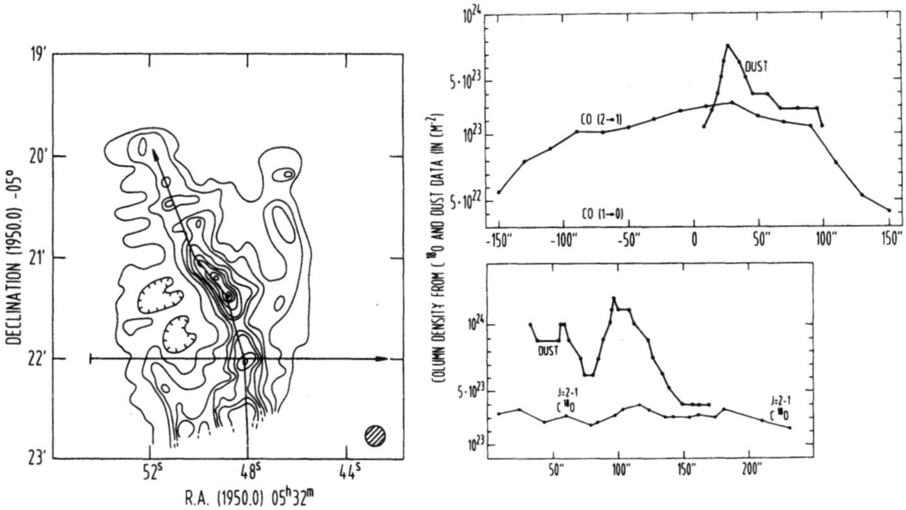

Fig. 4. To the left, a contour map of the region 3′ north of Orion KL in 1.3 mm dust continuum emission (Wink *et al.*in prep). To the right, a comparison of H_2 column densities from dust and $J = 2 - 1$ line emission for the two lines with arrows drawn on the contour map. Above is for the line in a roughly NS direction. Both data sets were taken with a 12″ angular resolution; the assumed T_{kin} and T_d are 30K over the whole map.

densities obtained from dust data would be much larger than those from $C^{18}O$, even for the regions away from the peaks. For the maxima, there is a large discrepancy between column densities. This may indicate internal heating sources, which would raise T_d and thus increase the dust intensities. For the compact peaks, high resolution data is needed to estimate sizes. There are results obtained from maps of H_2CO in the 2 cm line, using the VLA (Wilson & Johnston 1989). From unpublished H_2CO data and excitation studies, n(H_2) is less than $10^6 cm^{-3}$. When combined with source sizes, this gives N(H_2) estimates which are 10% of the N(H_2) values from dust data. Thus, it would appear that N(H_2) obtained from dust data is too large.

4.3 SgrB2

This is one of the most prolific sources of complex molecules, located at a projected distance of ~100 pc from the galactic center. The SgrB2 region has a total mass of ~ 10^6 $M_\odot$. Within an extended envelope are a number of dense molecular cores, H II regions, and maser centers. The determination of H_2 density and T_{kin} is complicated since the v_{LSR} of the envelope and dense cores are very nearly the same as that of the dense molecular cores. Thus, a separation of contributions is difficult. In the most determined effort so far, Hüttemeister *et al.* (1993a,b) have mapped 9 inversion lines of NH_3 and taken spectra of 25 additional inversion line transitions, with angular resolutions from 25 to 40″. The inversion lines are emitted from energy levels up to 2000K above ground. These data can be combined with Very Large Array (VLA) data taken with 2″ resolution (Vogel *et al.* 1987, Gaume & Claussen 1990) to produce a model of this region.

Of special interest are the properties of the envelope, which is extended over more than 4′. This has been mapped in NH_3 lines, in the $J = 1 - 0$ rotational transition of CO, and other low excitation species. The envelope is seen only in metastable (J=K) inversion lines of NH_3: in absorption toward the continuum sources, and in emission within one telescope beam of the peak of the continuum regions. That NH_3 is extended in the metastable inversion lines has been directly tested up to the (6,6) transition, which arises from energy levels ~500K above ground. Spectra toward the continuum maxima show absorption in the (14,14) transition, which arises from energy levels 2000K above ground. If the (14,14) line is distributed as the (6,6) line, the T_{rot} must be $\geq$200K. If a spectral line from an extended cloud is found in absorption, the excitation temperature, T_{ex}, must be lower than the main beam brightness temperature of the background source. For SgrB2(M) this is 23K, for a 40″ beam (Martín-Pintado *et al.*1990). A T_{ex} value which is 10% of T_{rot} indicates that the H_2 density is $\sim 10^3$ cm^{-3}. Thus, the envelope must be a hot, but not very dense gas. A more detailed analysis (Flower *et al.*1994) shows that for the low densities in the SgrB2 envelope, the NH_3 data are consistent with T_{kin} 600K, but that energetics requires that only 10% of the gas is this hot.

Closer to the compact sources, there is a region of size 1′ in α by 2′ in δ in which emission from the (2,1) and (3,2) non-metastable transitions is found. This is probably a region of moderate density. On a smaller scale, the compact cores are to be found.

The most complex species arise in the dense cores (Snyder *et al.*, this volume). The northern source, SgrB2(N), has two v_{LSR}'s, at 85 km s^{-1} and 65 km s^{-1}. The component at 65 km s^{-1} has a very large optical depth in some non-metastable NH_3 transitions, and a large T_{kin}. Combining these, one obtains an astoundingly large column density, N(NH_3)$\sim 10^{21}$cm^{-2} (averaged over the cloud but corrected for the beamsize), and high relative abundance of NH_3 to $H_2 \sim 10^{-4}$. Toward SgrB2(M), there are two major sources, at v_{LSR} 65km s^{-1} and 55 km s^{-1}. As with SgrB2(N), both have large column densities of NH_3, high T_{kin} values and relative abundances of NH_3. There is some indication that such large abundances are the result of evaporation from grains, but this is not completely certain.

There have been suspicions (Thronson & Harper 1986) that the SgrB2 cores have been 'hidden' by à large column density of intervening material. This intervening material is seen in absorption in the low excitation, metastable (J=K) inversion lines of NH_3. Since this lower density foreground gas is very extended, while the emission from the dense cores is very compact, so in larger radio telescope beams the absorption will dominate. With much higher angular reso-

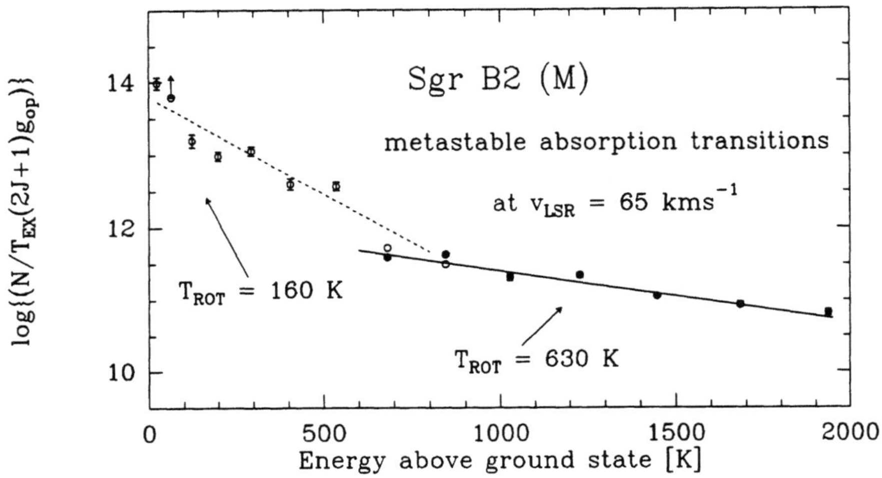

Fig. 5. A plot of the normalized column density of NH_3 for the envelope of SgrB2, obtained from metastable ($J = K$) inversion lines of NH_3 seen in absorption toward SgrB2(M). Data from Hüttemeister *et al.*(1994). The fact that absorption is seen toward this source, and emission one telescope beamwidth away shows that H_2 densities are low (Hüttemeister *et al.*1993).

lutions, one begins to find more detail in the molecular distribution. The advantage of the single dish measurements is the ability to measure many transitions; these must be supplemented by high resolution maps however. Measurements of the very highly excited transitions of NH_3, show emission toward SgrB2(M) and (N) at 65 km s^{-1}, and absorption at 55 km s^{-1}. This absorption is caused by a cloud in front of a bright continuum source; the VLA data of Gaume & Claussen (1990) shows that their components F1–3 are absorbed. A comparison with the distribution of other molecular species shows some similarities with NH_3, and generally supports a picture in which there is a moderately dense extended region toward SgrB2(M) and (N). However in addition, there is a strong differentiation of molecular species between SgrB2(M) and (N), as discussed by Snyder *et al.*and Welch (this volume).

5 Future Prospects

Necessary for progress in understanding cloud structure are a large number of high quality high resolution maps. There are still far too few 5″ to 25″ resolution maps of isotopomers of CO, or CS and NH_3 with $\geq 10^3$ beams. Models are still based on spherical geometries with constant T_{kin} and n(H_2), and making use of the Large Velocity Gradient approximation. The production of true 3 dimensional distributions of rather simple quantities such as $\Delta V_{1/2}$ and v_{LSR} without the use of idealized geometries is difficult. Interpreting the map data may require the application of advanced topological concepts (Scalo 1990, Adams

1992). The question of how much the pre-history of a cloud affects its structure is not clear.

The data so far show that maps in different species appear to be significantly different. This is *not* caused by excitation, and must be seen as one of the central problems of molecular cloud research. So far this problem shows no sign of being settled in the very near future. Solutions range from spatial variations in the radiation densities, the relative abundance of C, N or O, or the relative amounts of atomic versus molecular forms of N or O, or changes in the characteristics of dust grains. However, all of these quantities are unmeasured. Either new diagnostic tools must be found, or the chemical models developed to the stage that definite predictions about measurable quantities can be made with certainty.

In the last few years, estimates of column densities and masses from dust continuum emission, made with 12″ resolution have appeared. The dust continuum maps differ from all of the molecular line maps. For compact, dense regions the dust continuum data predict much larger masses than the spectral line data. There are two extreme interpretations: In the one it is supposed that in each region, there are a number of sources with different T_{kin}, $n(H_2)$, and sizes, but little variation in the relative abundance of species such as NH_3, CS or CO. The alternative is to suppose that the relative abundances of these molecules vary by orders of magnitude, and that in the densest regions, these molecules have condensed onto grains. This argument is not settled as yet. Only a few sources have been studied, and these show much more detail than previously thought. There is evidence that molecules have been liberated from dust grains in dense regions in Orion KL and SgrB2. Thus, it seems likely that at an earlier stage, molecules have condensed onto grains. The crucial question is whether this effect is a factor of a few or a factor of 100. In the past there have been a large number of surprises; we can reasonably expect many more in the near future.

References

Adams, F.H. 1992 Ap. J. 387, 572

Allen, D.A., Burton, M.G. 1993 Nat 363, 54

Bally, J., Langer, W.D., Stark, A.A., Wilson, R.W. 1987 Ap. J. 312, L45

Benson, p.J., Myers, P.C. 1989 Ap. J. Suppl. 71, 89

Bieging, J.H., Crutcher, R.M. 1986 Ap. J. 310, 853

Blitz, L. 1990 in IAU Symp. 144, 'The Interstellar Disk-Halo Connection in Galaxies' ed. H. Bloemen, Kluwer, Dordrecht, p. 41.

Bloemen, H. 1988 in 'Molecular Clouds in the Milky Way and Other Galaxies', eds. R.L. Dickman, R.L. Snell, J.S. Young, Springer, Heidelberg, p. 71

Castets, A., Duvert, G., Dutry, A., Bally, J., Langer, W.D., Wilson, R.W. 1990 A&A 234, 469

Cesaroni, R., Walmsley, C.M., Kömpe, C., Churchwell, E.B. 1991 A&A 252, 278

Cheung, A.C., Rank, D.M., Townes, C.H., Thornton, D.D., Welch, W.J. 1968 Phys. Rev. Lett. 21, 1701

Chièze, J.P. 1987 A&A 171, 225

Cesaroni, R., Wilson, T.L. 1994 A&A 281, 209

Crutcher, R. 1988 in 'Molecular Clouds in the Milky Way and Other Galaxies', eds. R.L. Dickman, R.L. Snell, J.S. Young, Springer, Heidelberg, p. 105
Dame, T., Elmegreen, B.G., Cohen, R.S., Thaddeus, P. 1986 Ap. J. 305, 892
Danby, G., Flower, D.R., Valiron, P., Schilke, P., Walmsley, C.M. 1988, MNRAS, 235, 229
Elmegreen, B.G. 1989 Ap. J. 338, L103
Falgarone, E., Phillips, T.G. 1991 in Proc. of IAU Symp. 147, 'Fragmentation of Molecular Clouds and Star Formation', eds. E. Falgarone, F. Boulanger, G. Duvert, Kluwer, Dordrecht
Fleck, R.C. 1981 Ap. J. 246, L151
Flower, D.R. 1990 'Molecular Collisions in the Interstellar Medium', Cambridge Astrophysics Series, CUP, Cambridge.
Flower, D.R., Pineau des Forêt, G., Walmsley, C.M., 1994, A&A (submitted) Fukui, Y. 1991 in IAU 'Reports on Astronomy', ed D. McNally, Kluwer, Dordrecht, p. 457
Gaume, R., Claussen, M. 1990 Ap. J. 351, 538
Gaume, R., Johnston, K.J., Wilson, T.L. 1992 Ap. J. 388, 489
Genzel, R., Stutzki, J. 1989, ARA&A, 27, 41
Genzel, R. 1993 in 'The Galactic Interstellar Medium', Saas-Fee Advanced Course 21, eds. W.B. Burton, B.G. Elmegreen, R. Genzel, Springer Verlag, Heidelberg, p. 275.
Goldsmith, P.F., Snell, R.L., Lis, D.C. 1987 Ap. J. 313, L5
Goldsmith, P.F. 1988 in 'Molecular Clouds in the Milky Way and Other Galaxies', eds. R.L. Dickman, R.L. Snell, J.S. Young, Springer, Heidelberg, p. 1
Goldreich, P. Kwan, J. 1974 ApJ 189, 441
Goss, W.M., Kalberla, P.M.W., Dickel, H.R. 1984 A&A 139, 317
Guélin, M., Cernicharo, J. 1988 in 'Molecular Clouds in the Milky Way and Other Galaxies', eds. R.L. Dickman, R.L. Snell, J.S. Young, Springer, Heidelberg, p. 81
Hasegawa, T., Kaifu, N., Inatani, J., *et al.* 1984, Ap. J. 283, 117
Hasegawa, T. 1986 in Proc. of IAU Symp 115, 'Star Forming Regions', ed M. Peimbert and J. Jugaku, Reidel, Dordrecht, p. 123
Hauschildt, H., Güsten, R., Phillips, T.G. *et al.* 1993 A&A 273, L23
Hermsen, W., Wilson, T.L., Walmsley, C.M., Henkel, C. 1988 A&A 201, 285
Hüttemeister, S., Wilson, T.L., Henkel, C., Mauersberger, R. 1993 A&A 276, 445
Hüttemeister, S., Wilson, T.L., Mauersberger, R., Lemme, C., Dahmen, G., Henkel, C. 1994 A&A (submitted)
Lada, C. 1985 Ann Rev A&A 23, 267
Lequeux, J., Allen, R.J., Guilloteau, S. 1993 A&A 280, L23
Maloney, P. 1988 Ap. J. 334, 761
Martín-Pintado, J., Rodriguez-Franco, A., Bachiller, R. 1990 Ap. J. 357, L47
Martín-Pintado, J., de Vincente, P., Wilson, T.L., Johnston, K.J. 1990 A&A 236, 193
Mauersberger, R., Wilson, T.L., Mezger, P.G., Gaume, R., Johnston, K.J. 1992 A&A 256, 640
Mangum, J., Wootten, A. 1993 Ap. J. Suppl. 88,
McKee, C. 1989 Ap. J. 345, 782
Mezger, P.G., Chini, R., Kreysa, E., Wink, J.E., Salter, C.J., 1988 AA 191, 44
Mezger, P.G., Wink, J.E., Zylka, R. 1990 A&A 228, 95
Mezger, P.G., Sievers, A.W., Haslam, C.G.T., Kreysa, E., Lemke, R., Mauersberger, R., Wilson, T. 1992 AA 256, 631
Migenes, V., Johnston, K.J., Pauls, T.A., Wilson, T.L. 1989 Ap. J. 347, 294
Mundy, L.G., Snell, R.L., Evans, N.J. II, Goldsmith, P.F. 1986 Ap. J. , 306, 670
Myers, P., Goodman, A. 1988 Ap. J. 326, L27

Norman. C., Silk, J. 1980 Ap. J. 238, 158
Olano, C., Walmsley, C.M., Wilson, T.L. 1986 A&A 196, 194
Sanders, D., Scoville, N.Z., Solomon, P.M. 1985 Ap. J. 289, 373
Scoville, N.Z., Solomon, P.M. 1974 ApJ 187, L67
Swade, D.A. 1989 Ap. J. 345, 828
Scalo, J. 1990 in 'Physical Processes in Fragmentation and Star Formation' eds. R. Capuzzo-Dolcetta, C. Chiosi, A. deFazio, Kluwer, Dordrecht, p. 151
Schwartz, U.J., Troland, T.H., Albinson, J.S., Bregman, J.D., Goss, W.M., Heiles, C. 1986 Ap. J. 301, 320
Schilke, P., Güsten, R., Schulz, A., Serabyn, E., Walmsley, C.M. 1992 A&A 261, L5
Shu, F.H., Adams, F.C., Lizano, S. 1987, ARA&A, 25, 23
Solomon, P.M., Rivolo, A.R., Barrett, J., Yahil, A. 1987 Ap. J. 730
Stutzki, J., Güsten, R. 1990 Ap. J. 356, 513
Tatematsu, K., Tomofumi, U., Osamu, K., *et al.* 1993, ApJ, 404, 643
Thronsen, H., Harper, D.A. 1987 Ap. J. 300, 396
van Dishoeck, E.F., Black, J. 1988 Ap. J. 334, 771
Vogel, S., Genzel, R., Palmer, P. 1987 Ap. J. 316, 243
Vogel, S.N., Bieging, J.H., Plambeck, R.L. *et al.* 1985 Ap. J. 296, 600
Walmsley, C.M., Hermsen, W., Henkel, C., Mauersberger, R., Wouterloot, J.G.A., Wilson, T.L. 1987 A&A 172, 311
Walmsley, C.M., Schilke, P. 1992 in 'Dust and Chemistry in Astronomy' ed. T.J. Millar and D.A. Williams (in prep)
Weinreb, S., Barrett, A.H., Meeks, M.L., Henry, J.C. 1963 Nature 200, 829
Wilson, T.L., Johnston, K.J. 1989 Ap. J. 340, 894
Wilson, T.L., Walmsley, C.M. 1989, A&AR, 1, 411
Wilson, T.L. 1990 in 'From Ground Based to space-borne sub-mm astronomy' Proc. of 29th Liege Int'l Astrophys. Symp., ESA SP 314, Paris, p. 279
Wilson, T.L., Henkel, C., Hüttemeister, S., Dahmen, G., Linhart, A., Lemme, C., Schmid-Burgk, J., 1993 A&A 276, L29
Wilson, T.L., Mauersberger, R. 1994 A&A 282, L41
Wolfendale, A., in in 'Molecular Clouds in the Milky Way and Other Galaxies', eds. R.L. Dickman, R.L. Snell, J.S. Young, Springer, Heidelberg, p. 76
Womack, M., Ziurys, L.M., Wyckoff, S. 1991, Ap. J. 370, L99
vanden Bout, P.A., Loren, R.B., Snell, R.L., Wootten, A. 1983, Ap. J. , 271, 161
Zhou, S., Wu, Y., Evans, N.J., II, Fuller, G., Myers, P. 1989 Ap. J. 346, 168

High Resolution Studies of Molecular Clouds

W. J. Welch

Radio Astronomy Laboratory University of California Berkeley, CA 94720

Abstract: High spatial resolution observational results for molecular clouds are reviewed. Results obtained with millimeter wavelength interferometers are emphasized.

1 Introduction

The highest resolution observations of molecular clouds have been made with interferometers, and this review will be restricted to that subset of molecular cloud studies. Before 1970, there were relatively few interferometric studies of galactic objects, either in the HI line or in the thermal continuum, largely because of the intrinsic low brightness of these objects. When it was learned in 1969 that there were polyatomic molecules in the interstellar medium and that there had to be dense clouds of hydrogen to excite these molcules into emission (Rank *et al.*1971), the view of the interstellar medium changed. It became clear that the ISM contained substantial amounts of dense gas, roughly organized into clouds, where stars formed and the evolution of such regions could be studied by the emission of the trace molecules. Furthermore, the line emision might be bright enough to observe at high resolution with interferometers. A number of groups began to plan the development of arrays at the short centimeter wavelengths and millimeter wavelengths where the molecular emission is strongest. Also, the VLA, whose construction was begun about 1972, was redesigned to accomodate spectral line observations.

The earliest observations with existing arrays were studies of the thermal continuum from HII regions in cloud cores. For example, the luminous core in the W49 region was studied with the Cambridge 5 kilometer telescope (Wynn-Williams, 1971), with the Greenbank Interferometer (Wink *et al.*1975) and with the Parkes interferometer (Radhakrishnan *et al.*1971). They found a distribution of sizes of HII regions, similar to what had been inferred from single antenna spectra of the region (Mezger *et al.*1967).

The first shortwavelength interferometer, a two element system, was constructed at Hat Creek, California, to operate at 1.3 centimeters wavelength. Its first spectral line observations were the measurement of the positions of a few interstellar water masers (Hills *et al.*1972) to an acuracy of about 2". The coincidence of one with a luminous infrared source, IRS5 in W3, suggested that the water masers might be signposts for massive star formation.

By about 1980, the VLA went into operation at wavelengths of 1.3 cm and longer, the Hat Creek Array moved on to three millimeter wavelength operation and was joined by a similar array built by Caltech in the Owens Valley. By 1985, both of these millimeter arrays were operating with three antennas and low temperature receivers. Results from these arrays in the early 1980's include, for example, an OVRO study of the massive outflows from O stars in W49, NGC 2071, and NGC 7538 (Scoville *et al.*1986) and the discovery of a young stellar object radiating in HCN near the water masers in W3OH (Turner and Welch, 1984).

The Nobeyama Millimeter Array became operational by about 1988, and the IRAM millimeter array at the Plateau de Bure in Eastern France began operations in 1990. Thus, at the present time four arrays at millimeter wavelengths, as well as the centimeter wavelength arrays, are engaged in studies of molecular clouds at resolutions of the order of one arc second or better. In the following we briefly summarize some of the work done with these instruments.

2 Cloud Observations

2.1 Clump Structure

The large scale mapping of molecular clouds has been principally carried out with single antennas, and it is typically found that the clump mass spectra follow a power law. For example, studies of the Rosette Nebula in the J=1-0 line of CO (Blitz, 1987; Blitz and Stark,1986) lead to a power law mass spectrum with spectral index -1.5 for clumps greater than about 0.7 pc diameter. An early interferometric search for structure by Brown and Padman (1988) with the Hat Creek array gave brightness upper limits of 0.02 to 0.04K for sizes in the range of 0.01 to 0.03pc in the nearby cold cloud L 1262. They used the interferometer as a spatial filter.

From a direct mapping of the high mass star forming region NGC 7538 at 3" resolution in HCN and HCO^+, Pratap *et al.*(1990) identified nearly 100 clumps over a fully sampled field of 120". These were found to have a number/diameter relation of the form $N(D) = 0.011\ D^{-3}$ for clumps larger than 0.08pc. If the clumps are all of the same density, the implied mass spectral index is -1.6. For densities inversely proportional to clump diameter, as is most often found for the larger clumps (Myers, 1985), the index is -2.0. This is closer to -2.5, the index of the Initial Mass Function for intermediate mass stars (Miller and Scalo, 1979), than for the larger clumps noted above, and it suggests that the smaller clumps may be precursors of individual stars.

2.2 Chemical Abundance Variations and Excitation

There is evidence of variations in the abundances of the different interstellar molecules on virtually all scales. On the largest scales, the maps of Sgr B2 by

Goldsmith *et al.*(1989) in several molecules show the prevalence of the long chain carbon molecules in the northern source, while the central OH source, at a projected distance of 2.5pc, is rich in sulfur compounds. On the other hand, the largest variations are found on the smallest scales.

The Orion-KL region of massive star formation is the closest such region at a distance of 500pc, and, with the relatively high linear resolution possible here, shows striking variations in chemical abundance. A young, highly imbedded and very luminous object IRc2 is in the core of this region of star formation and is evidently having a profound effect on its surroundings due to powerful winds and radiation. Several features are identified from their appearance and excitation: the 'hot core', dense material close to the star; the 'plateau source' with the broad spectrum of a wind; the 'ridge', a slightly more distant quiescent region. Figure 1 shows maps of a number of different molecules at two different radial velocities and at a resolution of 3.5" (Plambeck and Wright, 1988). In this, 7.4 km s^{-1} is the

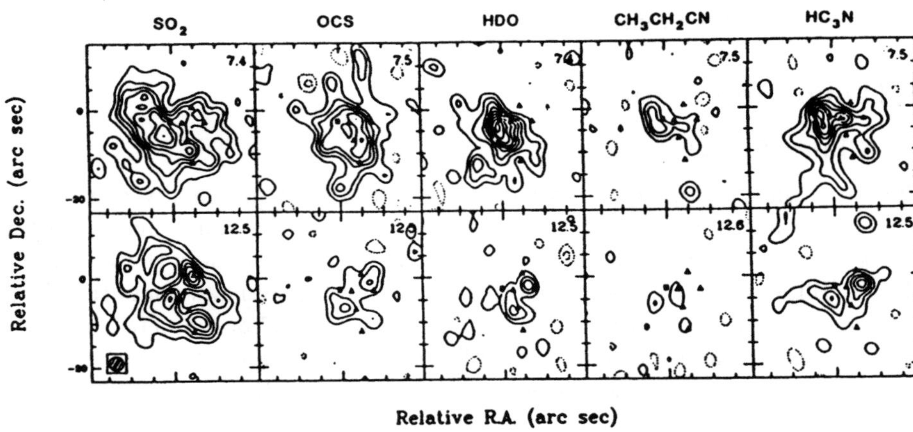

Fig. 1. BIMA line Maps at 7.4 and 12.5 km s^{-1} at 3.5" resolution toward Orion-KL (Plambeck and Wright, 1988). The square marks the position of IRc2, and the triangles mark the position of other prominent infrared sources.

'hot core' velocity. Whereas the sulfur compounds, such as SO_2, have the broad velocity of the 'plateau' gas, the other compounds shown are strongy enhanced in the 'hot core', particularly the long chain compounds. Both ammonia and deuterated water are enhanced here as well, possibly because they were once ices on grains that have sublimed due to the winds and high temperatures.

Understanding the excitation, and hence the physical conditions, of the gas requires the observation of more than one transition, and symmetric top molecules with closely spaced energy levels, such as methyl cyanide, are expecially valuable for this. Several nearby transitions can be observed at once, leading to well calibrated line ratios. Figure 2 shows maps of the Orion region in five transitions of methyl cyanide, along with a 450 μm map of the distribution of dust emission over the region (Wilner et al, 1994). The compactness of the map of the k=4 transition, with its excitation temperature of 127K, clearly identifies the 'hot core'. Plots of the relative strengths of the rotational transitions show that the

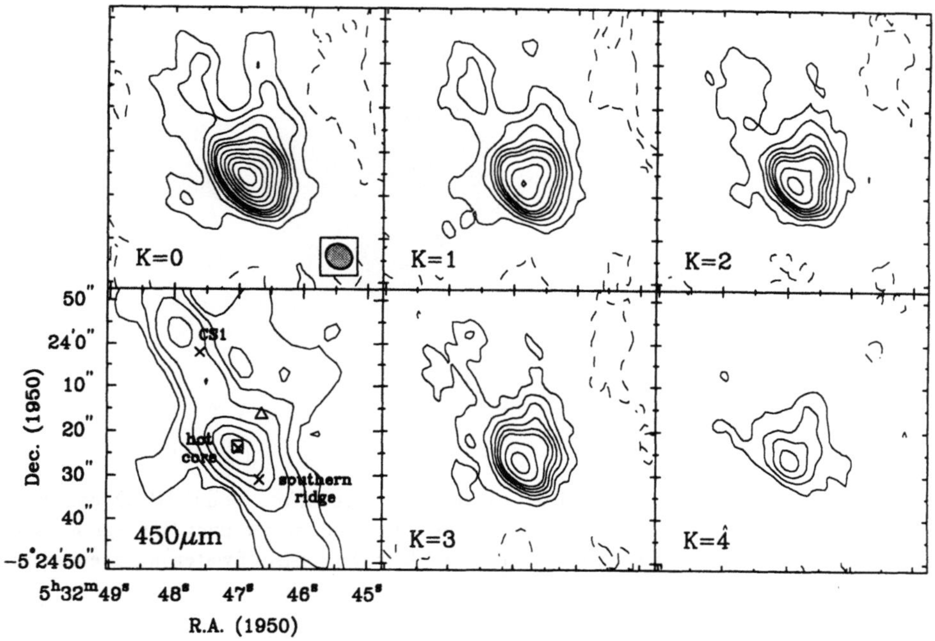

Fig. 2. Velocity integrated maps for each K component in the J=5-4 transition of Methyl Cyanide toward Orion-KL (Wilner *et al.*1994). The resolution beam is shown in the lower right hand corner. The lower left panel shows an 8" resolution 450 μm map.

levels are close to Local Thermodynamic Equilibrium, implying gas densities of 10^6 hydrogen molecules per cm^3 or greater, but at different temperatures. The temperature of the CS1 clump in the northwestern ridge lies in the range 52-83K, whereas the "hot core" temperature has a lower bound of about 280K.

2.3 Outflows

One of the most striking features of the star formation process is the existence of massive outflows from both the low and the high mass stars. For the more massive protostars, which are typically more distant, interferometric observations are essential to resolve them at all. The early OVRO observations of three massive outflows discussed above (Scoville *et al.*1986) found that for both W49 and NGC 2071, there were mass flow rates of the order of 0.02 solar masses per year. These are very large rates.

Figure 3, from Lis *et al.*(1993), shows images of blue shifted and red shifted emission from HC_3N toward SgrB2. The separation of the lobes is only a few seconds of arc, about 0.2 pc at the distance of 8 kpc. The blue shifted con-

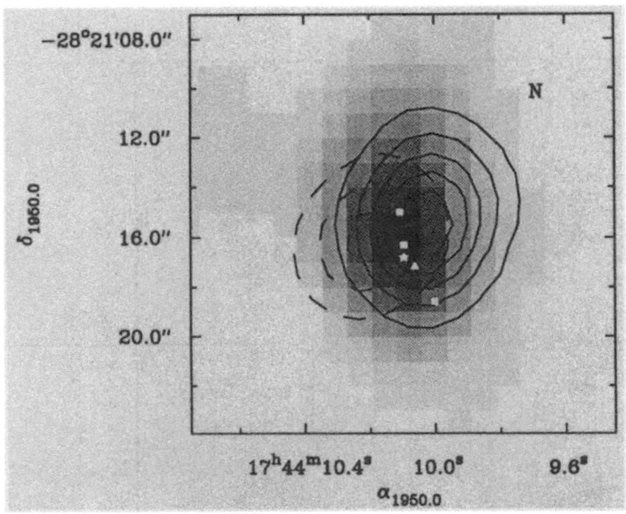

Fig. 3. Contour map of the blueshifted (dashed) and redshifted (solid) HC3N emission from SgrB2(N) showing bipolar structure. The squares show compact HII regions, the star is a compact ammonia source, and the triangle is the center of water maser emission (Lis *et al.*1993).

tours cover 28-55 $\mathrm{km\,s^{-1}}$ and the red shifted contours cover 75-108 $\mathrm{km\,s^{-1}}$. The inferred mass loss rate is about 0.2 solar masses per year and is probably the highest rate known. There are a number of compact HII regions and water vapor masers near the continuum peak in the center. It is unclear whether this flow is from one or several young protostars. Even higher resolution is needed for understanding of this source.

There is considerable evidence of clumpiness in the winds in the outflows from lower mass protostars (Kitamura *et al.*1992; Rudolph and Welch, 1988). In the case of L1448 (Guilloteau *et al.*1992) 2" maps in SiO(2-1) show highly collimated outflows made up of distinctive clumps with positive and negative velocities ranging up to more than 70 $\mathrm{km\,s^{-1}}$. This outflow is unresolved by the 2" beam in the transverse direction.

2.4 Infalling Material

Whereas the surprising outflows are seen coming from stars of all masses, there is relatively little evidence of the expected signatures of infall. This could be expected for the low mass stars,for which the gravitational infall velocities at distances corresponding to the present best resolutions of 1" - 2" are of the order of one $\mathrm{km\,s^{-1}}$ and are easily lost in the noise of random cloud velocities.

There is, however, fairly compelling evidence in the case of T Tauri, which has been studied interferometrically by Van Langevelde *et al.*(1993). Toward the continuum emission source at the star, they have detected weak red shifted absorption in the J=1-0 line of HCO^+. Because there is very little optical extinction toward this star, it is very likely that the HCO^+ is associated with the star and not a random forgrould cloud. Figure 4 shows a number of spectra taken with different instruments, and the sharply defined red shifted feature is clear. It is

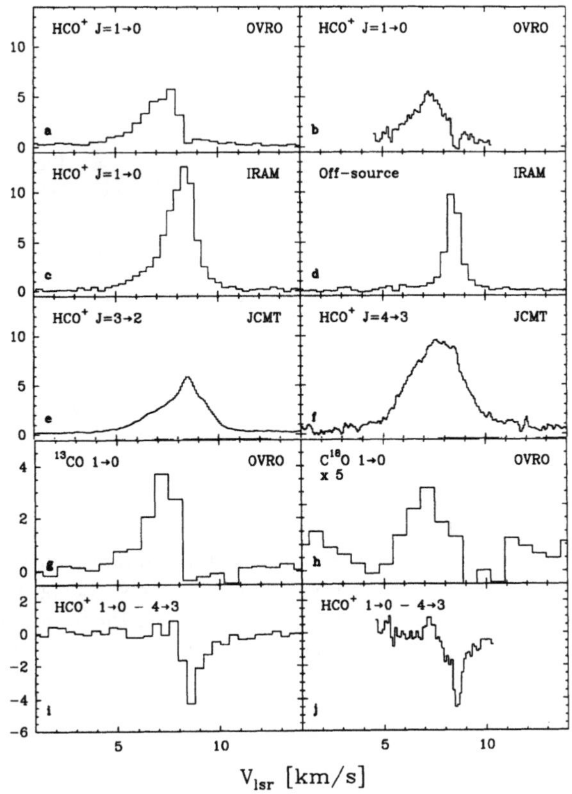

Fig. 4. Interferometer and single antenna spectra of T Tauri. The red-shifted absorption in HCO^+ implies present infall toward the young star (van Langevelde *et al.*1993).

often argued that infalling material is what must be driving the outflows, and for the massive star outflows discussed above we should expect to observe massive infalls at comparable rates. There is evidence of this for a few regions of massive star formation. Welch *et al.*(1988) reported inverse P-Cygni profiles (red shifted absorption and blue shifted emission) toward the very luminous core W49. The absorption is toward bright HII regions in the core, and can be understood as due to either the remnant infalling envelope that produced the multiple O associations as suggested by the authors, or separate infalls onto each of the HII regions (Dickel and Auer, 1994). Similar features have also been detected toward W51

(Rudolph *et al.*1990) and G34.2+0.2 (Carral and Welch, 1992). The most recent example is G5.89, where the effect is quite pronounced (Wilner *et al.*1994). Figure 5 shows the image spectral data in the form of a position/velocity diagram, with offset position from the center along an The implied rate of infall is 0.01, and the observed rate of outflow is 0.006 solar masses per year. There is thus evidence both for infalling material and that it is driving the outflow.

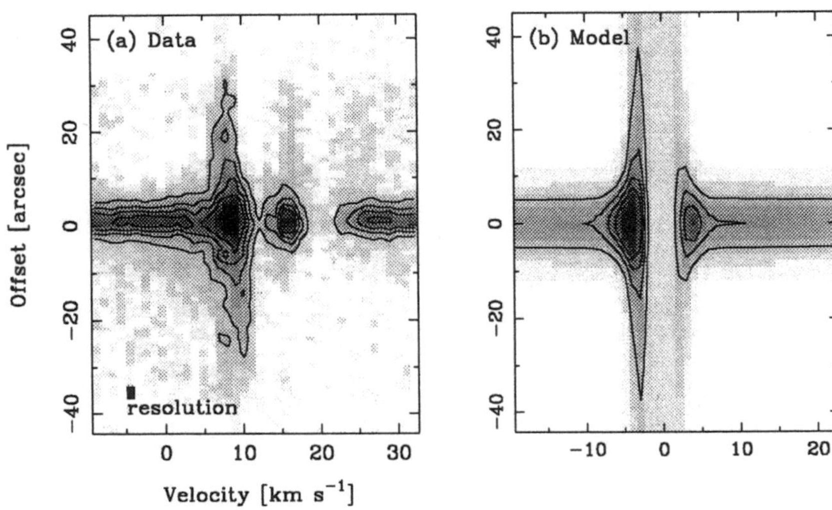

Fig. 5. $HCO^+(1-0)$ spectral map data shown as a position/velocity diagram toward the compact HII region G 5.89. The radial velocity is shown along an East-West line through the center of the source. The data is at the left and the 'inside-out' infall model is shown on the right. The right hand absorption is interpreted as a foreground cloud and not modelled (Wilner *et al.*1994).

2.5 Protostars and Protostellar Disks

Apart from the outflowing and infalling material, there is, of course, great interest in the structure and evolution of newly forming stars that are either embedded in the molecular clouds or, perhaps, just revealed. The hope that we may be able to understand the formation of stars is one of the main reasons for studying molecular clouds.

Optical and infra-red observations of the young T Tauri stars, which show thermal excesses but little visual extinction, argue that circumstellar dust must be organized into disks. Beckwith et al (1990), using the IRAM 30m telescope at 1.3mm wavelength, surveyed 86 pre-main-sequence-stars in the Taurus-Aurigae region, detecting thermal dust emission from nearly half, presumeably from disks. Then, Sargent and Beckwith (1991) mapped both the continuum and the

CO(1 − 0) emission from HL Tau, finding that the continuum was unresolved at their resolution of 3", but the CO distribution was a large flattened structure, possibly a large disk. They further found that there was evidence of Keplerian rotation in the gas close to the star, suggesting that some of the gas was bound to the star.

In an effort to understand the distributions of both gas and dust in such disks, Ohashi *et al.*(1991) mapped both the continuum and the CS(2-1) emission of 11 young stellar objects with resolutions of 3" - 9". One of their interesting results was that most of the imbedded stars showed no detectible thermal emission, whereas most of the revealed stars in their list did have continuum emission. In no case was the continuum resolved. Although their sample was small and probably subject to selection effects, they interpreted this intriguing result as posible evidence of disk evolution from the imbedded to the revealed state. In contrast, they detected CS emission more frequently toward the imbedded stars. Figure 6a shows their integrated CS maps of eight objects, including all of the invisible sources and two of the visible ones. What's striking is that the detected CS is typically not directly toward the associated infrared point source. This material is evidently not bound to the stars but rather may be the remnant of the star formation process. Blake *et al.*(1992) report that for HL Tau the CS in the disk must be depleted by at least a factor of 25-50 relative to its abundance in the nearby molecular cloud.

Perhaps the most interesting recent result is a 2" map which resolves the dust continuum and ^{13}CO emission from a disk around the binary pair GG Tau (Dutrey *et al.*1994). Because it surrounds a binary, this disk is evidently larger than what is expected for single stars, and it is clearly resolved. Figure 6b shows a continuous tone image of the dust continuum superposed on maps of the CO at three different radial velocities, showing rotation of the disk. The authors convincingly model the data with a gravitational interaction between the dusty disk and the binary system.

2.6 Distant Molecular Clouds

Whereas the large scale studies of molecular clouds in the Milky Way have been carried out entirely with single antennas, observations of these clouds in other galaxies at the same linear resolution requires the high resolution of arrays. Much of the excitement of high resolution observations of molecular material in other galaxies has focused on infrared luminous galaxies and Active Galactic Nuclei. Nevertheless, there have been important studies of spiral galaxies similar to our own which have enabled meaningful comparisons.

Wilson and Scoville(1990) mapped 19 fields in the Local Group galaxy M33 in CO(1−0) at a resolution of 7"-8" (30pc). Figure 7 shows maps in six typical fields. In these fields they identified 40 distinct clouds and found many similarities with those of our Milky Way. By comparing virial masses and CO fluxes, they found that the ratio of CO to H2 column densities was the same as in our galaxy. There were other similarities in the global properties. The cloud velocity dispersions are similar. The mean cloud gas density is approximately inversely proportional

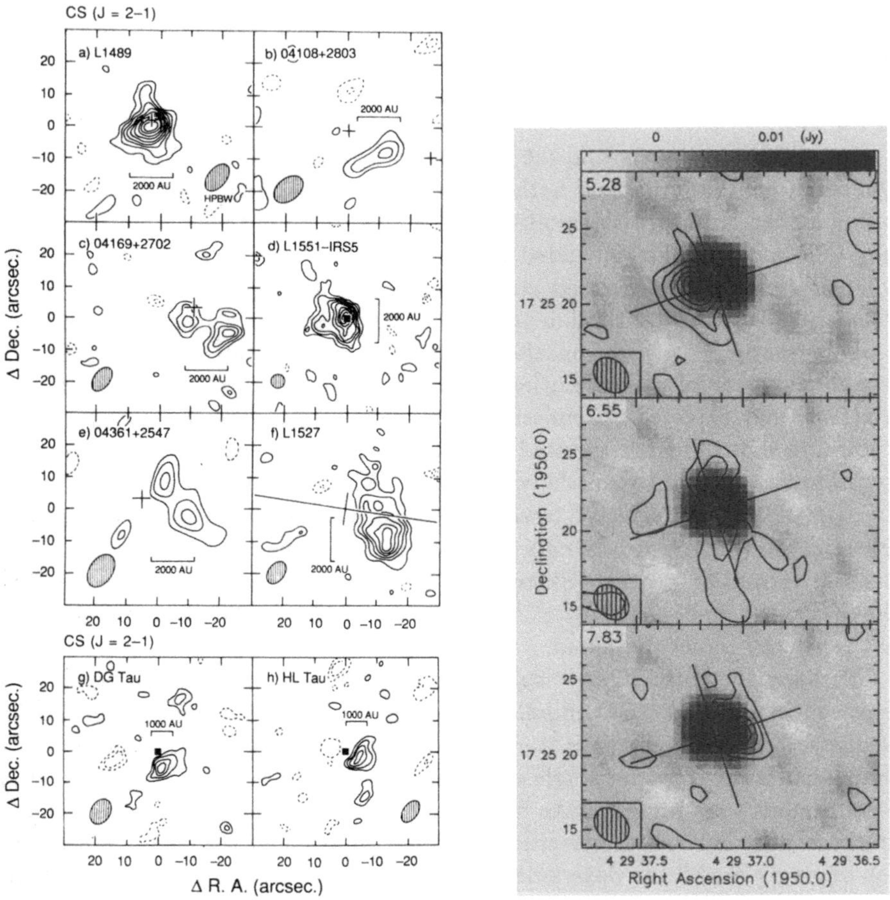

Fig. 6. (a) Nobeyama CS maps of 8 young stellar sources. The cross or square marks the IR or optical position. The resolution is shown as the hatched figure(Ohashi *et al.*1991). (b) The Plateau de Bure maps of the ring-like continuum disk around the GG Tau Binary, with three different radial velocities of the ^{13}CO emission, clearly showing rotation (Dutrey *et al.*1994).

to cloud diameter, and the cloud diameter varies approximately as the one third power of cloud mass. On the other hand, although the mass spectrum of M33 is similar to that of our galaxy, there seem to be no clouds with masses greater than about 4×10^5 as in the Milky Way. It is clear that with higher resolution an even greater range of cloud sizes could be studied, enabling a sharper comparison.

2.7 Boundary Layers

One effect of the presence of the dense clumpy molecular clouds is that there are large interstellar surface areas which separate the dense molecular mate-

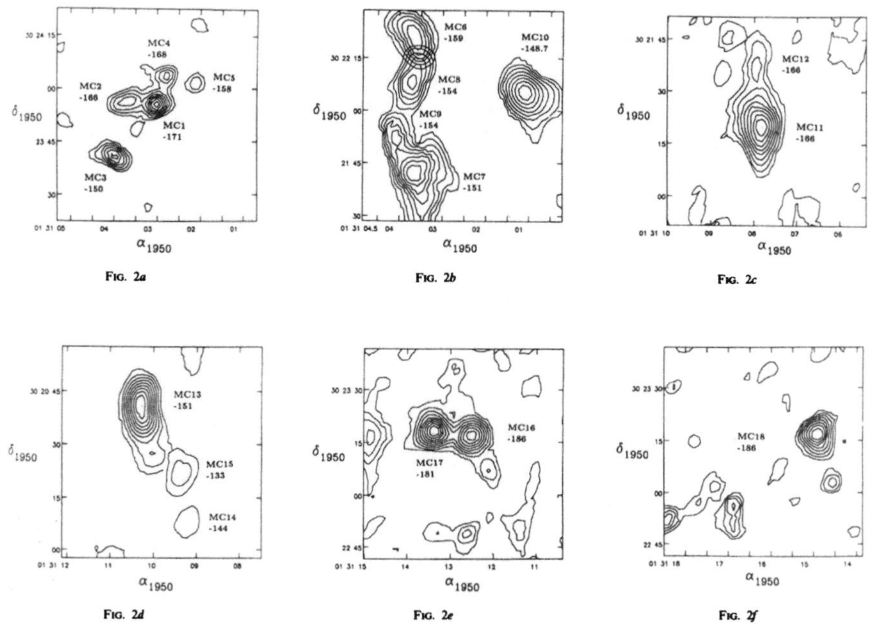

Fig. 7. CO maps from OVRO of six of the 16 fields mapped toward M33 showing molecular clouds (Wilson and Scoville, 1990).

rial from the diffuse ultra-violet of the hotest stars. These regions, known as PDR's (photodissociation regions or photon dominated regions), are thought to contain much interstellar gas and to play an important role in the physical and chemical evolution of the interstellar gas (Tielens and Hollenbach, 1985; Genzel *et al.*1989). An excellent example of the layering of different species in such a region is the Orion Bar, a sharp boundary between the ionized nebula in Orion and its surrounding molecular envelope. Figure 8 from Tielens *et al.*(1994) shows the variation of a few important species with depth into the cloud. Right at the boundary layer is the sharply defined flourescence of the 3.3 micrometer emission from PAH particles. Beyond this is the 2.1 micrometer emission from molecular hydrogen that has just recombined. Finaly, there is the emission from recombined $CO(1-0)$ mapped with the BIMA array. The heating is due to the ultraviolet from the HII region, and the recombination of species proceeds into the cloud as the ultraviolet is absorbed.

It is clear that the detailed study of such regions requires high resolution in the molecular emission.

3 Expectations

The previous paragraphs have shown some highlights of molecular cloud studies that have benefited from the high angular resolution of interferometer arrays. In

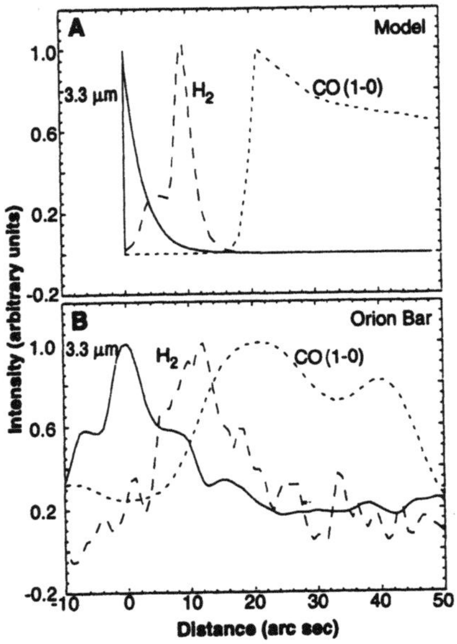

Fig. 8. The distribution of PAH (3.3 micron) emission, H_2 emission and CO$(1-0)$ emission (from BIMA) normal to the Orion Bar PDR. Data below and model above (Tielens *et al.*1994).

almost every area, having a more detailed picture is important. For the understanding of perhaps the most fundamental process, star formation, it is crucial. Until very recently, most of the arrays have had few elements and, as a result, have been very slow. The exception, of course, is the VLA which, with its 27 elements, is very fast. On the other hand it is the millimeter wavelength arrays which have the necessary sensitivity for studying processes that are basically thermal.

All the instruments are curently increasing in antenna number. The increased speed which is roughly proportional to the square of the number of elements, will bring a flood of new results, going quickly beyond the findings discussed above. The other new development will be the extension of baselines to one kilometer or more, and with angular resolutions at fractions of one arc second, the nearby protostars will give up thei secrets.

References

Beckwith, S. V. W., Sargent, A. I., Chini, R., and Gusten, R., 1990, AJ, 99, 924.

Blake, G. A., van Dishoeck, E., and Sargent, A. I., 1992, Ap. J., 391, L99.

Blitz, L., 1987, In 'Physical Processes in Interstellar Clouds', ed. G. Morfill, M. Scholer, Dordrecht:Reidel, pp. 35-58.

Brown, I. D., and Padman, R., 1988, 'Small Scale Structure in Small Clouds?' in "Molecular Clouds in the Milky Way and External Galaxies", R. L. Dickman, R.

L. Snell, and J. Young (eds), Proceedings, Amherst, Mass. Springer-Verlag, Berlin, 1988, p. 229.

Carral, P., and Welch, W. J., 1992, Ap. J., 385, 244.

Dickel, L., and Auer, L., 1994 (preprint).

Dutrey, A., Guilloteau, S., and Simon, M., 1994 (preprint)

Genzel, R., Harris, A. I., and Stutzki, J., 1989, in 'IR Spectroscopy and Astronomy', ed. B. H. Kaldeich, ESA, SP-290, p.115.

Goldsmith, P. F., Snell, R. L., Hasagawa, T., and Ukita, N., 1989, AP. J.,

Guilloteau, S., Bachiller, R., Fuente, A., and Lucas, R., 1992, A&A , 265, L49.

Hills, R., Janssen, M., Thornton, D. D., and Welch, W. J., 1972, Ap. J., 175, L59.

Kitamura, Y., Kawabe, R., and Ishiguro, M, PASPJ, 44, 407.

Lis, D. S., Goldsmith, P. F., Carlstrom, J. E., and Scoville, N. Z., AP. J., 402, 238-248.

Mezger, P. G., Scharml, J., and Terzian, Y., Ap. J., 1967, 150, 807.

Miller, G. E., and Scalo, J. M., 1979, Ap. J. Suppl., 41, 513-547.

Myers, P. C., 1985, In 'Protostars and Planets, II' ed. D. C. Black, M. S. Mathews, pp.81-103; Tucson, Univ. Ariz. Press.

Ohashi, N., Kawabe, R., Hayashi, M., and Ishiguro, M.,1991, AJ, 102, 2054.

Plambeck, R.L., and Wright, M.C.H., 1988, in 'Molecular Clouds in the Milky Way and External Galaxies', ed. R. L. Dickman, R. L. Snell, and J. Young, New-York, Springer-Verlag, p. 182.

Pratap, P., Batrla, W., and Snyder, L. E., 1990, Ap. J.. 351, 530-537.

Radhakrishnan, V., Goss, W. M., Murray, J. D., and Brooks, J. W., 1971, Ap. J. Suppl 24, 49.

Rank, D. M., Townes, C. H., and Welch, W. J., 1971, Science, 174, 1083-1101.

Rudolph, A., Welch, W. J., Dubrulle, B., and Palmer, P., 1990, Ap. J., 99, 1289.

Rudolph, A., and Welch, W. J., 1988, Ap. J., 397, L111.

Sargent, A. I., and Beckwith, S. V. W., 1991, Ap. J., 382, L31.

Scoville *et al.*1986, Ap. J., 303, 416-432.

Tielens, A. G. G. M., and Hollenbach, D., 1985, Ap. J., 291, 722.

Tielens, A. G. G. M., Meixner, M. M., van der Werf, P. P., Bregman, J., Tauber, J. A., Stutzki, J., and Rank, D. M., 1994, Science, 262, 86.

Turner, J. L., and Welch, W. J., 1984, Ap. J., 287, L81.

Van Langevelde, J., van Dishoeck, E., and Blake, G., 1993 (preprint).

Welch, W.J., Dreher, J. W., Jackson, J. M., Terebey. S., and Vogel, S. N., 1988, Science, 238, 1550.

Wilson, C. D., and Scoville, N., 1990, Ap. J., 363, L135.

Wilner, D. W., Forster, J. R., and Welch, W. J., 1994(preprint).

Wilner, D. W., Wright, M. C. H., and Plambeck, R., 1994, Ap. J., 422, 642.

Wink, J. E., Altenhof, W. J., and Webster, W. J., 1975, A&A , 38, 109.

Wynn-Williams, C. G., 1971, MNRAS,151,397.

Dense Molecular Gas in Star-Forming Regions – The Importance of Submillimeter Observations

Karl M. Menten

Harvard-Smithsonian Center for Astrophysics, 60 Garden Street, Cambridge, MA, USA

Abstract: An account of the prospects for astronomy in the submillimeter range is given, emphasizing spectral line measurements.

1 Introduction

The submillimeter band remains one of the last portions of the electromagnetic spectrum that have yet to be systematically explored by astronomers. Between wavelengths of 300 μm and 1 mm (1 THz $> \nu >$ 300 GHz) there exist several broad "windows" in which astronomical observations from the ground are possible from favourable sites such as high mountain tops or Antarctica under good weather conditions (Fig. 1). At far-infrared wavelengths, i.e. between 300 μm and 30 μm, comparable windows do not exist and ground-based observations are essentially impossible.

While submillimeter observations of spectral lines and dust continuum emission are providing significant contributions to many areas of astronomy from planetary science to cosmology (see Phillips & Keene 1992 for a recent review), we shall in the following mostly discuss molecular line observations of the warm ($T \gtrsim 100$ K), compact ($r \lesssim 0.05$ pc) , and dense ($n \gtrsim 10^6$ cm^{-3}) core regions of giant molecular clouds. These regions are of great astronomical interest since they are giving birth to massive stars. Furthermore, the high molecular column densities and excitation temperatures in these cores make them important laboratories for studying interstellar chemistry; indeed, many molecules, although also present in colder, lower density gas, are only observable toward dense cores. Most high-mass star-forming regions are at distances greater than 1 kpc, and the emitting regions have angular extents smaller than $\approx 10''$. Therefore, systematic studies in the submillimeter range had to await large ground-based telescopes with high precision surfaces, namely the 10.4 m telescope of the Caltech Submillimeter Observatory (CSO) and the 15 m James Clerk Maxwell Telescope (JCMT). These telescopes have angular resolutions of order $10''$ at their highest operating frequency, which is at the atmospheric cutoff near 1 THz. Both the CSO and the JCMT started operating in the late 1980's on the top of Mauna Kea, Hawaii. Before then, several of the most prominent spectral lines had been detected with the 0.9 m telescope aboard the Kuiper Airborne Observatory (KAO) and optical telescopes on Mauna Kea. A third large dish that will operate throughout the submillimeter range, the 10 m Heinrich Hertz Telescope, is currently nearing completion on Mt. Graham in Arizona. Even higher spatial resolutions will be obtained with the Submillimeter Array (SMA) currently under construction by the Smithsonian Institution (Moran 1994). Apart from the large telescopes now available, molecular line astronomy at submillimeter wavelengths is greatly benefitting from breathtaking new developments in heterodyne receiver technology (see Blundell & Tong 1992 for a recent review).

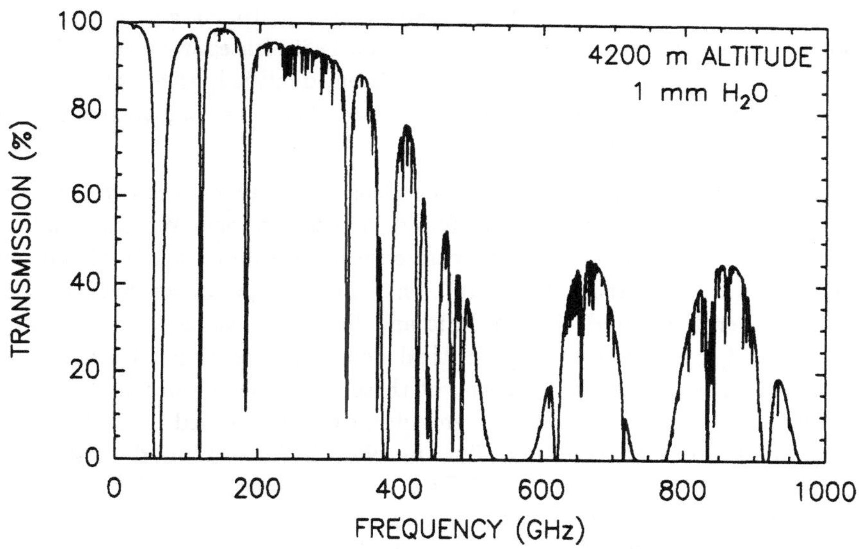

Fig. 1. Atmospheric transmission on Mauna Kea for 1 mm of precipitable water.

In section 2 of this article, we give examples for the unique information that submillimeter observations of molecular lines are providing on the physics and chemistry of dense molecular cores. We summarize some recent results and point out a number of open questions that will be addressed by future observations. In the third section we describe recent submillimeter observations of the particularly important water molecule.

2 Submillimeter Observations of Hot, Dense Cloud Cores

2.1 Hot Cores

It is clear from the association of hot, dense molecular cloud cores with ultracompact H II regions and luminous infrared sources that these regions are the sites of recent or on-going high-mass star formation (see, e.g., Wilson & Walmsley 1989). Because of its proximity, D = 460 pc, the dense molecular core coincident with the Kleinmann-Low (KL) nebula in the Orion Molecular Cloud (OMC-1) is the best studied example. Analyses of line profiles and interferometric observations indicate that the molecular emission from the Orion-KL region arises from a number of components with different kinematic properties and spatial distributions (see Blake et al. 1987 and Genzel & Stutzki 1989 for a review). A complex picture has emerged involving an emission component from relatively cool, quiescent cloud material, at least two distinct hot regions (the "hot core" and "compact ridge") heated to 150 – 200 K by one or more embedded young stellar objects, and high velocity emission arising from outflows driven by these young stars. Each of these regions has its chemical peculiarities, with high temperature "shock-chemistry" and other processes, such as injection of molecules off icy grain surfaces into the gas phase, playing important roles (Blake et al. 1987; Brown et al. 1988; Hasegawa et al. 1992). Hot cores in other giant molecular clouds have been identified, although systematic line surveys have only been published for the core(s) in the Sagittarius B2 cloud (e.g. Sutton et al. 1991).

Molecules are excellent probes of hot, dense molecular cores over the whole range of densities and temperatures found in these regions. Within a radius of ≈ 0.01 pc from an embedded star, rotational lines from a variety of molecules probe material of densities $10^6 - 10^9$ cm^{-3} and temperatures between a few tens and several hundred K. Diatomic and linear molecules with high dipole moments and, consequently, high critical densities, are of particular interest. As demonstrated in Fig. 2, for many of these molecules numerous transitions, spanning a wide range of excitation energies, can be observed from the ground, with the submillimeter wavelength lines probing the hottest and densest regions. Recent detections include the 797 GHz $J = 9 \rightarrow 8$ transition of HCN (Stutzki et al. 1988) toward the Orion hot core and the 490 GHz $J = 10 \rightarrow 9$ line of CS (Hauschildt et al. 1993) toward a number of star-forming regions; high-J CO data are reviewed by Stutzki (1989).

Systematic line surveys are necessary to study the chemistry in a given region and to analyse the excitation of various molecules, thus allowing an assessment of their importance for the energy balance. At millimeter wavelengths, wideband line surveys have been conducted for Orion-KL and Sgr B2 and are summarized by Hjalmarson (1989). Submillimeter wavelength line surveys to date have been published only for the same two sources in the 330 – 360 GHz frequency interval (Jewell et al. 1989, Sutton et al. 1991). A new sensitive survey of Orion (Fig. 3) covering these frequencies indicates a line density at least comparable to that found in the millimeter region (Groesbeck et al. 1994); essentially, the number of detected lines is a function of the limiting sensitivity of the survey. About half of the lines in the Orion spectrum shown in Fig. 3 are unidentified! The majority of these U-lines is likely to be emitted from various asymmetric rotor molecules with relatively high abundances in this source. Many of these species have low-lying vibrationally excited states that are easily populated under hot core conditions. Laboratory data on the rotational spectra of these states is very sparse. Recent major efforts include a thorough analysis of the spectrum of methanol (CH_3OH)

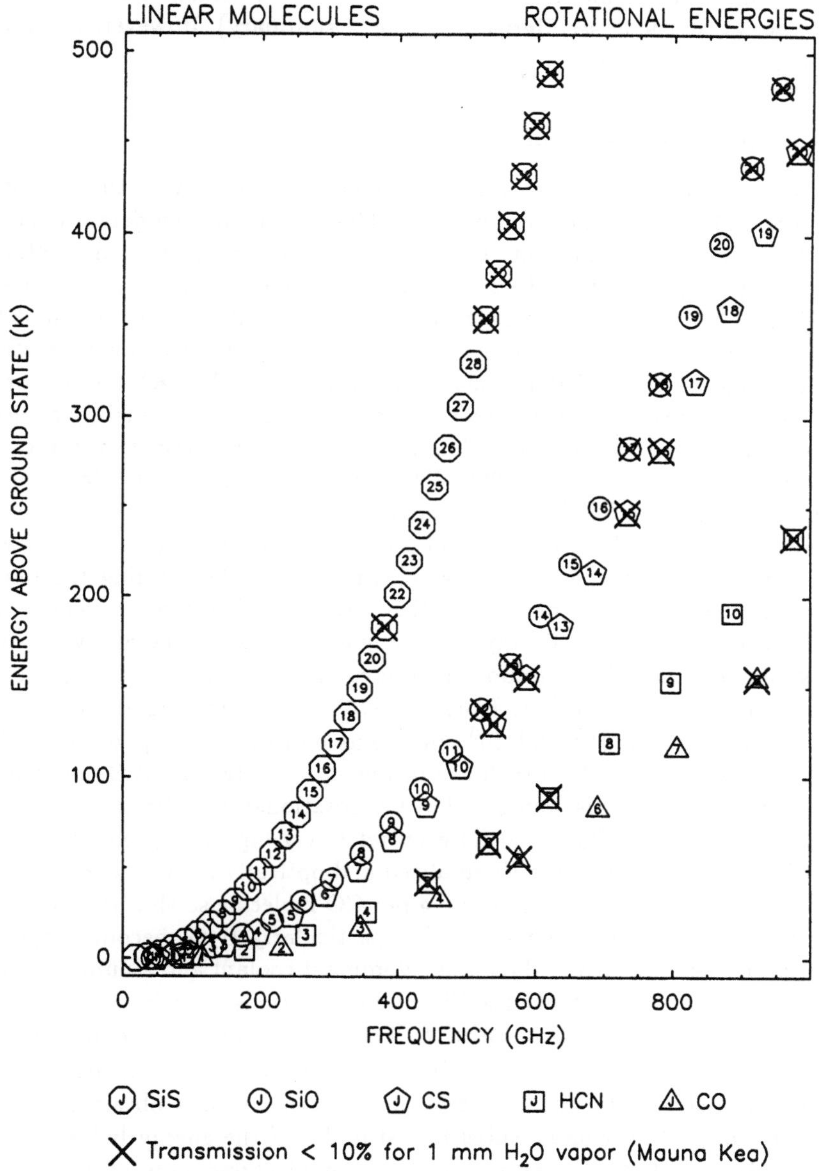

Fig. 2. Energies above ground state versus transition frequency for the mm- and submm-wavelength rotational lines with energies below 500 K of some important diatomic molecules and HCN. For each transition the number inside the respective symbol denotes the rotational quantum number of the upper state. If the entry for a transition is "crossed out", the atmospheric transmission at its frequency is less than 10% on Mauna Kea with 1 mm of precipitable water (cf. Fig. 1

in its ground state and first two torsionally excited states that resulted in reliable frequency information for all methanol lines of astronomical interest up to 1 THz (Anderson et al. 1990; see Herbst 1991 for a summary). However, much more work is needed in this area.

2.2 Cooling of Dense Cloud Cores

On a galactic scale the most important cooling transition of the interstellar gas is the 158 μm C II fine-structure line. This is impressively demonstrated by the COBE far-infrared absolute spectrometer results (Wright et al. 1991) and a rapidly growing number of extragalactic C II measurements (e.g. Stacey et al. 1991). C II emission arises both from the purely atomic medium and from the photon-dominated outer regions of molecular clouds, which are penetrated by UV radiation from H II regions. The warm ($T > 100$ K), mostly atomic parts near the surfaces of these photodissociation regions are mainly cooled by C II, O I, and other fine-structure lines. With increasing extinction the two C I fine-structure lines at 492 and 809 GHz, respectively, and CO rotational lines dominate the cooling (Tielens & Hollenbach 1985).

Submillimeter and far-infrared rotational lines from various molecules are the dominant cooling agents in dense cloud cores. Understanding the relevant cooling processes, which determine the rate at which gravitational energy, transformed into kinetic energy of the gas, can be radiated away during protostellar collapse, is an important prerequisite for understanding the star-forming process. Goldsmith & Langer (1978) studied molecular cloud cooling for densities between 10^2 and 10^6 cm^{-3} and temperatures up to 40 K. Their results, an example of which is shown in Fig. 4, indicate that at H_2 densities up to 10^5 cm^{-3} the cooling is dominated by CO rotational transitions. For moderate to high temperatures ($T_{\rm kin} \gtrsim 50$ K) even relatively high lying (e.g. $J = 6 \rightarrow 5$) transitions of $^{12}C^{16}O$ become thermalized and optically thick in dense ($n > 10^5$ cm^{-3}) cores, which reduces the cooling per CO molecule so that the rarer ^{13}CO and $C^{18}O$ isotopes and other molecules, particularly H_2O, become important. For densities near 10^6 cm^{-3} H_2O is the dominant coolant and a number of other less abundant molecules make significant contributions as well.

An extrapolation of Goldsmith & Langer's results to the temperatures ($T_{\rm kin} \gtrsim 150$ K) and densities ($n \gtrsim 10^7$ cm^{-3}) prevalent in hot, dense cores is by no means straightforward. Several observational results obtained in recent years have important consequences for estimates of the energy balance in such regions. Goldsmith & Langer assumed fractional abundances of 10^{-9} for the "rare" molecules they considered, i.e. species other than H_2, CO, H_2O, and O_2 and their isotopic varieties. The molecules in question (HCN, HCO^+, CN, CS, and others) do indeed have abundances of this order (or somewhat higher) in "quiescent" low temperature regions, such as TMC-1 (Irvine et al. 1985) or the extended Orion Ridge (Blake et al. 1987). However, as mentioned above, inside hot cores molecular abundances can be drastically different from cold cloud values as a result of various processes. As an example we note that for the Orion Hot Core ($T_{\rm kin} \approx 150 - 200$ K) Blake et al. (1987) estimate a very high

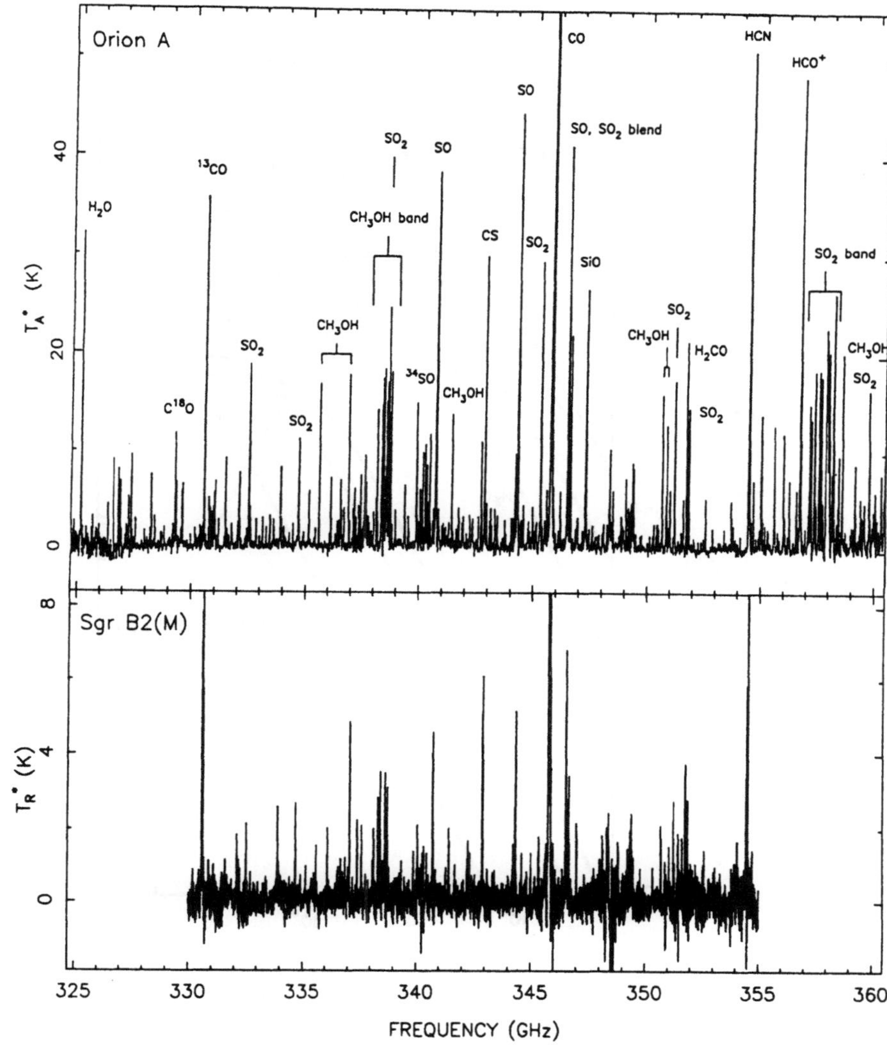

Fig. 3. Spectral-line surveys of the 345 GHz atmospheric window toward the Orion-KL (*top*) and Sgr B2(M) (*bottom*) molecular cloud cores. The figure combines the results of Groesbeck et al. 1994 (Orion, see also Phillips & Keene 1992) and Sutton et al. 1991 (Sgr B2), which were obtained with the CSO and the JCMT, respectively. Some of the stronger lines are labelled in the upper part of the figure. Many lines in the Orion spectrum remain unidentified. Some of the negative features in the Sgr B2(M) spectrum are spurious, although genuine absorption is observed toward some lines against the compact dust continuum emission of this source.

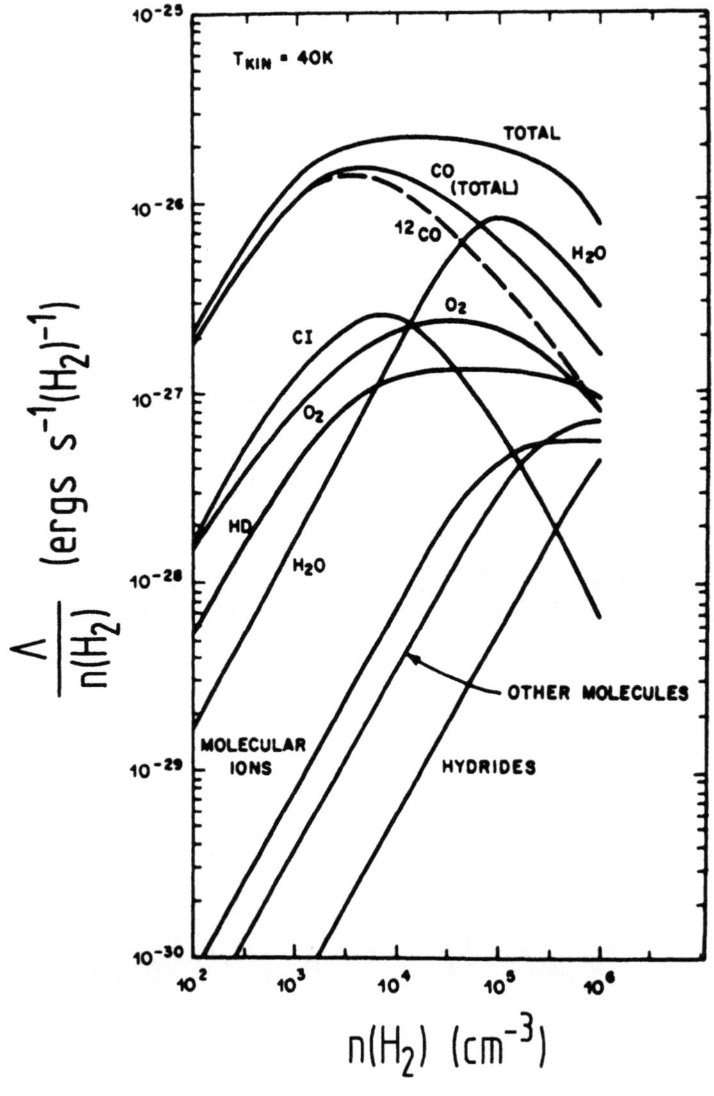

Fig. 4. Molecular cloud cooling per H_2 molecule for a kinetic temperature of 40 K and H_2 densities between 10^2 and 10^6 cm^{-3} (from Goldsmith & Langer 1978).

HCN to H_2 abundance ratio of 3×10^{-7}, possibly a result of high temperature chemistry in warm postshock gas. Modeling their observations of the 797 GHz HCN $J = 9 \rightarrow 8$ transition, Stutzki et al. (1988) estimate an optical depth of several 100 for this line, concluding that even the rare $H^{13}CN$ and $HC^{15}N$ isotopes will have strong high-J submillimeter lines. Other molecules, such as SO and SO_2, as well as more complex species like CH_3OH, all of which have rich (sub)millimeter spectra, also show high overabundances in the Orion core and are possibly significant coolants. Evaporation of molecular ices off dust grain mantles is a plausible scenario to explain the observed high gas phase abundances of some of these species (Menten et al. 1987; Brown et al. 1988; Hasegawa et al. 1992).

At present, the contribution of light hydrides to dense core cooling is poorly understood. These molecules have large rotational constants so that their rotational spectra are in the submillimeter or far-infrared wavelength bands. Phillips & Keene (1992) give a compilation of important transitions and their frequencies. Because of their wide level spacings, light hydrides are very effective coolants approaching the cooling efficiency of CO despite their much lower fractional abundance. Dalgarno et al. (1974) discuss the case of hydrogen chloride (HCl) which is thought to be a major reservoir of chlorine in molecular clouds and find that for a fractional abundance of 10^{-7}, which is 300 times lower than the fractional abundunce they assume for CO, the HCl cooling efficiency becomes comparable to that of CO for densities exceeding a few times 10^5 cm^{-3}. Observations of the $J = 1 \rightarrow 0$ ground state transition of HCl near 626 GHz using the KAO indicate somewhat lower abundances of $(0.5\text{–}5.0)\times10^{-8}$ toward OMC-1, where the size of the emitting region relative to the $2'$ beam size is a major source of uncertainty (Blake et al. 1985). Since ground-based observations of this line with resolutions of order $12''$ are now feasible, important new information on HCl toward Orion and other molecular cloud cores will become available soon.

3 Submillimeter Observations of Interstellar Water

3.1 Water in Molecular Clouds

The water (H_2O) molecule is an asymmetric rotor which has most of its pure rotational transitions at submillimeter and far-infrared wavelengths (Fig. 5) As discussed above, water is a dominant coolant in warm and dense molecular clouds and knowledge of the H_2O/H_2 ratio therefore of great importance.

For quiescent, cold molecular clouds, where water is formed by ion-molecule reactions, most theoretical models predict H_2O/H_2 abundance ratios between 10^{-6} and 10^{-5} (see Wannier et al. 1991). At the high temperatures (> 1000 K) found in shocked regions the neutral reactions

$$O + H_2 \rightarrow OH + H$$

$$OH + H_2 \rightarrow H_2O + H$$

proceed rapidly and may produce water abundances that are hundreds of times higher than quiescent cloud values in gas behind slow non-dissociative shocks (Draine et al. 1983). In the case of fast shocks, which dissociate molecular hydrogen and other molecules, the water abundance in the postshock gas stays comparatively low because the above reactions can only form H_2O effectively after significant amounts of H_2 have reformed. Furthermore, destructive UV radiation emitted from the region around the shock front keeps the H_2O concentration small (Neufeld & Dalgarno 1989). Evaporation of molecular ices

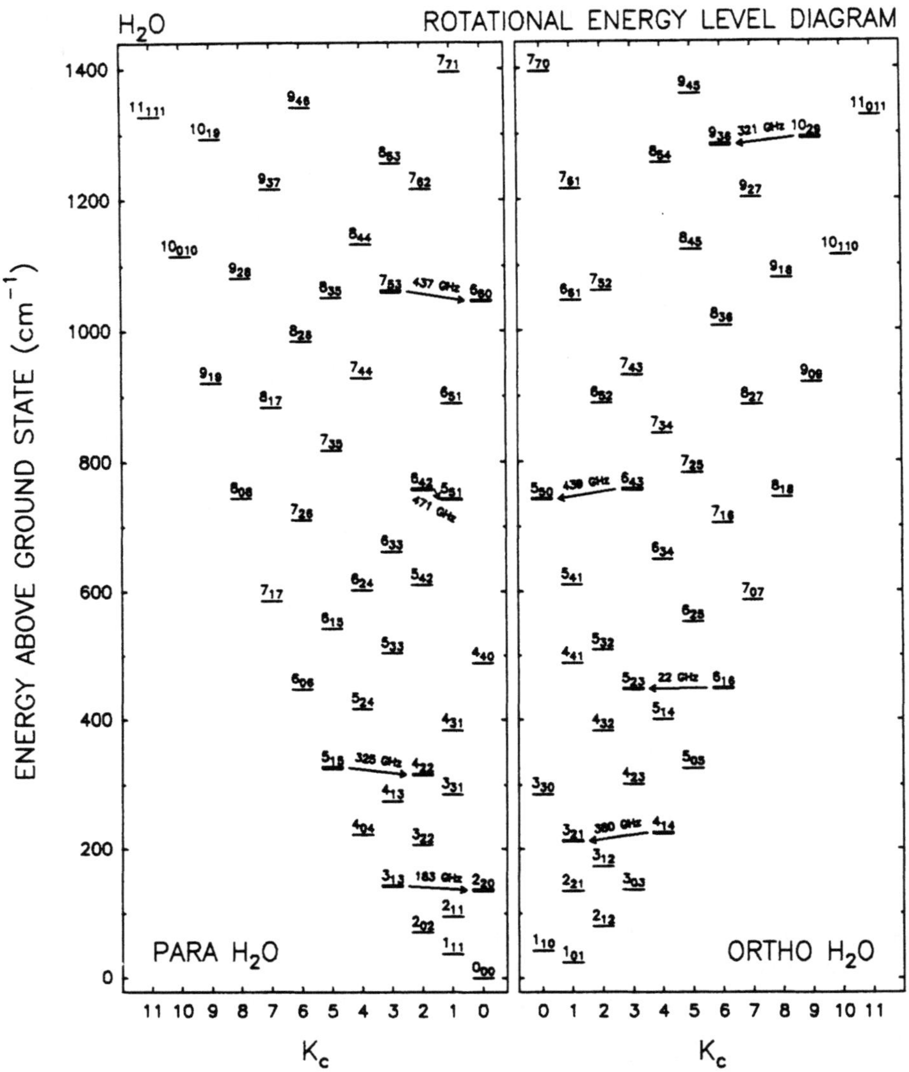

Fig. 5. Rotational energy levels of para-H_2O (*left*) and ortho-H_2O (*right*) with energies above the ground state below 1400 cm^{-1}. The arrows denote the observed interstellar maser lines, for which transition frequencies are also indicated. The 437 GHz $7_{53} \rightarrow 6_{60}$ transition has been detected in evolved stars, but not in star-forming regions.

off grain mantles heated by embedded stars is another way to produce high gas phase abundunces of water and other molecules in hot cloud cores.

Determinations of the interstellar water abundance are notoriously difficult, because observations of thermal (non-maser) emission and/or absorption of H_2O lines toward molecular clouds are practically impossible because of strong absorption from the same transitions in the Earth's atmosphere (see Fig. 1). At the frequencies of low excitation H_2O lines the atmosphere is totally opaque, so that studies of water in cold, quiescent molecular clouds will require observations from space. In particular, measurements of the $1_{10} \rightarrow 1_{01}$ ground-state transition of ortho-H_2O and its $H_2^{18}O$ isotope at 557 and 548 GHz, respectively, with the soon-to-be-launched Submillimeter Wave Astronomy Satellite (SWAS, Melnick et al. 1991) are expected to yield the first determinations of the water abundance in such regions.

Emission from a number of higher excitation H_2O transitions, with energies $\gtrsim 200$ K above the ground state, has been observed toward various star-forming regions and evolved stars (see Fig. 5). However, as discussed below, all these lines are inverted, which makes abundance estimates difficult. Observational approaches invoke the $H_2^{18}O$ and HDO isotopes, which have lines at frequencies with better atmospheric transmission. Since the molecular structure of the deuterated species is significantly different from that of H_2O, numerous HDO lines with a wide range of energies above the ground state occur throughout the centimeter-, millimeter-, and submillimeter-wavelength bands; in particular the $1_{01} \rightarrow 0_{00}$ ground-state transition near $\nu = 465$ GHz has been detected toward Orion-KL (Schulz et al. 1991). Unfortunately, the unknown degree of deuterium enrichment makes HDO an unreliable tracer of the water abundance. Interestingly, HDO observations provide strong evidence for the grain mantle evaporation scenario as the source for high water abundances in hot cores. The high HDO column densities and HDO/$H_2^{18}O$ abundance ratios measured in a number of hot regions (Henkel et al. 1987; Jacq et al. 1990) indicate that the HDO was formed at low temperatures where significant deuterium fractionation takes place, then froze out on dust grains and reentered the gas phase when newly formed stars started to heat their environments.

Observations of the $3_{13} \rightarrow 2_{20}$ and $5_{15} \rightarrow 4_{22}$ transitions of $H_2^{18}O$ at 203 GHz and 322 GHz, respectively, (Jacq et al. 1988, 1990; Menten et al. 1990b) indicate H_2O abundances around 10^{-5} in several molecular hot cores. These observations are plagued with contamination from neighbouring lines from other molecules. Furthermore, there are uncertainties about source geometries and molecular excitation, since by analogy with the main isotope, where the corresponding transitions at 183 and 325 GHz show maser emission, one expects these lines to be inverted. Airborne searches for the 548 GHz $H_2^{18}O$ ground-state ortho transition toward a number of giant molecular cloud cores were unsuccesful and imply very low upper limits on H_2O/H_2 between a few times 10^{-7} and 10^{-6} (Wannier et al. 1991).

3.2 Submillimeter Water Masers

Maser emission in the 22 GHz $6_{16} \rightarrow 5_{23}$ transition of H_2O was first discovered by Cheung et al. (1969) and has since been studied in hundreds of star-forming regions and circumstellar envelopes around oxygen-rich evolved stars. Because of their often very high intensities, 22 GHz masers are important signposts for high- and low-mass star formation that can be detected throughout the Galaxy. Milliarcsecond resolution, VLBI, proper-motion studies have firmly established that interstellar H_2O masers form in high velocity outflows driven by young

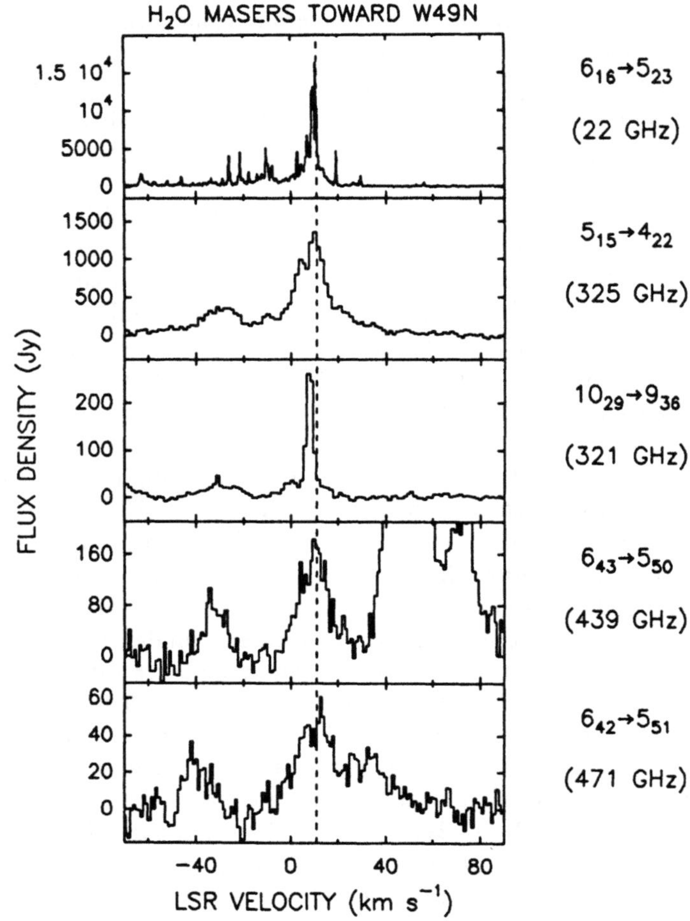

Fig. 6. Spectra of the 22 GHz line and various submillimeter water maser transitions detected toward W49N, the most luminous 22 GHz maser source in the Galaxy. The 325, 439, and 471 GHz transitions spectra were taken during the same observing run (see Melnick et al. 1993), while the 22 and 321 GHz spectra were observed at a different time. The dashed vertical line marks an LSR velocity of 11 km s^{-1}.

stellar objects (see, e.g., Moran, this volume). For a long time, the excitation of water masers remained poorly understood, since observations of a single line are insufficient to meaningfully constrain excitation models. Subsequent observations of the $3_{13} \rightarrow 2_{20}$ and $4_{14} \rightarrow 3_{21}$ transitions at 183 and 380 GHz with the KAO did little to change this situation, because only a single source, Orion-KL, was detected (Waters et al. 1980; Phillips et al. 1980).

Over the last few years the observational situation has changed markedly. Using the CSO, new water transitions, the $10_{29} \rightarrow 9_{36}$ and $5_{15} \rightarrow 4_{22}$ lines at frequencies of 321 and 325 GHz, respectively, were detected toward a number of star-forming regions (Menten et al. 1990a,b). Also, high quality observations of the 183 GHz line toward various regions have been made with the IRAM 30-m telescope (Cernicharo et al. 1990). Maser emission is observed in all of these lines. Model calculations by Neufeld & Melnick (1990, 1991) showed that these masers can be collisionally pumped in the warm ($T \gtrsim 500$ K) gas found in postshock regions, and they predicted maser emission in the $6_{43} \rightarrow 5_{50}$ and $6_{42} \rightarrow 5_{51}$ transitions near 439 and 471 GHz, which was subsequently detected by Melnick et al. (1993) using the CSO. These authors use contemporaneously measured spectra of the 325, 439, and 471 GHz transitions (see Fig. 6) to put constraints on the excitation conditions in the masing gas for W49N and five other star-forming regions and find that in all cases the kinetic temperature has to be higher than 900 K to explain the observed line ratios. This seems to suggest that the masers are found in the postshock gas behind non-dissociative C-shocks, which can have higher temperatures than the gas behind dissociative J-shocks. Another H_2O maser model (Elitzur et al. 1989) had placed the masers in the warm (300 – 400 K) molecular reformation zone behind J-shocks.

Obviously, multi-transition studies of water masers have great potential for providing interesting information on the shocked regions in which they are forming. At present, one inherent limit in using observed line ratios together with model calculations to derive source parameters is given by the fact that the emitting regions of the individual maser lines might differ in their small scall structure, which is unresolvable with the $15''$ – $20''$ beams afforded by single dish telescopes. This problem will be resolved by synthesis imaging with future interferometers such as the Smithsonian Submillimeter Array, which will produce maps of submillimeter water masers at resolutions of a few tenths of an arcsecond, comparable to the resolution attainable with the Very Large Array (VLA) for the 22 GHz line.

4 Conclusions

Submillimeter wavelength observations of dense molecular cloud cores have provided astronomers with important information on the chemistry and energy balance in the immediate neighborhood of newly formed stars. In the near future sensitive heterodyne receivers will allow single-dish observations at the highest submillimeter frequencies with resolutions better than $10''$ and interferometry in the subarcsecond regime.

New observations and theoretical studies have to address the problem of molecular cooling of hot, dense cores, many aspects of which remain poorly understood. Line surveys at the highest submillimeter as well as at far-infrared wavelengths are necessary to study the excitation of lines from abundant molecules and, thus, allow an assessment of their cooling efficiency. In particular, such observations will yield important information on various light hydrides, which might be significant coolants and are also of great astrochemical interest.

Rotational line emission from the water molecule is of central importance for the energy balance in warm, dense cores and determinations of the presently poorly known water abundance are therefore highly desirable. Multi-transition studies of the various submillimeter-wavelength H_2O maser lines are useful probes of the hot postshock gas in which these masers are forming.

I would like to thank Mark Reid for reading the manuscript and Ed Sutton for providing the Sgr B2 spectrum shown in Fig. 3 in computer-readable format.

References

Anderson, T., Herbst, E., De Lucia, F. C., 1990, ApJS, 82, 797
Blake, G. A., Keene, J., Phillips, T. G., 1985, ApJ 295, 501
Blake, G. A., Sutton, E. C., Masson, C. R., Phillips, T. G., 1987, ApJ, 315, 621
Blundell, R., Tong, C.-Y. E, 1992, Proc. IEEE, 80, 1702
Brown , P. D., Charnley, S. B., Millar, T. J., 1988, MNRAS, 231, 409
Cernicharo, J., Thum, C., Hein, H., John, D., Garcia, P., Mattioco, F., 1990, A& A, 231, L15
Cheung, A. C., Rank, D. M., Townes, C. H., Thornton, D. D., Welch, W. J., 1969, Nature, 221, 626
Dalgarno, A., De Jong, T., Oppenheimer, M., Black, J. H., 1974, ApJ, 192, L37
Draine, B. T., Roberge, W. G., Dalgarno, A., 1983, ApJ, 264, 485
Elitzur, M., Hollenbach, D. J., and McKee, C. F., 1989, ApJ, 346, 983
Genzel, R., Stutzki, J., 1989, ARAA, 27, 41
Goldsmith, P. F., Langer, W. D., 1978, ApJ, 222, 881
Groesbeck, T. D., et al., 1994, in prep.
Hasegawa, T., Herbst, E., Leung, C. M., 1992, ApJS, 82, 167
Hauschildt, H., Güsten, R., Phillips, T. G., Schilke, T. G., Serabyn, E., Walker, C. K., 1993, A& A, 273, L23
Henkel, C., Mauersberger, R., Wilson, T. L., Snyder, L. E., Menten, K. M., Wouterloot, J. G. A., 1987, A& A, 182, 299
Herbst, E., 1991, in Skylines (Proc. Third Haystack Observatory Meeting), eds. A. D. Haschick & P. T. P. Ho (San Francisco: ASP), 313
Hjalmarson, Å, 1989, in The Physics and Chemistry of Interstellar Molecular Clouds, eds. G. Winnewisser & J. T. Armstrong (Berlin: Springer), 73
Irvine, W. M., Schloerb, F. P., Hjalmarson, Å., Herbst, E., 1985, in Protostars & Planets II, eds. D. C. Black & M. S. Matthews (Tucson: The University of Arizona Press), 579
Jacq, T., Jewell, P. R., Henkel, C., Walmsley, C. M., Baudry, A., 1988, A& A, 199, L5
Jacq, T., Walmsley, C. M., Henkel, C., Baudry, A., Mauersberger, R., Jewell, P. R., 1990, A& A, 228, 447
Jewell, P. R., Hollis, J. M., Lovas, F. J., Snyder, L. E., 1989, ApJS, 70, 833
Melnick, G., et al., 1991, in Skylines (Proc. Third Haystack Observatory Meeting), eds. A. D. Haschick & P. T. P. Ho (San Francisco: ASP), 119
Melnick, G. J., Menten, K. M., Phillips, T. G., Hunter, T., 1993, ApJ, 416, L37
Menten, K. M., Melnick, G. J., Phillips, T. G., 1990a, ApJ, 350, L41
Menten, K. M., Melnick, G. J., Phillips, T. G., Neufeld, D. A., 1990b, ApJ, 363, L27
Menten, K. M., Walmsley, C. M., Henkel, C., Wilson, T. L., 1988, A& A, 198, 253
Moran, J. M., 1994, in Very High Angular Resolution Imaging, eds. J. G. Robertson & W. J. Tango (Dordrecht: Kluwer), 27
Neufeld, D. A., Dalgarno, A., 1989, ApJ, 340, 869

Neufeld, D. A., Melnick, G. J. 1990, ApJ, 352, L9
Neufeld, D. A., Melnick, G. J. 1991, ApJ, 368, 215
Phillips, T. G., Keene J., 1992, Proc. IEEE, 80, 1662
Phillips, T. G., Kwan, J., Huggins, P. J., 1980, in Interstellar Molecules, ed. B. H. Andrew (Dordrecht: Reidel), 21
Schulz, A., Güsten, R., Serabyn, E., Walmsley, C. M., 1991, A& A, 246, L55
Stacey, G. J., Geis, N., Genzel, R., Lugten, J. B., Poglitsch, A., Sternberg, A., Townes, C. H., 1991, ApJ, 373, 423
Stutzki, J., 1989, in The Physics and Chemistry of Interstellar Molecular Clouds, eds. G. Winnewisser & J. T. Armstrong (Berlin: Springer), 53
Stutzki, J., Genzel, R., Harris, A. I., Herman, J., Jaffe, D. T., 1988, ApJ, 330, L125
Sutton, E. C., Jaminet, P. A., Danchi, W. C., Blake, G. A., 1991, ApJS, 77, 255
Tielens, A. G. G. M., Hollenbach, D., 1985, ApJ, 291, 722
Wannier, P. G., et al., 1991, ApJ, 377, 171
Waters, J. W. et al., 1980, ApJ, 235, 57
Wilson, T. L., Walmsley, C. M., 1989, A& A Rev, 1, 141
Wright, E. L., et al., 1991, ApJ, 381, 200

Conditions in Regions of High Mass Star Formation

C.M. Walmsley and R. Güsten

Max-Planck-Institut für Radioastronomie, Auf dem Hügel 69, D-53121 Bonn, Germany

Abstract: A review is given of our present knowledge of the physical characteristics of the high density molecular gas associated with regions of high mass star formation. We give a brief summary of the methods used to derive the temperature and density of such regions and discuss the characteristics of the hot high density clumps known as "hot cores". A discussion is given of recent interferometric observations of the nearby HII regions NGC 2024 and W3(OH).

1 Introduction

High mass star formation regions tend to be more distant than their low mass brothers. They also usually present a more confused picture due partly to the interactions of the young newly formed O-B stars with their surroundings. Another factor may be that O-B stars tend to form in clusters whereas solar mass stars occasionally form as single objects. Whatever the reasons, the result is that a close study of the large body of observational data available for the Orion-KL nebula tends to lead to stupor on the part of the neophyte entering the field.

We offer no solution to this problem but nonetheless feel that some of the characteristics of high mass star formation regions are becoming clearer. One of the developments which has helped has been the ability to carry out detailed multi-transition studies of an appropriately chosen molecule in order to determine the characteristics (density and temperature) of the dense gas clumps where massive stars are currently forming. A description of some of the techniques used and of some recent results is given in section 2. In view of the growing awareness that many young O-B stars at an early phase in their evolution are surrounded by a hot cocoon of molecular gas similar to the Orion hot core (see Genzel and Stutzki (1989) for a discussion of the latter), we discuss hot core characteristics in section 3. Finally, the most important development of recent years has perhaps been the growing ability to perform high angular resolution studies using both millimeter interferometers and the VLA. In section 4, we discuss a few recent studies of this type.

2 Molecular Tracers of Density and Temperature

The "classical" approach to determine the density and temperature distribution in nearby star forming regions has been to observe a symmetric top molecule (usually ammonia) in order to derive the temperature, and a linear (or quasi-linear) species in order to estimate the density. The rationale behind this is that for symmetric top molecules, radiative transitions which change K (the quantum number giving the projection of angular momentum along the molecular symmetry axis) are forbidden. Thus measurements of the relative population of levels of different K (which are mainly sensitive to collisions) allow a reliable temperature estimate.

The temperature thus derived can be used to analyse the relative intensities measured in some species such as CS or HCN which has several observable transitions with widely differing radiative decay rates. In this case, one makes use of the fact that the Einstein A-value for the rotational transitions of such molecules varies with the cube of the frequency. In a situation where there is a balance between collisional excitation and radiative decay, one can thus define a "critical density" needed to excite a given rotational level J. This is $n_{cr}(J) = A_{J,J-1}/ < \sigma v >$ where $< \sigma v >$ is the collisional excitation rate from level J-1 to J and $A_{J,J-1}$ is the appropriate Einstein A value. Very roughly, one may say that this "critical density" is the minimum density required in order that the transition J to J-1 becomes observable. Since collisional rates have typically a slow dependence on J but the radiative decay rates are proportional to J^3, the critical density thus defined can be two orders of magnitude higher, say, for CS(5-4) than for CS(1-0). The variation of the critical density as a function of observing frequency and the excitation energy of the relevant levels is given in Figure 1 for HCN and CS, revealing that at typical interstellar temperatures (10-50 K), these tracers are sensitive to densities in the range 10^5 to 10^8 cm^{-3}. One should be aware however that the estimates shown in Figure 1 neglect the effects of trapping (high optical depth) and hence the effective "critical density" can be considerably lower.

Moreover, if attempting to estimate densities on the basis of observations of such transitions, one has to solve a multi-level statistical equilibrium problem in order to derive the populations expected at a given density and temperature. The first detailed studies of this type were carried out for CS and $C^{34}S$ by Snell et al. (1984) and by Mundy *et al.* (1986) (see also reviews by Cernicharo (1991) and Walmsley (1987)). They observed the regions M17, S140, and NGC2024 and concluded that much of the gas in the neighbourhood of these HII regions was in "clumps" with densities between $4\,10^5$ and $9\,10^5$ cm^{-3}. A more recent higher angular resolution study of M17 (Stutzki and Güsten 1990) not only confirms the earlier work but provides direct evidence for clumps with sizes down to 0.14 parsec and masses between a few and a few thousand solar masses. Thus there are about 10^4 $M_\odot$ of high density ($> 10^5$ cm^{-3}) gas in this region in the form of clumps which at least potentially are capable of collapsing to form stars. More recent work (e.g Cesaroni *et al.* 1991, Hauschildt *et al.* 1993) has shown that such high density complexes are common in the neighbourhood of HII regions.

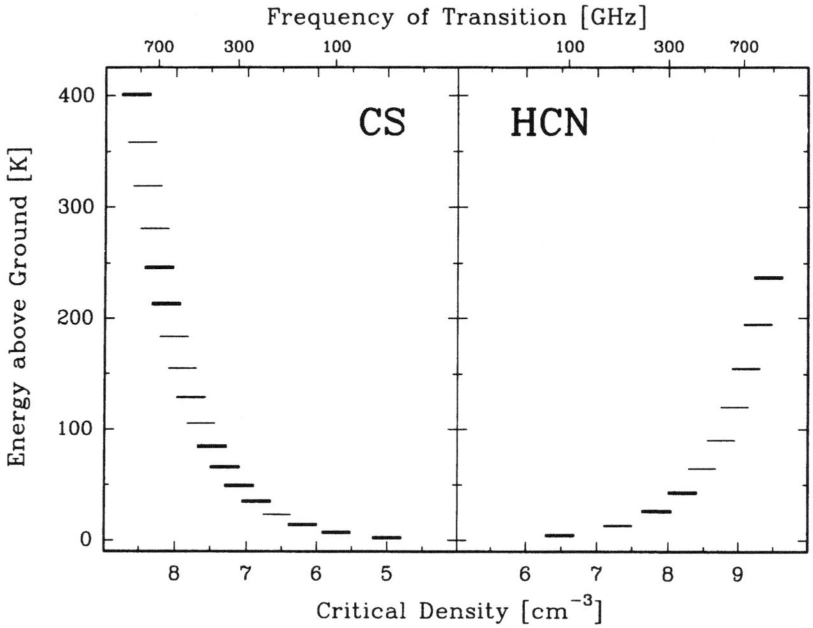

Fig. 1. The logarithm of the critical density ($n_{cr}(J)$, see text) is shown as a function of the excitation above ground (Kelvin) of the upper levels of the rotational transitions of CS (left panel) and HCN(right panel)

Excitation studies of the type discussed above have a number of weaknesses. One, which was mentioned before, is that trapping within optically thick lines has a marked effect upon level populations and has to be taken account of. Doing this requires some knowledge of the geometry and velocity field of the observed region and this is often poorly known. One approach to this problem is to avoid it by using observations of less abundant species whose transitions (hopefully) are optically thin. Hence, for example, one observes $C^{34}S$ rather than CS. This has the advantage of shifting the difficulties onto the shoulders of the receiver engineers who are asked to construct more sensitive mixers.

Another problem is that the basic assumptions of isothermality and homogeneity are rarely satisfied. This causes the well known effect that observers measuring transitions of varying critical density find that they derive higher densities from the higher frequency lines with larger decay rates. The ratio of the intensities of a given pair of lines is typically only sensitive to density within a given range and can give no information outside that range. Worse still, it is becoming clear that in the hot core clumps which we discuss below, something close to LTE (a Boltzmann population distribution) holds at least for the lower frequency lines which are more readily observable from the ground. Clearly, if LTE holds, one can learn nothing about the density from excitation studies. This is one reason among others to obtain measurements of high J submillimeter transitions which may be sub–thermal. With this in mind, Hauschildt *et al.* (1993) have observed several sources in CS(10-9) and $C^{34}S$(10-9) at roughly 490 GHz. Their results indicate very high densities (~10^7cm−3) if excited by collisions, but the interpretation is complicated by the possibility that such high J transitions might also be pumped by near infrared radiation which can excite the first vibrational state.

Sorting out this latter problem almost certainly requires direct observations of transitions within the first vibrational state. It will also require detailed modelling of the transfer of the infrared radiation. While much of this is in the future, emission from the vibrationally excited states of several species has been observed both towards Orion-KL and several other sources (see e.g Schilke *et al.* 1990,1992). Thus the question of the influence of vibrational excitation upon ground state populations is certainly of interest. Carroll and Goldsmith (1981) provide a useful discussion of the physics of the problem.

Even if radiative pumping via vibrationally excited states can be neglected, there remains the problem that most of the models mentioned above assume homogeneity. In regions of star formation on the other hand, we expect to find centrally condensed objects. In fact (see Shu *et al.* 1987 and references therein), in many situations the density distribution should approach that of a singular isothermal sphere with the density $\rho(r)$ as a function of radius varying as r^{-2}, or as $r^{-1.5}$ if undergoing collapse. As a consequence, attempts have been made to model the molecular line data for sources in regions of star formation with the assumption of plausible density and temperature distributions. Examples are the work of Zhou et al. (1991,1993) for B335 and NGC2071, and of Serabyn et al. (1993) for IC1396N. In this type of model, the high frequency lines which sample the high density regions of the clouds clearly play a crucial role. The problems of comparing the predictions of such models with observations are at least threefold. Firstly, all such candidate collapsing objects appear to be associated with outflows and the emission from the outflowing gas is blended with that of the "collapsing" gas. Also, the models have many free (or partially free) parameters which renders it difficult to decide if a "good fit" has been reached. Finally , the models are sensitive to the presence within typical single dish beams of the "hot core" type regions discussed in the next section. The lines formed in hot cores where the molecular abundances are anomalous and both temperature and density are extremely high confuse the interpretation of the "well behaved" molecular cloud gas. Sorting this problem out probably requires interferometry .

3 Hot Cores

It has been known for some time that low mass stars lose mass rapidly during the early stages of their evolution towards the main sequence (see e.g. Lada 1985). For high mass star forming regions, the situation is less clear but there is evidence for bipolar flows in regions such as NGC2024 (Richer 1990) and DR21

(Garden *et al.* 1991). There is also evidence for a high mass outflow in the Orion-KL region (see Genzel and Stutzki) with a mass loss rate of around 10^{-3} $M_\odot$ per year.

In the case of Orion-KL, the outflowing gas appears to be interacting with the high temperature gas seen in ammonia interferometer maps and known as the Orion hot core. The hot core phenomenon is relatively common in regions of high mass star formation (see Mauersberger et al. 1986a, Cesaroni et al. 1992) and has recently been reviewed by Walmsley and Schilke (1993). Walmsley and Schilke define a hot core as having a dimension of roughly 0.1 parsecs, a mass of a few $M_\odot$ and a temperature of at least 100 K. The cores are most easily observable in the emission of high excitation transitions such as NH_3(4,4) which is roughly 200 K above the ground level. There is some evidence that the hot cores seen in ammonia coincide spatially with groups of water masers which are partaking in outflows but the nature of the link between the two is unclear. What does seem certain is that the hot cores are powered by very luminous objects which must be at a very early stage of their evolution. It seems plausible also that the hot core phase occurs during the evolution of a young massive object prior to the development of an ultra-compact HII region (UCHII).

The latter are often associated with hot core regions but have dispersed their surrounding circumstellar material to an extent which suffices to allow them to ionize their surroundings. UCHII regions appear to be relatively long lived with a lifetime which is a substantial fraction of an O-star lifetime (see Churchwell 1991). In the case of hot core regions, the statistics on their presence or absence is rather poor. However, combining recent 100-m surveys by Olmi et al. (1993) and Cesaroni et al.(1992) with earlier work by Mauersberger et al.(1986a) , we find that there are now 25 objects known with luminosities in the NH_3(4,4) line as large or greater than that of Orion-KL (2.5 Jy kpc^2 MHz ; see discussion of Olmi et al.). This clearly greatly underestimates the total number of hot cores in the galaxy. With the 100-m, one would only detect a source of the strength of Orion out to a distance of roughly 3 kpc at the present sensitivity. Nevertheless, it is interesting to compare with the estimate by Wood and Churchwell (1989) that there are 1650-3300 embedded O-stars in the Galaxy and that the lifetime of this UCHII phase is of order $5\,10^5$ years. If hot cores evolve into ultracompact HII regions, their lifetime is at least 5000 years but probably much more. An improvement of these statistics would be certainly worthwhile.

Measuring the mass and density of hot cores is difficult for a number of reasons. One of these is the profusion of lines which they emit at millimeter and submillimeter wavelengths. An example of this is the spectrum shown in Figure 2. Clearly, an appreciable contribution to the "continuum" is due to blending of weak spectral line features. This renders an estimate of the dust emission at millimeter wavelengths extremely difficult and consequently our estimates of dust (and implicitly gas) masses become extremely inaccurate. It is also difficult to use tracers such as CO isotopomers because the low-J transitions of species such as $C^{18}O$ are confused by surrounding lower density material. Finally, the abundances of many species measured in these hot regions are radically different from "normal" molecular cloud abundances. Ammonia for example is typically enhanced in abundance by two orders of magnitude. While this facilitates the observations, it makes gas mass estimates based upon the observed line intensities more difficult.

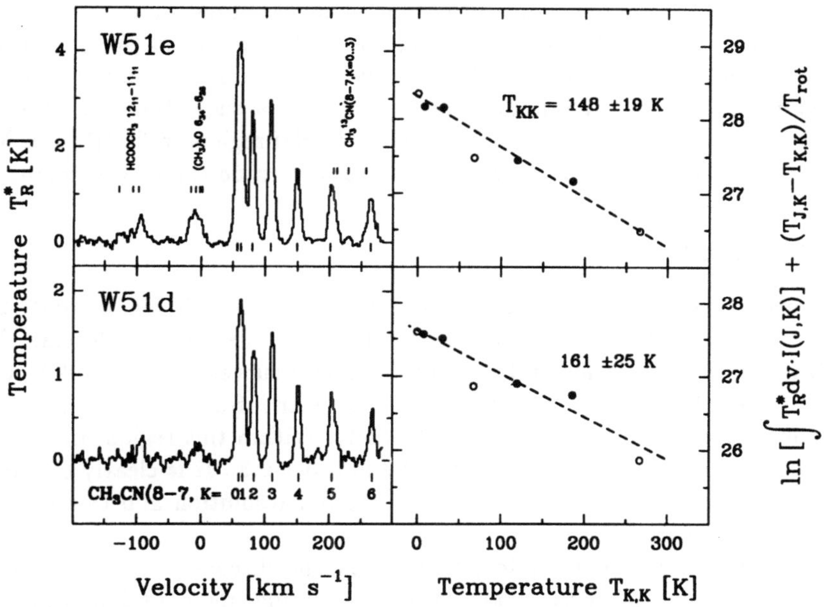

Fig. 2. The left hand panels show spectra of the J=8-7 transition of Methyl Cyanide towards the star forming regions W51d and W51e (Güsten et al. 1994 in preparation). The positions of some "interfering" lines of methyl formate and dimethyl ether are shown on the W51e spectrum. Also shown are the positions of the different K-components of CH_3CN as well as the positions of the lower K-components of ^{13}C CH_3CN. The rotation temperature diagrams to the right are derived using the analytic approach described by Hollis(1982). The slope of the fit to the observed points on these diagrams allows an estimate of the kinetic temperature. These estimates are given as T_{KK} on the rotation diagrams. The emitting gas appears to be at temperatures of order 150 K.

One might think that a solution to the above problems lies in excitation analyses of the type mentioned in section 2. However, one of the reasons that one observes a "line forest" towards sources such as Orion-KL and Sgr-B2N is that one is observing many transitions of vibrationally excited species. These are presumably at least in part radiatively pumped and it is the combination of radiative and collisional effects which makes the interpretation difficult (see the analysis of HDO by Jacq et al. 1988). On top of the strong abundance gradients thus observed, temperature and density gradients within hot core regions almost certainly cause extra complications. So one can conclude that interpretation of the vast amount of information contained in hot core spectra will be a challenge.

4 Interferometric Studies of Regions of High Mass Star Formation

Detailed study of high mass star formation regions requires interferometry both at millimeter wavelengths as well as with the VLA and indeed (for the masers) with VLBI. Much of the early work in this field is summarised in the review by Wilson and Walmsley (1989). Two obvious targets for such studies are the high density molecular complexes associated with the HII regions W3(OH) and NGC 2024. In this section, we summarise some recent work on these two regions.

4.1 NGC 2024

The dense molecular condensation associated with the HII region NGC 2024 is the most productive star forming center within the giant molecular cloud L1630 in Orion. This is shown by a recent infrared survey (E.Lada 1992) which demonstrates the presence of a rich cluster of near infrared sources coincident with the molecular clump observed in CS(2-1). Within this region, the ratio of mass in stars to mass in gas appears to be of order 0.3 . It is clearly of interest to examine the gas distribution at higher angular resolution and this has been done by a number of groups.

One result of particular interest has been the discovery by Mezger et al. (1988) of a number of dense condensations bordering upon the HII region and emitting strongly at submm wavelengths. Mezger et al. concluded that they were observing dust emission from dense clumps of temperature 16 K, mass between 10 $M_\odot$ and 60 $M_\odot$, and densities of a few times 10^8 cm^{-3}. There has been considerable controversy about the parameters quoted by Mezger et al. as well as about the question of whether a considerable fraction of the heavy elements have condensed out upon grain surfaces (see Walmsley and Schilke 1992 for a discussion of the freeze out problem). While the controversy has not died out (see e.g. Moore and Chandler (1989), Schulz et al. 1991, Gaume et al. 1992, Mezger et al. 1992, Mauersberger et al. 1992, Ho et al. 1993), there is clearly evidence both for very high density gas clumps and young newly formed stars. The debate mainly centers upon the degree of evolution of the "protostellar condensations" embedded within the dense clumps seen by virtue of their dust emission. Are there in fact self-luminous objects within these dense clumps ?

The answer may be that in some there are and in some there are not. There seems to be good evidence for a bipolar flow emerging from the submm condensation FIR6 and a unipolar flow from FIR5 (Richer 1990, Richer et al. 1992). Presumably, the objects which power these flows are stars or protostars in the sense that their ultimate energy source is the gravitational potential of the central mass. Then, they should presumably heat up their surroundings and much of the discussion is about whether there is evidence for this or not. Getting better answers to these questions requires interferometry and of most importance is spectroscopy since only with kinematic information can we hope to understand the processes going on in the high density clumps. In this context, it is important to note that *even if* most of the heavy elements have disappeared onto dust grain surfaces as suggested for example by Mauersberger et al. (1992), it is likely that observable amounts of some molecules will remain in the gas phase (Walmsley and Schilke 1992). However, the chemistry under these circumstances is hard to predict.

There are already some interferometric measurements of NGC 2024 available (Gaume et al. 1992, Ho et al. 1993). The results of Gaume et al. show a reasonable correspondence between the submm condensation FIR3 and the ammonia contours as seen with the VLA suggesting that at least in this condensation, an observable amount of molecules is present in the gas phase. More recent high-resolution images in the $C^{34}S$(2-1) transition obtained with the Plateau-de-Bure interferometer, however reveal a striking anti-correspondance between the dust continuum emission towards FIR5 and its associated molecular core (Wiesemeyer et al. 1993). A first analysis suggests an underabundance of CS by at least an order of magnitude. This suggests molecular depletion in this high-density environment but more molecular line measurements are needed to decide whether the underabundance is due to gas phase processes or depletion.

4.2 W3(OH)

W3(OH) can perhaps lay claim to the title of the most studied ultracompact HII region. This can partly be attributed to the fact that the OH masers and associated OH absorption lines towards this object are exceptionally strong (see Baudry et al. (1993) for a recent discussion). The HII region itself (see e.g Guilloteau et al. 1983) shows a shell like structure suggestive of the effects of a stellar wind emanating from the exciting star. The gas surrounding the compact HII region has been sampled in a number of lines which in general are red shifted relative to the velocities of the surrounding molecular emission lines as well as to the velocity of the ionized gas (see Wilson et al. 1991 for a tabulation of measured velocities). This was at first interpreted as evidence for collapse. However recent proper motion studies of the OH masers (Bloemhof *et al.* 1992) on the contrary show evidence for expansion and it seems that the kinematics of the region is complex.

In addition to the star or stars associated with the compact HII region, there appears to be a second star formation center in W3(OH) associated with the grouping of water masers 5 arc seconds east (0.05 parsec in projected distance) of the HII region. A very recent proper motion study of the water masers (Alcolea et al. 1992) shows evidence for a bipolar outflow in this region with an expansion velocity of 20 $km\,s^{-1}$. This seems likely to be associated to the hot clump found by Turner and Welch (1984) observing HCN(1-0) with the Hat Creek Interferometer. Turner and Welch derived a temperature of at least 130 K, a density of at least 10^6 cm^{-3} and a mass of at least 1 $M_{\odot}$for the clump. This suggests the presence of a hot core and indeed Mauersberger et al. (1986b) found that the center of hot ammonia (NH_3(5,5)) emission peaks on the water masers.

More recent observations of this region with the Plateau de Bure Interferometer show evidence for a continuum source at 110 GHz coincident with the HCN clump (Wink et al. 1993). It appears as if this is due to dust emission but this interpretation combined with the normal estimates for dust emissivity implies a column density of H_2 of $3\,10^{25}$ cm^{-2} which is extremely high. The mass of the clump is then 30 $M_\odot$. Wink et al. also observed $C^{18}O$, HCO^+, and CH_3CN. Their integrated intensity maps are displayed in Figure 3, which shows that the main mass of gas seen in $C^{18}O$ ($\sim$60 $M_\odot$) is centred upon the grouping of water masers rather than upon the compact HII region. Also, it is notable that the CH_3CN emission is compact and coincident with the HCN clump. Taken together, these facts suggest though do not prove that the exciting star of the UCHII is older than the star or stars responsible for the hot core seen in the dust continuum and HCN. An implication of this would be that the water masers associated with hot molecular cores are the outward manifestation of the earliest known stage in high mass star formation.

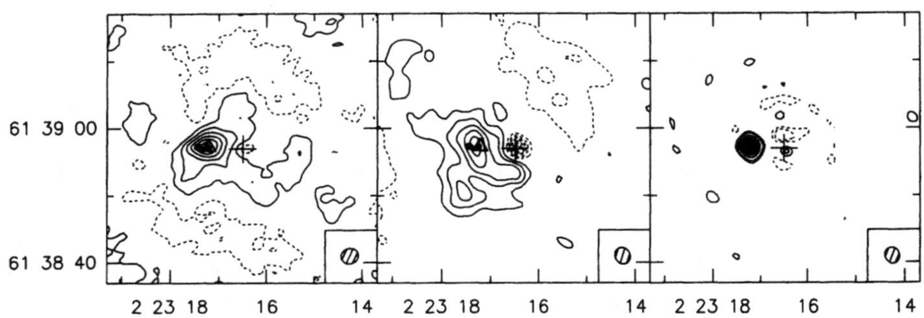

Fig. 3. Integrated intensity maps of (left) $C^{18}O$(1-0), (centre) HCO^+(1-0) , and (right) $CH_3CN(5_3-4_3)$ towards the compact HII region W3(OH) taken from the work of Wink et al. (1993) . The filled circles are water maser positions from Dreher and Welch (1981) and the open triangle is the HCN maximum found by Turner and Welch(1984).

5 Future Prospects

Progress in understanding high mass star formation is likely to be linked to progress in the understanding of the formation of bound galactic clusters. The mass distribution of the stars in those clusters is presumably linked to the mass distribution of the gas clumps or cores within complexes such as NGC 2024, M17, or W3(OH). This suggests a number of questions. In particular, is the initial mass function (IMF) in such high mass star forming regions similar to that in regions such as Taurus where apparently only low mass stars are presently forming. And is the mass distribution in clumps similar ? Is star formation in regions such as NGC2024 being "triggered" in a manner analogous to that proposed by Elmegreen and Lada (1977) more than a decade ago ? The morphology of regions such as NGC 2024 and M17 is rather suggestive of this. Answering these questions will require sampling the high density concentrations in high mass star forming regions and this in turn implies observations with high angular resolution, observations at high frequency or both. The new generation of sub-millimeter telesopes and millimeter interferometers should be capable of providing at least some answers to these questions.

References

Alcolea J., Menten K.M., Reid M. 1992 in *Astrophysical Masers*, (editors A.Clegg, G.Nedoluha ; publ. Springer Verlag).
Baudry A., Menten K.M., Walmsley C.M., Wilson T.L. 1993 A&A **271**,552.
Bloemhof E.E., Reid M.J., Moran J.M. 1992 Ap. J. , **397**,500.
Carroll T.J., Goldsmith P.F. 1981 Ap. J. **245**,891.
Cernicharo J. 1991 in *The Physics of Star Formation and Early Stellar Evolution*, edited C.J.Lada , N.Kylafis, NATO ASI Vol 342, publ. Kluwer.
Cesaroni R., Walmsley C.M., Kömpe C., Churchwell E. 1991 A&A **252**,278.
Cesaroni R., Walmsley C.M., Churchwell E. 1992 A&A **256**,618.
Churchwell E. 1991 in *The Physics of Star Formation and Early Stellar Evolution* , (edited by C.J. Lada, N.Kylafis ; publ. Kluwer), NATO ASI Series Vol. 342.
Dreher J.W., Welch W.J. 1981 Ap. J. ,**245**,857.
Elmegreen B.G., Lada C.J. 1977 Ap. J. , **214**,725.
Garden R.P., Hayashi M., Gatley I., Hasegawa T., Kaifu N. 1991 Ap. J. **374**,540.
Gaume R.A., Johnston K.J., Wilson T.L. 1992 Ap. J. **388**,489.
Genzel R., Stutzki J. 1989 Ann Rev A&A **27**,41.
Guilloteau S., Stier M.T., Downes D. 1983 A&A **126**,10.
Hauschildt H., Güsten R., Phillips T.G., Schilke P., Serabyn E., Walker C.K. 1993 A&A **273**,L23.
Ho P.T.P., Peng Y-L., Torrelles J.M., Gómez J.F., Rodruigez L.F., Cantó J. 1993 Ap. J. **408**, 565.
Hollis J.M. 1982 Ap. J. ,**260**,159.
Jacq T., Walmsley C.M., Henkel C., Baudry A., Mauersberger R., Jewell P.R. 1990 A&A **228**,447.
Lada C.J. 1985 Ann Rev A&A **23**,267.
Lada E. 1992 Ap. J. **393**,L25
Mauersberger R., Henkel C., Wilson T.L., Walmsley C.M. 1986a A&A **162**,199.
Mauersberger R., Wilson T.L., Walmsley C.M. 1986b A&A **166**,L26.
Mauersberger R., Wilson T.L., Mezger P.G., Gaume R., Johnston K.J. 1992 A&A **256**,640.
Mezger P.G., Chini R., Kreysa E., Wink J.E., Salter C.J. 1988 A&A **191**,44.
Mezger P.G., Sievers A.W., Haslam C.G.T., Kreysa E., Lemke R., Mauersberger R., Wilson T.L. 1992 A&A **256**,631.
Moore T.J.T., Chandler C.J. 1989 M.N.R.A.S. **241**,19P.
Mundy L.G *et al.* (1986) Ap. J. **306**,670.
Olmi L., Cesaroni R., Walmsley C.M. 1993 A&A (in press)
Richer J.R. 1990 M.N.R.A.S. **245**,24P.

Richer J.R., Hills R.E., Padman R. 1992 M.N.R.A.S. **254**,525.
Schilke P.,Mauersberger R., Walmsley C.M., Wilson T.L 1990 A&A **227**,220.
Schilke P. *et al.* 1992 A&A **261**,L5.
Schulz A., Güsten R., Zylka R., Serabyn E. 1991 Ap. J. **246**,570.
Serabyn E., Güsten R., Mundy L.G. 1993 Ap. J. **404**,247.
Shu F.H, Adams F.C., Lizano S. 1987 Ann Rev A&A **25**,31.
Snell R.L., Mundy L.G., Goldsmith P.F., Evans N.J. II, Erickson N.R. 1984 Ap. J. **276**,625.
Stutzki J., Güsten R. 1990 Ap. J. **356**,513.
Turner J.L., Welch W.J. 1984 Ap. J. **287**, L81.
Walmsley C.M. 1987 in *Physical Processes in Interstellar Clouds*; ed. G.E.Morfill, M.Scholer, NATO ASI Vol. 210, publ. Kluwer.
Walmsley C.M., Schilke P. 1992 p251 in *Astrochemistry of Cosmic Phenomena*, IAU Symposium 150, (edited P.D.Singh , publ. Kluwer).
Wiesemeyer H., Güsten R. , Wink J.E. 1993, in preparation.
Wilson T.L., Johnston K.J., Mauersberger R. 1991 A&A **251**,220.
Wilson T.L., Walmsley C.M. 1989 *The Astron. and Astrophys. Review* 1,411.
Wink J.E., Duvert G., Guilloteau S., Güsten R., Walmsley C.M., Wilson T.L. 1993 A&A (in press)
Zhou S., Evans N.J. II, Kömpe C., Walmsley C.M. 1993 Ap. J. **404**,232.
Zhou S., Evans N.J. II, Güsten R., Mundy L.G., Kutner M.L. 1991 Ap. J. **372**,518.

H_2O Masers and Star Formation

Marcello Felli

Osservatorio Astrofisico di Arcetri, Largo E. Fermi 5, 50125 Florence, Italy

Abstract: An overview of the connection between water vapor masers and star formation is given. A survey of near infrared sources and H_2O-masers is presented.

1 Introduction

Although H_2O masers in star forming regions (SFRs) have been widely studied in the last 25 years and their general properties are fairly well understood (for recent reviews see e.g. Cohen, 1989; Elitzur, 1992) some basic questions still remain partly unanswered. For instance, we know that the primary source of energy supply for the maser emission comes from massive young stellar objects (YSO) which form at the centre of a molecular cloud. However, there are only few cases where such a YSO has been directly observed (e.g. IRc2 in the Orion-KL region, Gezari, 1992). One would also like to know how close the source of energy has to be to the H_2O maser, what is the precise mechanism that converts stellar energy into maser pumping energy, in which phase of the evolution of a SFR the H_2O maser occurrs and how long does it last. Knowledge of these properties would enable one to use the H_2O maser as an accurate clock to estimate the age of a SFR.

Since mass outflow phenomena, such as stellar winds, jets and CO outflows are also associated with the early stages of star formation, it is likely that they have some relationship with H_2O masers, most probably through the formation of shocks. The large velocity extent of the H_2O emission, up to a hundred km s^{-1}, by itself suggests that high velocity outflows are present in the region where H_2O masers are formed. In a few cases, such as Orion-KL and SgrB2, VLBI proper motion studies suggest that the H_2O masers are entrained in an outflow from a central point of expansion (Reid and Moran, 1988). It has been conjectured that H_2O masers are created and driven by supersonic mass outflows from YSO (Genzel, 1985). The fact that the mechanical luminosity of the CO outflow correlates with the luminosity of the maser and that the velocity ranges of CO and H_2O emission show a general correspondence, provide additional support for this hypothesis (Felli *et al.*1992).

A theoretical model proposed by Elitzur, Hollenbach, McKee (1989) involves fast ($v_s \sim 100$ km s^{-1}) J–type shock propagating in a dense molecular gas ($n \sim 10^7$ cm^{-3}). In the postshock region there should be the proper density and temperature conditions to enhance the H_2O fractional density and to pump the maser transition. The geometry is considered to be planar (Elitzur *et al.*1992) in order to produce the required amplification and beaming angle. However, the details of this interaction are still obscure, even observationally, since the scales involved are very small ($\leq 10^{13}$ cm for the individual component and 10^{16} cm for the cluster size).

Interstellar water masers are sources with angular size less than a few milliarcseconds, with brightness temperatures far above thermal, in the range of 10^{11} to 10^{15}, and with luminosities in the 22 GHz line that, in the extreme cases, can be up to one solar luminosity. It has been known for many years that they are signposts of SFRs. In fact, single-dish surveys in the 1970's of the $6_{16} \rightarrow 5_{23}$ rotational transition of H_2O at 22.23507985 GHz have shown that water masers are found in diffuse HII regions, which are one of the best indicators of recently formed massive stars (Genzel & Downes 1977). Subsequently, interferometric radio continuum maps have evidenced a class of higher density, small diameter HII regions, called ultra-compact (UC) HII regions, related to an earlier evolutionary phase of a SFR and more directly correlated to the YSO. UCHII regions can be located either within more diffuse HII (in which case they indicate a continuing star forming process in the same region) or deeply embedded in molecular clouds (representing the first generation of newly formed massive stars in a collapsing molecular cloud). The H_2O maser spots are preferentially found in the proximity of the UCHII regions. This fact linked even closer the occurence of H_2O maser to the youth of a SFR. However, the masers do not necessarely coincide with UCHII regions, the separation being 0.01 pc or more (Forster and Caswell, 1989).

In this presentation we want to investigate the precise connection between H_2O maser and UCHII region as a function of their evolution and see if the lack of a strict association between H_2O masers and well developed UCHII regions may come from the fact that the maser emission occurs in even earlier phases of evolution, preceeding that in which a fully developed UCHII region becomes observable.

The H_2O maser emission is highly variable (much more than that of OH masers), with time scales of the order of several years and dynamic range of about an order of magnitude, with occasional bursts of shorter duration and larger variation. This is another indication that H_2O masers are formed in regions where strong activity is taking place, i.e. in the immediate surrounding of massive YSO.

Near infrared images are the most suitable tools for investigating the powering sources of the maser because one can penetrate deep inside the molecular clouds and detect embedded stars and protostars. Near and mid infrared observations of maser regions have been carried out since the late seventies, using monopixel InSb photometers and typical diaphragms of about 10 arcsecond in diameter (Evans *et al.*1979; Morwood and Salinari, 1981; Epchtein and Lepine, 1981 and Braz and Epchtein, 1982). Correlations between infrared and maser sources observed with single dish were found to be very strong at these low resolutions.

After the successful completion of the IRAS mission far infrared data are also available to be compared with the radio measurements. Using the IRAS Point Source Catalogue many authors have found a high correlation between masers and FIR point sources. Since the FIR emission comes from cool dust cocoons that surround YSOs, the association further confirms the connection of H_2O masers and the first stages of the evolution of a SFR. The maser and the infrared fluxes seem to be correlated, although differing by many orders of magnitude (see for example Palagi *et al.*1993 and references therein).

The complex morphology of UCHII regions, found from VLA observations and the unclear relationship between this and the maser spots has opened many foundamental questions that the old low resolution and low sensitivity NIR observations could not approach. For instance, Wood and Churchwell (1989a, 1989b) in an extensive study of UCHII regions found that these objects were near or coincident with OH and, less frequently, with H_2O masers. On the other hand the studies of Gaume and Mutel (1987) and Forster and Caswell (1989) give a large catalogue of accurate positions (absolute position better than 0.5 arcsec) of OH and H_2O masers in the "classical" lines (ground state OH and the 6_{16}–5_{23} H_2O transition). The observations show that the maser spots tend to be grouped in compact well separated groups, UCHII regions tend to be related more to OH main and satellite lines, while H_2O masers are usually not closely associated with them. When both masers are associated with an UCHII regions the H_2O masers are usually closer to the peak of the radio continuum emission.

The advent of modern infrared arrays makes them the most suitable devices to be used for a new NIR survey of regions that show H_2O and OH masers. The comparison between this new generation of infrared images and arcsec resolution radio observations (line and continuum) will enable investigation of the maser location with respect to the YSO and its occurrence in the SFR life-time. For these reasons we have begun a systematic survey in the J, H, and K broad band filters of all the masers for which accurate positions are available using a modern NIR camera. The aim is to find if there is any source of NIR emission close to the maser spots, with a positional accuracy of $\lesssim 1''$ and, eventually, to study the nature of these sources and their role in the creation of the maser effect.

In the following, some of the questions outlined above are addressed by means of two distinct observational approaches: 1) a detailed study of a particularly representative SFR, the W75N complex, which presents all the typical aspects of a region where star formation just occurred and 2) a search in the near IR of a large sample (17 objects) of H_2O masers for possible intimately connected YSO.

2 The W75N region

Water emission in W75N was first detected in 1972 at four separate velocity components (Johnston *et al.*1973). A mid-infrared source was found in this region by Wynn-Williams, Becklin & Neugebauer (1974) while a far-infrared source was detected about 30″ to the south by Harvey, Campbell, & Hoffman (1977). Later VLA observations detected 2 HII regions: the more extended one (HII region (A)) associated with the mid-infrared source, and the more compact one (HII region (B)) associated with the far-infrared source (see figures 6 and 10 of Haschick *et al.*1981). VLBI and MERLIN observations of the OH maser emission find a series of spots coincident with and mostly aligned along the N-S direction of elongation of HII region (B) (Haschick *et al.*1981, Baart *et al.*1986). Zeeman splitting of the OH masers indicates a magnetic field of 3 mG. The 2.2μm continuum map of Moore *et al.*(1988) shows 3 clustered sources, the brightest of which (IRS 1) lies 3″ south and 6″ east of HII region (B) and, based on infrared colors, suffers approximately 25 magnitudes of visual extinction. The same authors present a map of the v=1→0 transition of H_2 which shows extended emission and strong peaks at the positions of IRS 1 and HII region (A), a possible indication of shock interaction between the molecular outflow and the surrounding cloud (alternatively the H_2 emission could be due to fluorescence). Submillimeter continuum at 350μm clearly peaks on HII region (B) and not IRS 1 (Moore *et al.*1991a), indicating that the source of energy (the total luminosity is estimated to be $\sim 10^5$ $L_\odot$) must be searched in HII (B). Imaging polarimetry in J, H and K of the W75N region with 2 ″resolution (Moore *et al.*1991b) indicates that the IRS1 source is an extended reflection nebulosity, whose source of illumination (not detected at K band) coincides with HII region (B). The obscuration implied in the direction of HII region (B) is $A_v \geq 90$ mag. Thus the main source of energy supply to the maser as well as to the entire region (whose only trace is the weak radio continuum source HII (B)) is so deeply absorbed that it is not observable even at K band. A cluster of YSO was found in the W75N area, with a K luminosity function similar to that found in the Orion cluster (Moore *et al.*1991b). The CO J=1→0 maps of Fischer *et al.*(1985) and Moore *et al.*(1991b) suggest a large-scale bipolar outflow oriented in a roughly east-west direction, centered on HII (B).

Our observations of the W75N star forming region (Hunter *et al.*1993) comprize: 1) high-resolution VLA radio continuum and H_2O maser maps, 2) Caltech Submillimiter Observatory molecular maps in the CS and CO lines, 3) results from a six-year monitoring of the H_2O maser emission with the 32-m Medicina telescope.

Five H_2O maser spots were found in W75N, which spatially coincide with three very weak, small diameter (< 1″) radio continuum sources (all placed within HII (B)) at the center of a more extended high density molecular cloud core, as illustrated in Figures 1 and 2. The continuum sources are thermal and self absorbed. Given the small size (they are almost unresolved with a HPBW of 0.3 ″) and weak emission (the most intense component has a flux density of 4 mJy) they may represent the first manifestation of an UCHII region, in

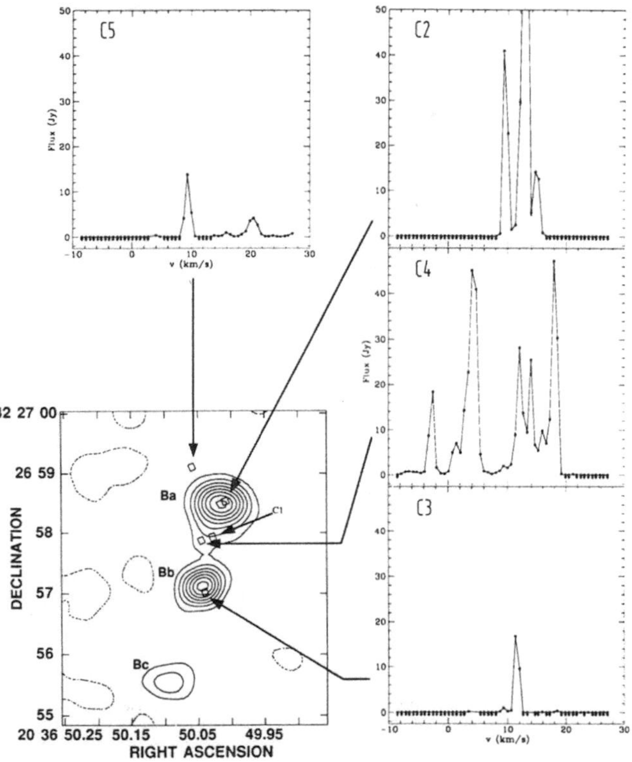

Fig. 1. The three compact continuum sources (8.4 GHz, contours), with the five H_2O maser (diamonds). Levels are drawn at -0.3,0.3,0.6,0.9,1.2,1.5, and 1.8 mJy/beam. The corresponding spectra of the four brightest maser spots (C2 to C4) are also shown (from Hunter *et al.*1993).

which the continuum flux density is very strongly reduced with respect to that of an optically thin HII region with the same exciting star, due to self absorption effects of the radio emission and possible strong absorption of dust in the uv. The (proto-)UCHII region occurs at the centre of a 1400 $M_\odot$ molecular cloud observed in CS, with an H_2 density of the order of 10^7 cm^{-3}. The very high optical obscuration ($A_v \geq 90$ mag), the high total luminosity of the source and the high electron density of the (proto-)UCHII region suggest that in W75N we are witnessing the first appearance of an UCHII region produced by an early type star still deeply embedded in the progenitor molecular cloud. In this early phase the H_2O maser and the UCHII region are closely connected. However, this association may last only for the short time ($\sim 10^3$ y) needed by the UCHII region to expand up to the distance where the masers are. By the time the UCHII region will have expanded and increased in intensity, the enviroment where H_2O maser can form will be overcome by the ionization front and destroyed. However, if other stars of the cluster will, at later times, produce (proto-)UCHII and H_2O

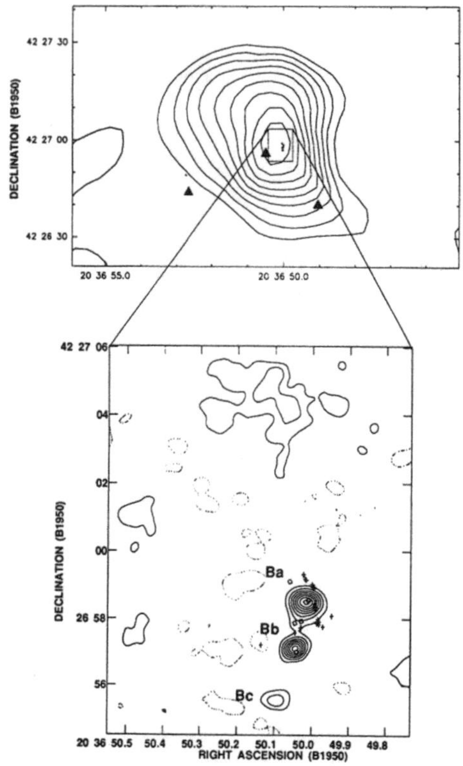

Fig. 2. The top panel shows the map of CS J=7→6 transition integrated from 5 to 15 km s^{-1}. Contours are drawn at 12, 18, 24, 30, 36, 42, 48, 54, and 60 K km s^{-1}. The infrared sources from Moore *et al.*(1989), IRS 1 (top), 2 and 3 (bottom right and left), are indicated by filled triangles and the H_2O maser positions from this work are indicated by dots within the inset box. The lower panel shows a larger field 8.4 GHz continuum map with the H_2O maser positions indicated by diamonds and the OH maser spots of Baart *et al.*(1986) indicated by crosses (from Hunter *et al.*1993).

masers, these will be found in the general area of the previous UCHII region, as observed. Other similar examples of (proto-)UCHII regions are IRc2 in the KL-Orion nebula (Felli *et al.*1993) and the H_2O maser in W3OH (Guilloteau *et al.*1985). In fact, in both cases a very weak continuum source is found close to the H_2O maser.

As traced by the CO J=3→2 transition, the origin of the outflow is at the position of the UCHII regions, suggesting them as the powering source. The H_2O spectrum extends over ~3 km s^{-1}, with many distinct velocity components, part of which can be unambigously identified with individual spatial components of the VLA H_2O map. The variability of each velocity component is very high, with time scales of the order of a few years and dynamic range up to almost two orders of magnitude, as shown in Figure 3. For the bluest velocity component, well isolated in the velocity and space domains, there is evidence of

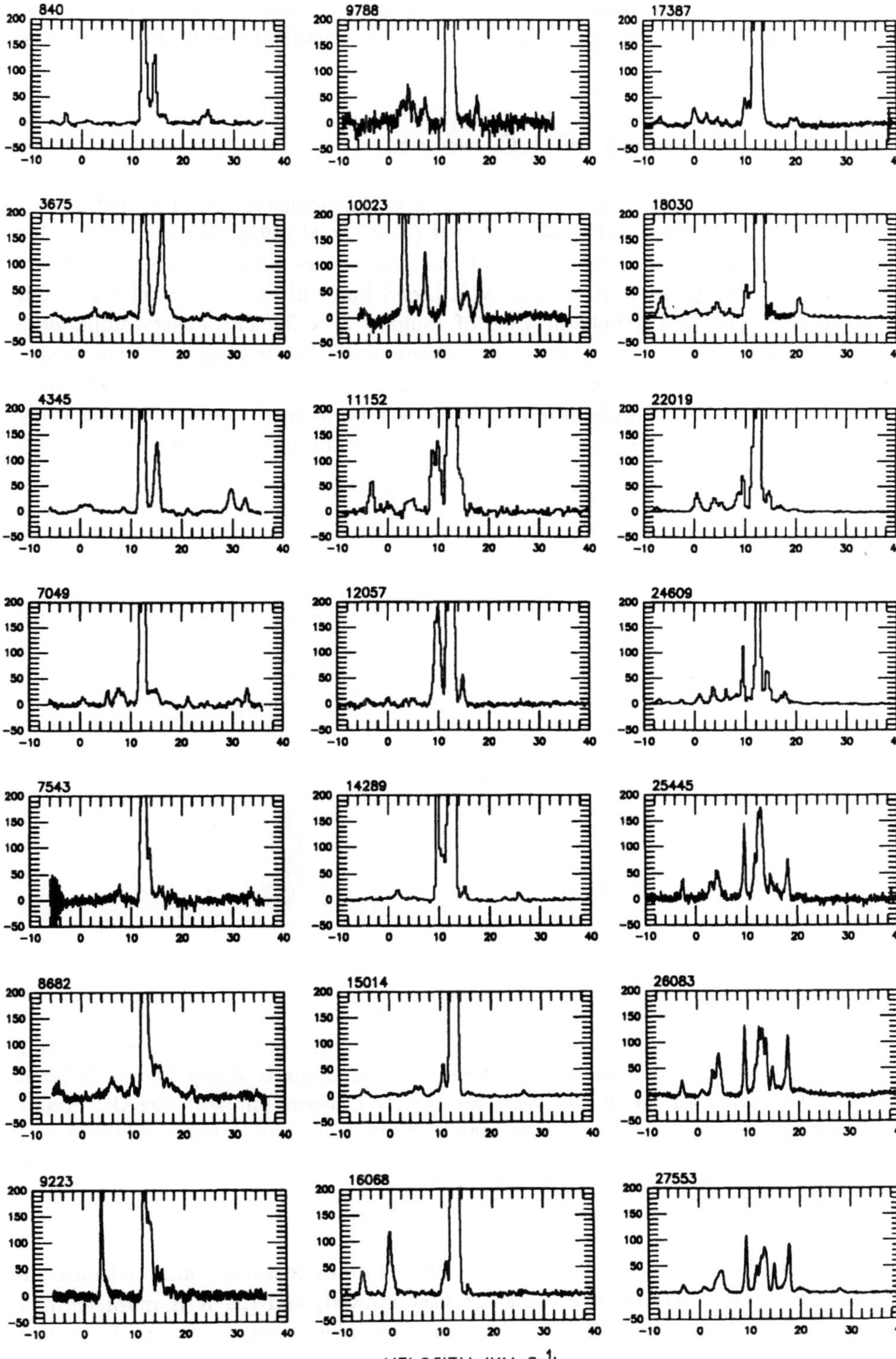

Fig. 3. The Medicina spectra of the H_2O maser in W75N over a six year interval. Time runs from top to bottom and from left to right (from Hunter *et al.*1993).

a continuous velocity drift of ∼4 km s^{-1}over 2 years. The energy required for this acceleration is amply compatible with the mechanical energy input from the molecular outflow.

3 Near IR images of H_2O masers

In this section we present the first results of an extensive near infrared survey of Galactic H_2O and OH masers in SFR (Testi *et al.*1993). Seventeen fields for which accurate VLA observations of the maser lines were available have been imaged in the three J, H, and K NIR broad band filters with pixel resolution of 1.34″/pixel and a field of view of roughly 3′ × 3′. The observations were carried out during June 1991 at Las Campanas Observatory, using the Near–Infrared Camera mounted on the 1 m telescope. Accurate astrometry with field star from the HST Guide Star Catalogue was performed to obtain an accuracy between radio and IR positions of 1 ″. As an example of the results Figure 4

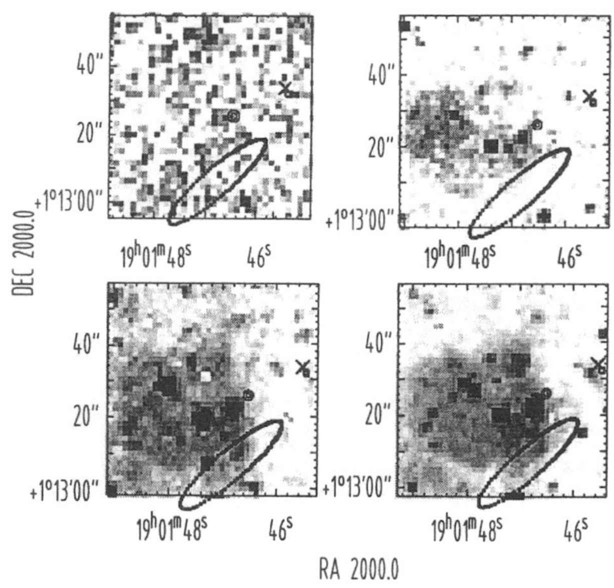

Fig. 4. From upper left to lower right: optical, J, H, and K images of the field (035.20 − 1.74). Only a small part of the coadded frame is shown for the infrared data. Overlayed on the images there are the H_2O masers (crosses), the OH masers (open squares), the radio continuum peak found by Wood and Churchwell (1989b, two concentric circles), and the error box of the IRAS source (ellipse). The coordinates are correct for epoch 2000.0 (from Testi *et al.*1993).

shows the field (035.20 − 1.74) which contains many features common to all the surveyed fields. Usually the fields are very crowed, and the NIR maps show a lot of pointlike and extended sources. The unresolved sources are usually stellar objects, while the extended emissions could be reflection nebulae or diffuse HII regions. Most of these infrared sources are not visible in the optical map, which implies heavy extinction. The masers are located near a faint source seen only in the K image. The brightest NIR sources, often coincident with the IRAS peak and the UCHII peak, are well separated from the masers and are embedded in a large nebulosity that becomes brighter in the 2.2 μm image.

All the regions observed show a high density of faint sources, completely undetected in previous surveys. For this reason it is not straightforward to decide whether a near infrared source is associated with the maser emission or not. From numerical simulations we find that the distributions of the observed "first–neighbor" for the two maser species are very unlikely to be due to random association from uniformly distributed field sources. Consequently, the first nearest infrared source to the maser position ($\lesssim 10''$) is considered to be associated with the maser.

All these sources have peculiar characteristics: they are weak and detected only in K, or if the H magnitude is measurable or the signal to noise ratio is large enough, they show an H$-$K color index >2.

In Figure 5a the sources detected are plotted in the (H$-$K, J$-$K) color–color diagram. We decided to use such plot instead of the standard $(H-K, J-H)$, because in most cases the source is detected only in the K map and only upper limits are available for the J and H magnitudes, and for the color index $(J-H)$ it is impossible to give even a lower/upper limit. Of course this choice does not alter in any sense the informative contents of the graph, but simply makes it more easily readable. The solid line labelled MS represent the colors of main sequence stars, a reddened main sequence star should lie inside the band delimited by the two dashed lines. As can be easily seen, many of them have near infrared colors very different from those of main sequence stars even if heavily reddened. For comparison, in Figure 5b the color–color diagram of all the sources detected in the K band images, without those near to the masers, is presented. More than 300 sources are plotted and only a few of them have colors similar to those found near to the masers, implying that these are somewhat peculiar.

After considering several plausible alternatives for the NIR emission associated with the H_2O maser, we find that the observed emission is probably produced by a massive YSO and a surrounding dusty circumstellar envelope. We developed a qualitative model in which the near infrared flux is composed of the sum of the fluxes of two black bodies, one at 10^4 K, representing the star's contribution, and the other at temperatures ranging from 100 to 1000 K, representing the hot dust envelope. In Figure 6 the curves plotted correspond to the colors of the model. The colors vary along the curve as the ratio of the fluxes of the two black bodies in the K band is varied. A curve is plotted for each temperature of the coldest black body from 100 K to 1000 K each 100 K. As can be easily seen such qualitative description can explain the observations fairly well, so we conclude that the objects detected are probably massive stars deeply embedded in dusty envelopes. A (proto-)UCHII region might also be present, but it is likely to be in a very early phase of its formation and it may not be detectable in the radio depending on its evolutionary phase. In fact, in a few cases a radio

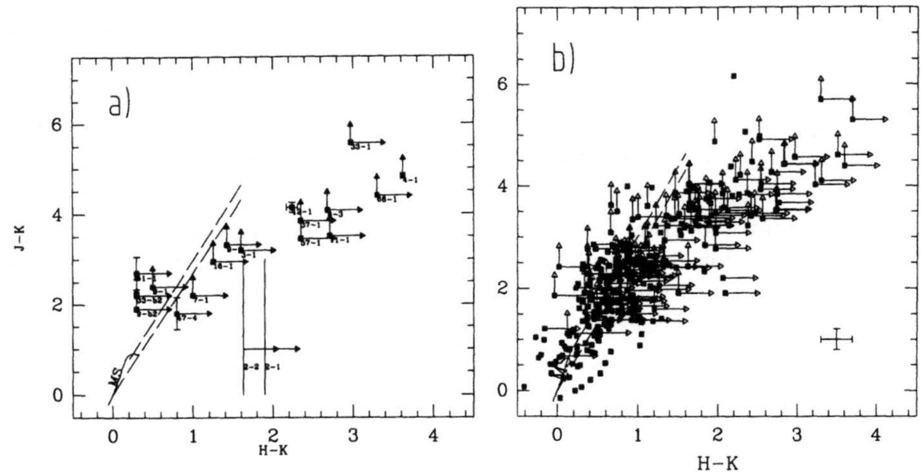

Fig. 5. (a) Diagram for the sources associated with the masers. (b) All the sources detected in the K-band images except those associated to the masers, the arrows indicate that the point is a lower limit, in the lower right corner of the graph the mean error bars are reported.

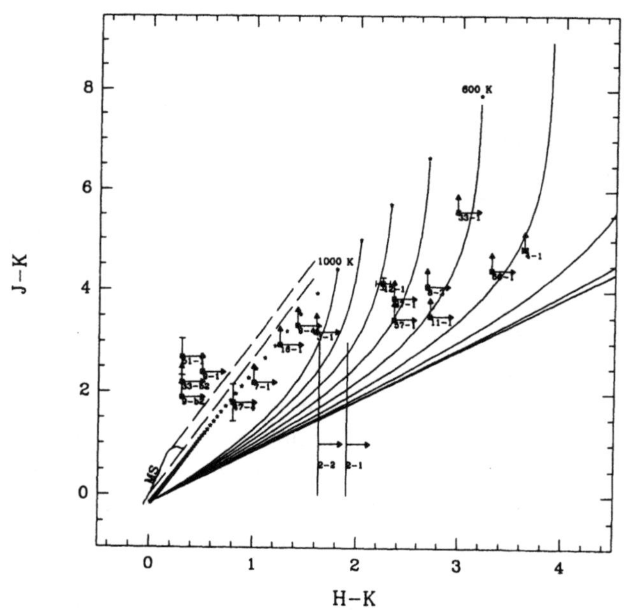

Fig. 6. The two black bodies model (curves) and the sources detected near the masers are plotted on the $(H-K, J-H)$ color-color diagram.

continuum source is coincident with the NIR source associated with the masers. However, since in most cases there is no detected radio continuum emission, many of the YSO are likely to be in such a young phase that they are still invisible with the current radio sensitivities. As already noted, evolved UCHII regions eem to be associated with OH rather than with H_2O masers. This view is in complete agreement with the expectation that OH masers form at larger distances from the YSO and is confirmed by the W3(OH)/(H_2O) source, where Guilloteau *et al.*(1985) and Baudry *et al.*(1993) found that a barely detected radio continuum emission (1.5 mJy at 4.7 GHz) is found near the strong H_2O masers, a few arcseconds away form the OH maser and the strong UCHII (447 mJy at 4.7 GHz). Molecular observations of the same region (Wilson *et al.*1991) seem to show that the two complexes are within the same cloud environment, but separated and probably in different stages of evolution, the youngest being associated with the H_2O masers group. Near infrared observations of the region would be very important in order to test our interpretation of the infrared emission from sources near to maser activity.

The most important results of this preliminary sample is the detection of peculiar NIR sources that seem to be closely related to the maser activity itself. The sources discovered are usually faint in K ($m_K \sim 13$ to 14) and are usually undetectable in J and H images. Consequently the NIR colors of the sources assume high values corresponding to large infrared excesses. The high rate of detection and the fact that many of the detected sources are near the magnitude limits of our images, induces to predict that a more sensitive survey should discover a near infrared source from almost all the maser activity sites. This fact also implies that maser emission needs a close-by luminous YSO.

4 Conclusions

The picture that emerges from the above new observations is that: 1) H_2O maser activity is characteristic of the youngest phases of the life of a massive star, 2) they form very close to the exciting star, which can be seen either in the NIR (if not too heavily absorbed) or in the radio continuum as weak unresolved thermal sources, 3) the most likely process that accelerates and provides the energy are shocks created by the winds in the immediate surrounding of the YSO, 4) the life time of H_2O maser might be very short, determined by the time required to the UCHII region to expand up to the masers location.

To test these ideas and to really understand the relationship between NIR sources, UCHII regions and masers, more NIR and radio observations with higher sensitivity and high positional accuracy are needed.

References

Baart, E.E., Cohen, R.J., Davies, R.D., Norris, R.P., & Rowland, P.R. 1986, MNRAS, 219, 145
Baudry, A., Menten, K.M., Walmsley, C.M., & Wilson, T.L. 1993, A & A, 271, 552
Braz, M.A., & Epchtein, N. 1982, A & A, 111, 91
Cohen, R.J. 1989, Rep. Prog. Phys., 52, 881
Elitzur, M., & de Jong, T. 1978, A & A, 67, 323
Elitzur, M. 1992, Annual Reviews of Astronomy and Astrophysics, 30, 75
Elitzur, M., Hollenbach, D.J., & McKee, C.F. 1989, ApJ, 346, 983
Elitzur, M., Hollenbach, D.J., & McKee, C.F. 1992, ApJ, 394, 221
Epchtein, N., & Lepine, J.R.D., 1981, A & A, 99, 210
Evans, N.J., Beckwith, S., Brown, R.L., & Gilmore, W. 1979, ApJ, 227, 450
Felli, M., & Panagia, N. 1981, A & A, 102, 424
Felli, M., Palagi, F., & Tofani, G. 1992, A & A, 255, 293
Felli, M., Taylor, G.B., Catarzi, M., Churchwell, E., & Kurtz, S. 1993, A & A Suppl, in press
Fischer, J., Sanders, D.B., Simon, M., & Solomon, P.M. 1985, ApJ, 293, 508
Forster, J.R., & Caswell, J.L. 1989, A & A, 213, 339
Gaume, R.A., & Mutel, R.L., 1987, ApJS, 65, 193
Genzel, R., & Downes, D. 1977, A & AS, 30, 145
Genzel, R., Reid, M.J., Moran, J.M., & Downes, D. 1981, ApJ, 244, 884
Genzel, R. in Masers, Molecules and Mass Outflows in Star Forming Regions, 1985, ed. Haschick, A.D., p. 233
Gezari, D.Y. 1992, ApJ, 396, L43
Guilloteau, S., Baudry, A., Walmsley, C.M. 1985, A & A, 153, 179
Harvey, P.M., Campbell, M.F., & Hoffman, W.F. 1977, ApJ, 211, 786
Haschick, A.D., Reid, M.J., Burke, B.F., Moran, J.M., & Miller, G. 1981, ApJ, 244, 76
Hunter, T.R., Taylor, G.B., Felli, M., & Tofani, G., 1993, in preparation
Johnston, K.J., Sloanaker, R.M., & Bologna, J.M. 1973, ApJ, 182, 67
Moore, T.J.T., Mountain, C.M., Yamashita, T., & Selby, M.J. 1988, MNRAS, 234, 95
Moore, T.J.T., Mountain, C.M., & Yamashita, T. 1991,a, MNRAS, 248, 79
Moore, T.J.T., Mountain, C.M., Yamashita, T. & McLean, I.S. 1991,b, MNRAS, 248, 377
Morwood, A. & Salinari, P. 1981, A & A, 94, 299
Palagi, F., Cesaroni, R., Comoretto, G., Felli, M. & Natale, V. 1993, A & A Suppl, in press
Reid, M.J., & Moran, J.M., 1988, in *Galactic and Extragalactic Radio Astronomy*, Ed. G.L. Vershuur and K.I. Kellermann (2nd ed; New York, Springer), chap 6.
Testi, L., Felli, M., Persi, P., & Roth, M. 1993, in preparation.
Wilson, T.L., Johnston, K.L., & Mauersberger, R. 1991, A & A, 251, 220
Wood, D.O.S., & Churchwell, E., 1989a, ApJ, 340, 265
Wood, D.O.S., & Churchwell, E. 1989b, ApJ Suppl, 69, 831
Wynn-Williams, C.G., Becklin, E.E., & Neugebauer, G. 1974, ApJ, 187, 473

Where Is the Heavy Molecule Heimat in Sgr B2?

L.E. Snyder, Y.-J. Kuan, and Y. Miao

Astronomy Department, University of Illinois, 1002 West Green Street, Urbana, Illinois 61801, U.S.A.

Abstract: We review some of the early history of molecular radio detections and show how the Sgr B2 region came to be recognized as an important molecular source. We show the usefulness of millimeter array maps in distinguishing the various chemical processes in the different regions of Sgr B2. We suggest that the differing chemical processes may be indicators of different evolutionary states in the Sgr B2 clouds, which may explain why we have found Sgr B2(N) to be the heavy molecule heimat.

1 Historical Introduction to Radio Molecular Detections

1.1 1968 and 1969: The Beginning of Molecular Radio Astronomy

The dense molecular cloud complex in the galactic center was brought to our attention 25 years ago in a dramatic way with the first detection of an interstellar polyatomic molecule, ammonia (NH_3), at 1.26 cm wavelength in Sgr A by Cheung, Rank, Townes, Thornton, and Welch (1968). Next, water (H_2O) was detected in Sgr B2 (as well as Ori A and W49) at 1.35 cm by the same group in 1969 (Cheung et al. 1969). The next molecule to be found was an organic polyatomic, formaldehyde (H_2CO), detected and reported in 1969 March by Snyder, Buhl, Zuckerman, and Palmer (1969) at 6.2 cm wavelength. Formaldehyde was detected in absorption in the direction of 15 continuum sources, including Sgr A and Sgr B2. At the time, Sgr B2 didn't appear to be a very remarkable formaldehyde source (later, however, the analysis of Zuckerman et al. (1970) showed that Sgr B2 had a higher value of H_2CO integrated projected density divided by excitation temperature [$N_L/T_e = 66$] than any other H_2CO source then known.)

1.2 1970: The Beginning of Millimeter Wavelength Molecular Astronomy

The first molecule detected in 1970 was observed in the far ultraviolet, rather than in the radio region of the spectrum, when interstellar molecular hydrogen (H_2) was found in March by Carruthers (1970) in the far ultraviolet toward the star ξ Persei. In 1970 April, interstellar CO and the first radio lines of CN, both near 2.6 mm wavelength, were found by Wilson, Jefferts, and Penzias (Wilson et al. 1970; Jefferts et al. 1970). This was the beginning of millimeter wavelength molecular radio astronomy. Even though CO and radio CN were detected in other sources such as Sgr B2 and Sgr A, the intense CO spectrum from Orion attracted the most attention. Probably, this CO emission was responsible for stimulating other molecular studies of Orion which led to its current status as the most heavily studied source in our galaxy. The next two interstellar molecular detections were in 1970 June: hydrogen cyanide (HCN) at 3.38 mm, found in 6 sources, and formylium (HCO^+) at 3.36 mm, found in 5 sources (Snyder and Buhl 1971a; Buhl and Snyder 1970). But in the galactic center region, these molecules were detected only in Sgr A(NH_3A). In fact, Sgr B2(OH) produced a negative result in the initial HCN survey, probably due to the rather limited filter bank spectrometers available at the time. In 1970 July, interstellar cyanoacetylene (HC_3N) was detected at 3.3 cm wavelength by Turner (1971), but only in Sgr B2; 17 other "standard" molecular clouds produced negative results for HC_3N. Clearly the detection of HC_3N provided a dramatic demonstration of the importance of Sgr B2 as a source of polyatomic organic interstellar molecules. In 1970 September, this was shown again with the detection of methyl alcohol (CH_3OH) at 35.5 cm wavelength in Sgr B2 and Sgr A by Ball, Gottlieb, Lilley, and Radford (1970). Immediately afterward (in 1970 October), Zuckerman, Ball, and Gottlieb (1971) found formic acid (HCOOH) at 18.3 cm wavelength in Sgr B2 only (see also Winnewisser and Churchwell 1975).

1.3 1971: An Active Year for Discovery

Carbon monosulfide (CS) was detected in 1971 February in Orion and 3 other sources at 2.04 mm wavelength by Penzias, Solomon, Wilson, and Jefferts (1971), but Sgr B2 was not observable during the time of those observations (see also Solomon et al. 1971a). Formamide (NH_2CHO) was detected at 6.49 cm in 1971 March in Sgr B2 by Rubin, Swenson, Benson, Tigelaar, and Flygare (1971). Silicon monoxide (SiO) was detected in 1971 April only in Sgr B2 at 2.30 mm by Wilson, Penzias, Jefferts, Kutner, and Thaddeus (1971). Interstellar carbonyl sulfide (OCS) was reported at 2.74 mm in Sgr B2 only and methyl cyanide (CH_3CN) at 2.72 mm in both Sgr B2 and Sgr A in 1971 May (Jefferts et al. 1971; Solomon et al. 1971a; 1971b). Between 29 April and 6 May, 1971, Snyder and Buhl detected isocyanic acid (HNCO) at 3.41 mm and methyl acetylene (CH_3CCH) at 3.51 mm in Sgr B2; during the same period, hydrogen isocyanide (HNC) was discovered at 3.31 mm in W51 and DR21 (Snyder and Buhl 1971b; 1972; 1973). In 1971 August, interstellar acetaldehyde (CH_3CHO) was reported at 28.15 cm in both Sgr B2 and Sgr A by Ball, Gottlieb, Lilley, and Radford (Ball et al. 1971; Gottlieb 1973). The last interstellar radio molecule detected in 1971 was thioformaldehyde (H_2CS), found in 1971 September at 9.55 cm by Sinclair, Fourikis, Ribes, Robinson, Brown, and Godfrey (1973).

1.4 1972 and After: The Molecular Detection Rate Begins to Stabilize

In 1972, the molecular detection rate began to stabilize, but Sgr B2 remained an important source. In 1972 March, Thaddeus, Kutner, Penzias, Wilson, and Jefferts (1972) detected hydrogen sulfide (H_2S) at 1.78 mm in 7 galactic sources, including Sgr B2. In 1972 May, methylenimine (H_2CNH) was observed in Sgr B2 at 5.67 cm by Godfrey, Brown, Robinson, and Sinclair (1973). Finally, in 1972 November, Gottlieb and Ball (1973) found sulfur monoxide (SO) in 7 galactic sources, including Sgr B2. Subsequently, many other species were found, and

Table 1. The 93 Reported Interstellar and Circumstellar Molecules (December 1993)

2	AlF AlCl C_2 CH CH^+ CN CO CP CS CSi HC1 H_2 KC1 NH NO NS NaCl OH PN SO SO^+ SiN SiO SiS
3	C_2H C_2O C_2S HCN HCO HCO^+ HCS^+ H_2O H_2S HNC HNO MgNC N_2H^+ OCS SO_2 c-SiC_2 NH_2
4	c-C_3H 1-C_3H C_3N C_3O C_3S C_2H_2 $HCNH^+$ H_2CO H_2CN H_2CS H_3O^+ HNCO HNCS $HOCO^+$ NH_3
5	HCOOH C_4H C_4Si c-C_3H_2 CH_2CN CH_4 HC_3N H_2C_2O H_2CHN H_2NCN SiH_4
6	C_5H C_5O C_2H_4 CH_3CN CH_3NC CH_3OH CH_3SH $HCONH_2$ H_2C_4
7	C_6H CH_2CHCN CH_3C_2H HC_5N $HCOCH_3$ NH_2CH_3
8	CH_3C_3N $HCOOCH_3$
9	CH_3C_4H CH_3CH_2CN $(CH_3)_2O$ CH_3CH_2OH HC_7N
10	CH_3C_5N $(CH_3)_2CO$
11	HC_9N
13	$HC_{11}N$

today the Sgr B2 region has yielded more detections than any other source.

Of the 93 interstellar and circumstellar molecular species identified to date (Table 1), about half were initially detected in Sgr B2. Many of the largest and heaviest interstellar molecules are found in the Sgr B2 region, but the detection observations rarely show their precise location in the Sgr B2 complex. In other words, exactly where is the heavy molecule heimat in Sgr B2? Without this knowledge, it is impossible to make correct astrophysical and astrochemical models of this region. Hence, it is necessary to make maps of the heavy molecules using the highest spatial resolution available.

2 Previous Radio Maps of Large and Heavy Molecules in Sgr B2

Up to now, most of the maps of large and heavy molecules in Sgr B2 were made with single-element telescopes. This includes HNCO (isocyanic acid) (Buhl et al. 1973; Churchwell et al 1986; Goldsmith et al 1987; Lis and Goldsmith 1991), OCS (carbonyl sulfide), HC_3N(cyanoacetylene) (Goldsmith et al. 1987), the $HOCO^+$ ion (Minh, Irvine, and Ziurys 1988), and NH_3 (ammonia) (Hüttemeister et al. 1993) in Sgr B2(N), Sgr B2(M), and Sgr B2(S). Observations with single-element telescopes often don't have high spatial resolution, which is compounded by the large distance (7.1 kpc) of Sgr B2 from us (Reid et al. 1988). Clearly, this problem calls for high spatial resolution interferometric observations. Interferometric studies of recombination lines, molecular maser lines, and HII region radio continuum in Sgr B2 have been made at centimeter and millimeter wavelengths (e.g., Benson and Johnston 1984; Carlstrom and Vogel 1989; Gaume and Claussen 1990; Mehringer et al. 1994), but only a few interferometric studies of the neutral molecular gas in the Sgr B2 region, such as NH_3 and H_2CO, have been made in the Sgr B2 cloud (e.g. Vogel, Genzel, and Palmer 1987; Gaume and Claussen 1990; Martin-Pintado et al. 1990). As a consequence, the temperatures, velocity structures, and chemical abundances of the Sgr B2 molecular clouds are still poorly understood. Millimeter wavelength molecular line observations with high spatial resolution are thus needed in order to advance molecular and astrochemical studies in this important galactic region.

3 Locating the Heavy Molecule Heimat: Some Current Millimeter Wavelength Interferometric Observations of Large Molecules in Sgr B2

We are utilizing the high spatial resolution of the BIMA Array to study the exact location of the origin (or heimat) of the most interesting large and heavy interstellar molecular species in Sgr B2. Table 2 lists the Sgr B2 molecular maps that we are working on. These are listed by rest frequency in MHz, molecular name, chemical formula, rotational quantum numbers (Q. N.), and dipole moment in Debye. When the rotational transition is an "a" type, only the a component of the dipole moment is listed followed by an "a". As indicated by

Table 2. Some Current BIMA Array Sgr B2 Maps

Frequency (MHz)	Molecule	Formula	Q. N.	μ (D)
78,479.407	methyl formate	$HCOOCH_3$	$7_{17}-6_{16}E$	1.63a
78,481.394	" "	" "	$7_{17}-6_{16}A$	" "
81,477.563	nitroxyl hydride	HNO	$1_{01}-0_{00}$	1.03a
81,505.173	dicarbon sulfide	CCS	7_6-6_5	2.81
81,534.125	cyanoacetylene	$HC^{13}CCN$	9-8	3.724
86,846.998	silicon monoxide	SiO	2-1	3.0982
104,391.65	sulfur dioxide	$^{34}SO_2$	$10_{19}-10_{010}$	1.6a
104,408.903	vinyl cyanide	CH_2CHCN	$11_{57}-10_{56}$	3.68a
104,408.904	" "	" "	$11_{56}-10_{55}$	" "
104,410.48	methanol	CH_3OH	10_4-11_3A+	0.885a
104,411.262	vinyl cyanide	CH_2CHCN	$11_{48}-10_{47}$	3.68a
104,411.485	" "	" "	$11_{47}-10_{46}$	" "
107,502.426	ethyl cyanide	CH_3CH_2CN	$12_{58}-11_{57}$	3.85a
107,502.474	" "	" "	$12_{57}-11_{56}$	" "

the frequencies, the lines of $HCOOCH_3$ and CH_3CH_2CN are blended doublets. The CH_2CHCN lines in Table 2 are blended with CH_3OH. Some maps are complete and others are not. The HNO, CCS, and $HC^{13}CCN$ maps of Sgr B2 are full synthesis maps constructed from 20 nonredundant BIMA array baselines, ranging from 6.1 m to 100 m (1.6 to 28kλ), plus short spacings provided by the NRAO 12-m telescope. The $HCOOCH_3$ maps have 20 nonredundant baselines from the BIMA array, but no short spacing observations from the 12-m telescope. SiO was mapped in 15 baselines with the new BIMA C array. The remaining molecules were done with in the old BIMA B Array with 3 baselines. We note that the largest molecule on the list, CH_3CH_2CN, was identified 15 years ago by Johnson et al. (1977), but prior to this work no Sgr B2 synthesis maps of CH_3CH_2CN have been made.

The total column density, N_{tot}, for each molecular species in the case of low optical depth can be calculated from

$$N_{\rm tot} = \left(\frac{4\ln 2}{\pi\theta_a\theta_b}\right)\left[\frac{3hWS_f}{8\pi^3\, S\mu^2\, D_E}\right]\left[\frac{Q_{\rm rot}}{B_\nu(T_{\rm rot})-B_b}\right]\times 10^{-18}{\rm cm}^{-2} \qquad (1)$$

where $W=\int I_\nu dv$ in Jy beam^{-1} km s^{-1} is the integrated line intensity (with I_ν the spectral line intensity in Jy beam^{-1}) and S_f is the scaling factor which corrects the primary beam response. $B_\nu(T) = (2h\nu^3/c^2)[\exp(h\nu/kT)\text{-}1]^{-1}$ is the usual Planck function, and $D_E = [\exp(-E_L/T_{rot}) - \exp(-E_u/T_{rot})]$ where E_L and E_u are the lower and upper rotational energy levels of the transition. Q_{rot} is the partition function, S is the line strength, and μ the relevant component of the permanent dipole moment for the transition. B_b is the total background continuum contribution from the source and from the microwave background. T_{rot} is the estimated rotational temperature for each source. In equation 1 it has been assumed that the line excitation temperature is equal to the rotational temperature. We expect that this assumption can be improved as more molecular maps with high spatial resolution are obtained at other frequencies. The first multiplier in equation 1 corrects for a Gaussian beam shape where θ_a and θ_b are the HPBW dimensions in radians. The projected density values change with location in the Sgr B2 complex, but some representative values in the denser regions of Sgr B2 would be: N_{tot} ($HCOOCH_3$) $\approx 10^{15}cm^{-2}$; N_{tot}(HNO) $\approx 4 \times 10^{14}cm^{-2}$; N_{tot}(CCS) $\approx 4 \times 10^{14}cm^{-2}$; N_{tot} ($HC^{13}CCN$) $\approx 8 \times 10^{13}cm^{-2}$; N_{tot}($^{34}SO_2$) $\approx 3 \times 10^{14}cm^{-2}$; N_{tot}(CH_2CHCN) $\approx 4 \times 10^{13}cm^{-2}$; and N_{tot} (CH_3CH_2CN) $\approx 6 \times 10^{13}cm^{-2}$.

Mehringer et al. (1994) made a two-field mosiac map of Sgr B2 at X-band with the VLA and, at this meeting, Gaume showed how complicated the Sgr B2 region is at centimeter wavelengths. Sgr B2(N) has 7 detected ultracompact HII regions and Sgr B2(M) has 25-30, but Sgr B2(S) has only 1. Figure 1 shows our 20 baseline BIMA array map of the 3.8 millimeter continuum emission from the Sgr B2 region. Note the positions of Sgr B2(N) ($\alpha(1950) = 17^h44^m10.2^s$, $\delta(1950) =$

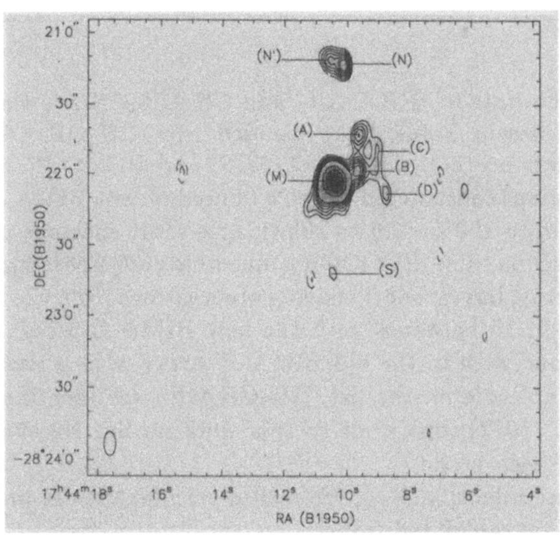

Fig. 1. Our 20 baseline BIMA array continuum map at 78.5 GHz. Contour units: -0.5, -0.4, 0.4, 0.5, 0.6, 0.7, 0.8, 0.9, 1.0, 2.0, 3.0, 4.0, 4.5 Jy beam^{-1}. Rms noise level = 0.11 Jy beam^{-1}. Solid lines denote positive contours and dashed lines are negative contours. The synthesized beam, $10'' \times 5''$, is shown in the lower-left. The map is cen- tered on Sgr B2(OH) and is not primary-beam corrected.

-28°21'13"), Sgr B2(M) ($\alpha(1950) = 17^h44^m10.4^s$, $\delta(1950) =$ -28°22'03"), and Sgr B2(S) ($\alpha(1950) = 17^h44^m10.5^s$, $\delta(1950) =$ -28°22'42"). Also note that there is nothing detected at the Sgr B2(OH) position ($\alpha(1950) = 17^h44^m11.0^s$, $\delta(1950) =$ -28°22'30"), which has been the pointing center for most millimeter wavelength molecular bandscans in Sgr B2. The maximum intensity is 5.05 Jy beam^{-1} at Sgr B2(M), but the peak molecular emission comes from the westward extension.

Sgr B2(N) is related to sources K1, K2, and K3 of Gaume and Claussen (1990) and the eastward extension corresponds to K5.

Figure 2 shows a full synthesis CCS map of Sgr B2 where natural weighting has been used. There are no negative contours displayed because none has an intensity exceeding -0.5 Jy beam^{-1}. Note that the widespread CCS distribution tends to hide any of the underlying structure. Figure 3 shows a naturally

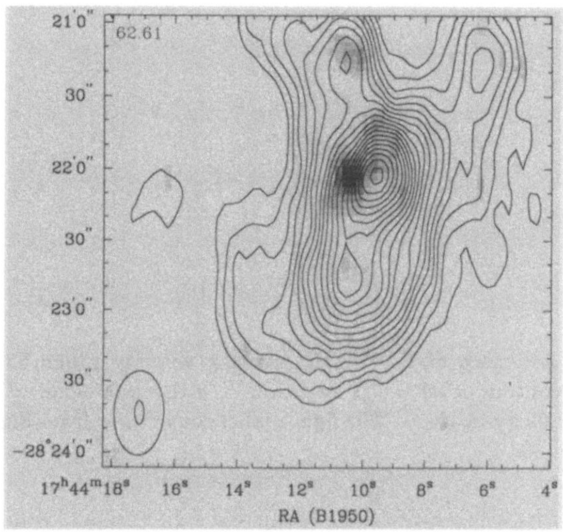

Fig. 2. The integrated CCS map averaged over the LSR velocity range 42.38 to 82.84 km sec^{-1}. The lowest contour level is 0.5 Jy beam^{-1} with increments of 0.1 Jy beam^{-1}. Rms noise level = 0.13 Jy beam^{-1}. The large ellipse in the lower-left is the synthesized beam for the CCS contour maps and the small ellipse is the synthesized beam for the grey scale continuum map.

weighted full synthesis line map of $HC^{13}CCN$ in Sgr B2. Negative contours are too low to appear in the figure. Even though this molecule is only one of three isotopomers of HC_3N, it is so abundant that, just as for CCS, few details of the underlying structure can be seen.

Figure 4 shows a full synthesis map of HNO in Sgr B2, with underlying 3.8 mm continuum. Four major HNO gas clumps were detected. The two strongest are associated with Sgr B2(N) and Sgr B2(M); the third is Sgr B2(NW), about 90″ northwest of Sgr B2(M); and the fourth is a cloud which extends more than 100″ and is located about 30″ south of Sgr B2(S). Kuan and Snyder (1994) have shown that several measured properties of this cloud agree with the model parameters for a young rotating object with mass $\approx 10^4 M_\odot$. A weak HNO emission source about 80″ east of Sgr B2(M) was also observed. Sgr B2(NW) is a relatively unexplored source which is seen in a few other molecules. The westward extension of HNO in Sgr B2(M) has also been observed in the 2,1 of

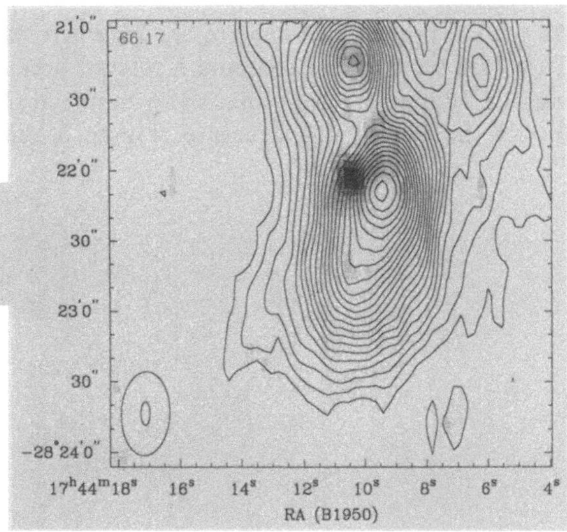

Fig. 3. The integrated map of $HC^{13}CCN$ covering velocity range 53.30 to 79.05 km sec^{-1}. The lowest contour level is 0.3 Jy $beam^{-1}$ with increments of 0.1 Jy $beam^{-1}$. Rms noise level = 0.09 Jy $beam^{-1}$. The figure label convention from Figure 2 is adopted here.

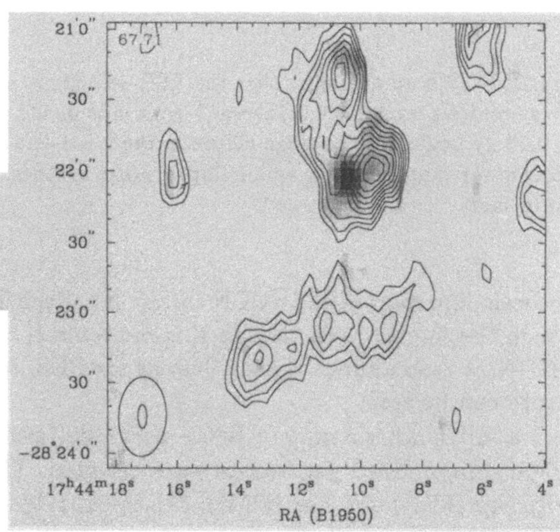

Fig. 4. The full synthesis HNO map in the Sgr B2 region, integrated over LSR velocity range 51.2 to 84.3 km sec^{-1}. The lowest con- tour level is 0.26 Jy $beam^{-1}$ with increments of 0.04 Jy $beam^{-1}$. No negative con- tours stronger than -0.26 Jy $beam^{-1}$ are seen. Rms noise level = 0.09 Jy $beam^{-1}$. Continuum emission at 3.8 mm is the underlying grey scale map. The figure label convention from Figure 2 is adopted here.

NH_3 by Hüttemeister et al. (1993). Note that no HNO is observed at the Sgr B2(OH) position.

Figure 5 shows a methyl formate ($HCOOCH_3$) map made with 20 nonredundant baselines from the BIMA array. Just as in Figure 4, the underlying grey scale is the continuum map. Note that the compact source of $HCOOCH_3$ emission from the Sgr B2(N) region dominates everything; no $HCOOCH_3$ emission is seen in the vicinity of Sgr B2(S) or Sgr B2(M) (although weak line emission from an extended source would require full synthesis mapping to detect). We found a similar distribution around Sgr B2(N) in our maps of the large molecules vinyl cyanide (CH_2CHCN) and ethyl cyanide (CH_3CH_2CN).

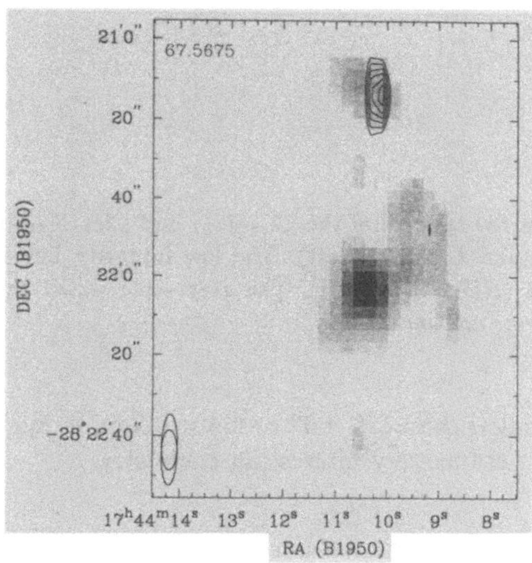

Fig. 5. A BIMA array map of $HCOOCH_3$ in the Sgr B2 region. The map is integrated over LSR velocity range 58.6 to 76.5 km sec^{-1}. The lowest contour level is 0.6 Jy $beam^{-1}$ with increments of 0.1 Jy $beam^{-1}$. Rms noise level = 0.19 Jy $beam^{-1}$. The figure label convention from Figure 2 is adopted here.

Figure 6 shows the results of a three baseline map of CH_3CH_2CN made with a synthesized beam of $11'' \times 6''$ using the BIMA B-array. The grey scale shows the continuum associated with Sgr B2(N) (upper) and Sgr B2 (M) (lower). The blended spectra of the strong Sgr B2(N) emission lines from the 12_{58}-11_{57} and 12_{57}-11_{56} transitions of ethyl cyanide are on the right and U107520, which may be another ethyl cyanide transition, is on the left. The inner vertical scale is in Jy $beam^{-1}$ and the horizontal scale is LSR velocity for the frequency center of U10752. We note that the 0.5 Jy $beam^{-1}$ emission feature between the two stronger lines (at 90 km sec^{-1} on the U107520 LSR velocity scale) is at the rest frequency of the J=18-17 line cluster of the small amino acid glycine (NH_2CH_2COOH) for an LSR velocity near 70 km sec^{-1}. No spectral lines above the noise level are seen in Sgr B2(M). Therefore, we find Figure 6 to be a striking display of the concentration of large, heavy molecules in an

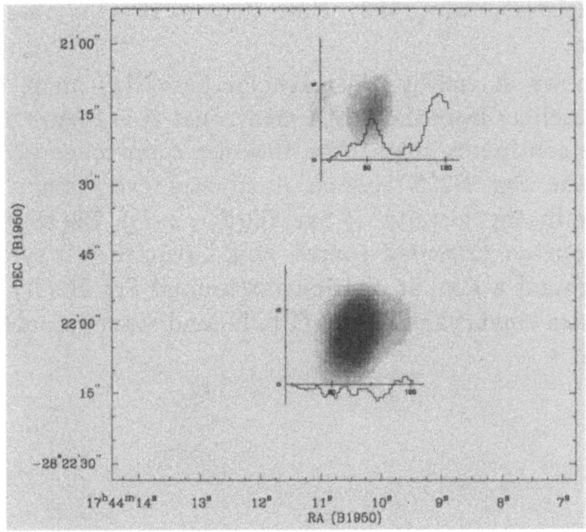

Fig. 6. The BIMA array spectra of the 12_{58}-11_{57} and 12_{57}-11_{56} blended transitions of CH_3CH_2CN (right) and U107520 (left). The low intensity line in the middle is at a frequency of glycine (NH_2CH_2COOH). The grey scale shows the Sgr B2(N) (upper) and Sgr B2(M) (lower) continuum.

emission region smaller than $11'' \times 6''$ or 0.4×0.2 pc in Sgr B2(N). Clearly this is a region which contains very interesting chemistry.

4 Summary

We believe that the our data show that the gas cloud associated with Sgr B2(N) is a dustier, less evolved source than the cloud associated with Sgr B2(M) because the Sgr B2(N) cloud contains a source of intense emission from heavy, complex molecules which are indicative of a chemistry dominated by dust grain-surface chemistry (see, for example, Caselli et al. 1993; Hasegawa and Herbst 1993; Hasegawa, Herbst, and Leung 1992). This view is supported by the smaller number of detected ultracompact HII regions in Sgr B2(N). The dominance of grain chemistry makes the Sgr B2(N) cloud the heavy molecule heimat in the Sgr B2 region. The cloud associated with Sgr B2(M) may have evolved to the point where both ion-molecule and grain-surface chemistry are occurring, but the radiation fields from ultracompact HII regions and lower shielding cause the net production rate for the more complex species produced by grain-surface reactions to fall below that for the simpler products of ion-molecule chemistry. The cloud associated with Sgr B2(S) may be dust deficient, with ion-molecule reactions and perhaps neutral-neutral reactions producing the simpler species observed there.

We acknowledge partial support from NASA grant NAGW-2299, NSF grant AST 90-24603, and the Laboratory for Astronomical Imaging at the University of Illinois.

References

Ball, J. A., Gottlieb, C. A., Lilley, A. E., & Radford, H. E. 1970, ApJL, 162, L203

Ball, J. A., Gottlieb, C. A., Lilley, A. E., & Radford, H. E. 1971, IAU Circ. No. 2350

Benson, J.M., & Johnston, K.J., 1984, ApJ, 277, 181

Buhl, D., & Snyder, L. E. 1970, Nature, 228, 267

Buhl, D., Snyder, L.E., Schwartz, P.R., & Edrich, J., 1973, Nature, 243, 513

Carlstrom, J.E., & Vogel, S.E., 1989, ApJ, 337, 408

Carruthers, G. R. 1970, ApJL, 161, L81

Caselli, P., Hasegawa, T. I., & Herbst, E. 1993, ApJ, 408, 548

Cheung, A. C., Rank, D. M., Townes, C. H., Thornton, D. D., & Welch, W. J. 1968, Phys. Rev. Lett., 21, 1701

Cheung, A. C., Rank, D. M., Townes, C. H., Thornton, D. D., & Welch, W. J. 1969, Nature, 221, 626

Churchwell, E., Wood, D., Myers, P.C., & Myers, R.V., 1986, ApJ, 305, 405

Gaume, R.A., & Claussen, M.J., 1990, ApJ, 351, 538

Godfrey, P. D., Brown, R. D., Robinson, B. J., & Sinclair, M. W. 1973, ApL, 13, 119

Goldsmith, P.F., Snell, R.L., Hasegawa, T., & Ukita, N., 1987, ApJ, 314, 525

Gottlieb, C. A. 1973, in Molecules in the Galactic Environment, eds. M. A. Gordon & L. E. Snyder (John Wiley and Sons: New York) 181

Gottlieb, C. A., & Ball, J. A. 1973, ApJL, 184, L59

Hasegawa, T. I., & Herbst, E. 1993, MNRAS, 261,83

Hasegawa, T. I., Herbst, E., & Leung, C. M. 1992, ApJS, 82, 167

Hüttemeister, S., Wilson, T.L., Henkel, C., & Mauersberger, R., 1993, A&A, 276, 445

Jefferts, K. B., Penzias, A. A., & Wilson, R. W. 1970, ApJL, 161, L87

Jefferts, K. B., Penzias, A. A., Wilson, R. W., & Solomon, P. M. 1971, ApJL, 168, L111

Johnson, D. R., Lovas, F. J., Gottlieb, C. A., Gottlieb, E. W., Litvak, M. M., Guelin, M., & Thaddeus, P. 1977, ApJ, 218, 370

Kuan, Y.-J., & Snyder, L. E. 1994, ApJ, submitted

Lis, D.C., & Goldsmith, P.F., 1991, ApJ, 369, 157

Martin-Pintado, J., de Vicente, P., Wilson, T.L., & Johnston, K.J., 1990, A&A, 236, 193

Mehringer, D.M., Palmer, P., Goss, W.M., & Yusef-Zadeh, F., 1994, ApJ, in prep.

Minh, Y.C., Irvine, W.M., & Ziurys, L.M., 1988, ApJ, 334, 175

Penzias, A. A., Solomon, P. M., Wilson, R. W., & Jefferts, K. B. 1971a, ApJL, 168, L53

Reid, M. J., Schneps, M. H., Moran, J. M., Gwinn, C. R., Genzel, R., Downes, D., & Rönnäng, B. 1988, ApJ 330, 809

Rubin, R. H., Swenson, G. W., Benson, R. C., Tigelaar, H. L., & Flygare, W. H. 1971, ApJL, 169, L39

Sinclair, M.W., Fourikis, N., Ribes, J. C., Robinson, B. J., Brown, R. D., & Godfrey, P. D. 1973, Aust. J. Phys. 26, 85

Snyder, L. E., & Buhl, D. 1971a, ApJ, 163, L47
Snyder, L. E., & Buhl, D. 1971b, BAAS, 3, 388
Snyder, L. E., & Buhl, D. 1972, ApJ, 177, 619
Snyder, L. E., & Buhl, D. 1973, Nature Phys. Sci. 243, 45
Snyder, L. E., Buhl, D., Zuckerman, B., & Palmer, P. 1969, Phys. Rev. Lett., 22, 679
Solomon, P. M., Jefferts, K., Penzias, A. A., & Wilson, R. W. 1971a, IAU Circ. No.2322
Solomon, P. M., Jefferts, K., Penzias, A. A., & Wilson, R. W. 1971b, ApJL, 168, L107
Thaddeus, P., Kutner, M. L., Penzias, A. A., Wilson, R. W., & Jefferts, K. B. 1972, ApJL, 176, L73
Turner, B. E. 1971, ApJL, 163, L35.
Vogel, S.N., Genzel, R., & Palmer, P., 1987, ApJ, 316, 243
Wilson, R. W., Jefferts, K. B., & Penzias, A. A. 1970, ApJL, 161, L43
Wilson, R. W., Penzias, A. A., Jefferts, K. B., Kutner, M., & Thaddeus, P. 1971, ApJL, 167, L97
Winnewisser, G., & Churchwell, E. 1975, ApJL, 200, L33
Zuckermam, B., Ball, J. A., & Gottlieb, C. A. 1971, ApJL, 163, L41
Zuckerman, B., Buhl, D., Palmer, P., & Snyder, L. E. 1970, ApJ 160, 485

The Relation Between OB Stars, H II Regions & Molecular Clouds

Ralph A. Gaume

Code 7213, Remote Sensing Division, Naval Research Laboratory,
Washington, DC 20375-5351

Abstract:
Observational and theoretical results over the last few years have highlighted the interrelationship between OB stars, ultracompact H II regions and molecular clouds. We review recent observational data and models for ultracompact H II regions, and show where the data fit and do not fit the models which have been designed to explain them.

1 Introduction

The properties of ultracompact H II regions are consistent with their being small photoionized nebulae produced by O and B stars embedded in their natal molecular clouds. These regions have electron densities $\geq 10^4$ cm^{-3}, emission measures $\geq 10^7$ pc cm^{-6}, and diameters < 0.1 pc (Wood & Churchwell, 1989). In their few times 10^6 year lifetimes, massive stars form from and ionize molecular cloud material. Massive stars have powerful stellar winds, and are believed to play a major role in the evolution and eventual destruction of the molecular clouds from which they have formed. The study of the ultracompact H II region phase of star formation leads to a better understanding of the star formation process and the interaction of newly formed stars with their surroundings.

Based on Strömgren theory, an H II region quickly forms around a new OB star. Pressure from the hot ionized gas causes the nascent H II region to expand into a spherical volume of diameter ≈ 0.3 pc in only a few times 10^4 years (Wood & Churchwell 1989). This is $\approx 2\%$ of the lifetime of the central star(s). Based on a knowledge of the number of O stars in the Galaxy, Wood & Churchwell (1989) predict that a total of 300 ultracompact H II regions should be detectable in the entire Galaxy with the VLA. Since there seem to be significantly more than 300 regions, some mechanism must be responsible for lengthening the ultracompact H II region phase; ultracompact H II regions cannot be as young as simple theory predicts.

Several models have been proposed to explain the large number and observed morphologies of ultracompact H II regions. At least one of these models, if correct, limits the role played by OB stars in molecular cloud evolution and destruction. We review these models in the context of current observations.

2 Discussion

2.1 The Moving Star Bow Shock Model

Recent high resolution ultracompact H II region surveys have classified most observed ultracompact H II regions into one of several morphological types. One of the most interesting morphological classes of ultracompact H II region is the "cometary" class, which makes up 20% of the observed ultracompact H II regions in the survey of Wood & Churchwell (1989) and a somewhat smaller percentage (16%) in the survey of Kurtz, Churchwell, & Wood (1993). The characteristic cometary morphology is defined by Wood & Churchwell (1989) as having an ionization front that is parabolic, and shaped like a comet, with a well defined, sharp leading edge (often unresolved), and exhibiting a surface brightness along the parabola's axis of symmetry which decreases gradually from the "head" to the "tail". The prototype of this class and most studied representative is G34.3+0.2 component C (see Fig. 1) A possible resolution to the ultracompact

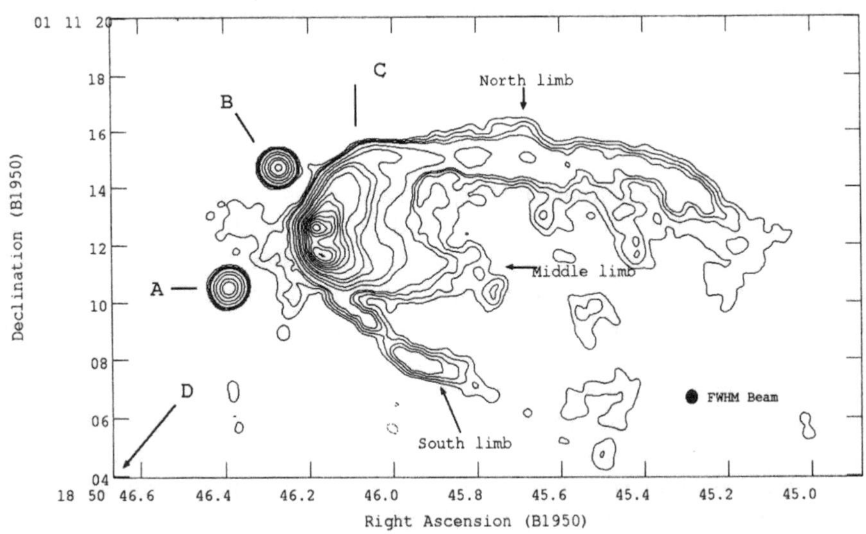

Fig. 1. Contour map of a 2 cm image of the G34.3+0.2 region. Contours are −0.3, 0.3, 0.5, 0.75, 1.0, 2.5, 5, 10, 15, 25, 35...95% of the peak flux density of 0.303 Jy per beam. Components A & B are hypercompact H II regions. Component C is the prototype of cometary H II regions. The direction to component D, a large shell source, is indicated. This deep image shows, for the first time, that the cometary H II region has a tri-limbed morphology. This morphology is inconsistent with the moving star bow shock model and the champagne flow model.

H II region age problem, and a model for cometary H II regions was suggested in a series of papers by Van Buren *et al.* (1990), Mac Low *et al.* (1991) and Van Buren & Mac Low (1992), who have proposed that cometary H II regions are stellar wind bow shocks. In this model, which we call the moving star bow shock model, an O/B star with a strong ionized stellar wind, moves supersonically (with a velocity of order 10 km s^{-1}) through a molecular cloud. Ram pressure from the ambient molecular material prevents expansion of the H II region. In the moving star bow shock model the size of a cometary H II region has little to do with its age. The ionized region itself is formed in a thin layer between the shocked molecular gas and the free-flowing stellar wind. The effect of the stellar wind is limited to behind the bow shock. As energy and momentum input from stellar winds are believed to play a significant role in the evolution and destruction of molecular clouds, by limiting the effect of stellar winds to behind the bow shock the moving star bow shock model limits the interaction between stellar winds and surrounding molecular cloud material. Besides modeling the continuum morphology, the radio recombination line velocity structure is predicted for the moving star bow shock model at various viewing angles (the angle between the line of sight and space velocity vector of the central star) by Van Buren & Mac Low (1992). Application of the moving star bow shock model has not been limited to cometary H II regions. Mac Low *et al.* (1991) propose that the moving star bow shock model accounts for most of the observed morphologies of ultracompact H II regions, excluding only those objects with spherical shells. Furthermore, Mac Low *et al.* (1991) suggest that simple time scale arguments imply that 85%–90% of *all* ultracompact H II regions are bow shocks.

G34.3+0.2, component C is the prototype of cometary H II regions. The high resolution continuum structure (see Fig. 1) shows a well defined core component, with a well developed tail to the west. Although G34.3+0.2 C has been demonstrated to exhibit characteristic cometary structure with a variety of observational parameters, we note that very few of the H II regions classified as cometary by Wood & Churchwell (1989), or Kurtz, Churchwell, & Wood (1993) have been shown to have long, well developed tails. Most H II regions in the cometary class exhibit an arc-like core structure, but do not possess well-developed tails. Fey *et al.* (1992) suggest that at least some of the arc-like H II regions in the cometary class, when observed with better sensitivity and greater dynamic range, are actually asymmetrically bright circular shell-like sources, i.e. shells with one portion a factor of ≈ 10 greater in brightness than the rest of the shell.

The moving star bow shock model predicts parabolic shaped continuum morphologies, with relatively narrow opening angles. The model produces continuum sources with a shuttlecock appearance. However, new deep 2 cm continuum imaging of G34.3+0.2 C (Fig. 1) shows a more complicated morphology. The core, north limb and middle limb of G34.3+0.2 C have been previously modeled by the moving star bow shock model (Mac Low *et al.* 1991). Taken by themselves, the north and middle limb, which trail to the west suggest that the exciting star of the G34.3+0.2 C H II region has a due east motion on the plane of the sky (Mac Low *et al.* 1991). However, the high sensitivity of the image contoured in Fig. 1 reveals, for the first time, a third limb, trailing to the southwest. This tri-limbed morphology is inconsistent with the moving star bow shock model.

The moving star bow shock model predicts that characteristic velocity gradients will be observed in the radio recombination lines from cometary H II regions (Van Buren & Mac Low, 1992). Velocity gradients are predicted to be observed both parallel and perpendicular to the symmetry axis, which passes through the core, and along the tail. The characteristic velocity structures predicted by the moving star bow shock model are axially symmetric for all viewing angles. However, Garay, Rodríguez, & van Gorkom (1986) and Gaume, Fey, & Claussen (1994) have found an asymmetric, north-south velocity gradient seen perpendicular to the continuum symmetry axis of G34.3+0.2 C. Again, this asymmetric velocity gradient is inconsistent with the moving star bow shock model.

The G34.3+0.2 molecular envelope/core has been well studied (see references in Gaume, Fey, & Claussen 1994) and ranges in temperature and density from 9–185 K and 5×10^3– 7×10^7 cm^{-3} respectively, and in size from 3.7 – 0.03 pc. The parameters from the model fit suggest that G34.3+0.2 has a velocity on the plane of the sky of 10 km s^{-1}, which is $6''$ of motion per 10^4 yr (Mac Low *et al.* 1991). The moving star bow shock model requires that the central star of G34.3+0.2 C is moving, by chance, into the $200''$ diameter G34.3+0.2 molecular envelope, and again by chance into the $\approx 2''$ diameter, dense, ultracompact, hot core of the envelope, which is also a site of pre-existing and on-going star formation as traced by continuum components A & B. The probability of this scenario seems extremely small. One could claim that observing a bow shock cometary near a dense core is a selection effect, i.e. bow shock cometary H II regions are only easily observed when they are near the dense cores of molecular clouds, as is G34.3+0.2 C. But this argument would require the existence of a much larger population of dense cores, moving O/B stars, or both. Therefore it seems unavoidable that the moving star bow shock model, as applied to G34.3+0.2 C, requires the extremely unlikely occurrence of an O star moving directly into a $2''$, pre-existing ultracompact, hot, dense, molecular core.

2.2 The Champagne Flow Model

The champagne flow model has been suggested by numerous authors to explain the nature of cometary and arc-like H II regions such as G34.3+0.2 C (Andersson & Garay 1986, Gaume & Mutel 1987, Gaume & Claussen 1990, Fey *et al.* 1992, Carral & Welch 1992). The champagne flow model (see Yorke *et al.* 1983 and references therein), was developed with the notion that stars form near the edges of molecular clouds. This model answers the question of what happens when an expanding H II region breaks through the edge of a molecular cloud, into the lower density inter-cloud region. Results of the model, with predicted continuum morphologies at various viewing angles, show that a cometary-like H II region is formed. However, the champagne flow model produces centrally condensed continuum sources, not limb brightened sources as are seen in the cometary and arc-like H II regions. This is because the champagne flow models have a constant density throughout the H II region, except near the opening. Similarly, the champagne flow model predicts that the largest velocity gradients in the ionized material should be observed near the opening nozzle, as the ionized gas flows into the region of lower density. Very little velocity structure should be observed near the core of the H II region. This again does not agree with our observations of G34.3+0.2 C or those of other cometary and arc-like H II regions. Finally, the champagne flow model predicts the rapid evolution of an H II region, as it expands into the surrounding molecular cloud. Thus, the champagne flow model does not help to resolve the question of why so many small H II regions are observed (see S1).

One improvement to the champagne flow model would be the inclusion of stellar winds, and more complicated ambient density distributions in the surrounding molecular cloud. The new model would perhaps be better labeled a photoevaporation model, as ionized gas photoevaporates (or boils) off of the surrounding molecular material and escapes into regions of lower pressure. This photoevaporation model, with the inclusion of stellar winds and more realistic irregular density distributions, might produce model H II regions similar to the cometary and arc-like H II regions, but this remains to be seen.

2.3 The Disk Photoevaporation Model

Whereas ultracompact H II regions are generally defined as H II regions with sizes < 0.1 pc ($\approx 20{,}000$ A.U.), recent deep radio wavelength imaging of some well studied star formation complexes have revealed a significant population of H II regions with sizes an order of magnitude smaller, i.e., $\lesssim$ a few thousand A.U. We define this subclass of H II regions as being hypercompact. Recent observations of the W51 d & e complex (Gaume, Johnston & Wilson, 1993), show 5 hypercompact H II regions with sizes $\lesssim$ 3000 A.U. The upper limit to the diameter of one of these, component d2, is 300 A.U. Recent 1.3 cm observations of Sgr B2 detail nearly 50 individual H II regions, 12 of which have Gaussian FWHM sizes of $\lesssim$ 2000 A.U. (see Fig. 2). The existence of a significant number of H II regions of this size greatly exacerbates the age/lifetime problem. Meanwhile, recent high spatial resolution radio recombination line observations of the hypercompact H II regions in Sgr B2 and NGC 7538 (see Fig. 3) show very wide line profiles.

Work is currently underway on a model that may be able to explain the hypercompact H II regions. Hollenbach, Johnstone, & Shu (1993) have proposed that many H II regions form by the photoionization and photoevaporation of neutral accretion disks that orbit newly-born OB stars. In this model, which we call the disk photoevaporation model, an accretion disk serves as a dense reservoir of material. The central star photoevaporates material off of the surface of the disk. The hot, overpressured ionized material then flows away from the central star and surrounding neutral disk. The model incorporates the case of both weak and strong stellar winds.

In the context of Strömgren theory, hypercompacts must be H II regions that have evolved little beyond the formation of the initial Strömgren sphere. These H II regions should rapidly evolve into larger H II regions, and should be observed only rarely. However, the disk photoevaporation model suggests

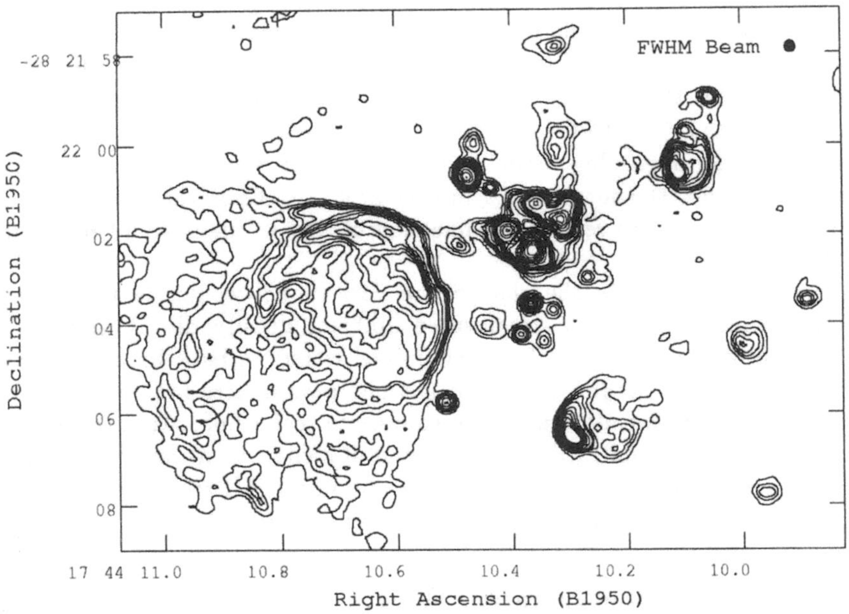

Fig. 2. Contour map of a 1.3 cm image of the inner portion of the core of the Sgr B2 complex. Contours are -1, 1, 2, 3, 4, 5, 7.5, 10, 12.5, 15, 17.5, 20, 30...90% of the peak flux density of 0.206 Jy per beam. We have identified nearly 50 individual H II regions in Sgr B2, 12 of which have Gaussian FWHM sizes of $\leqq$ 2000 A.U.

that H II regions of this size may be quasi-stable, as the ionized material which expands away from the star is constantly replenished by fresh material from the disk. Initial calculations suggest that the lifetime of an H II region produced from photoevaporation of a stellar disk is on the order of $10^{5.5}$ yr (Hollenbach, Johnstone, & Shu 1993) . This is about 10% of the lifetime of an O star. As material is photoevaporated from the surface of the disk, the resulting wind may attain speeds of 10–50 km s^{-1}. The broad radio recombination lines that have been detected toward some hypercompact H II regions (see Fig. 3) may be tracing the wind flowing away from the surrounding disk.

2.4 A Possible Model for G34.3+0.2

As an example, we suggest that the following scenario may explain many of the current observations of the G34.3+0.2 region (Gaume, Fey, & Claussen 1994). In addition to the cometary H II region G34.3+0.2 component C, the G34.3+0.2 region also contains two hypercompact H II regions, components A and B (see Fig. 1). We propose that the exciting stars of components A, B, & C have formed from the same dense core. The remnant of this dense core has been traced by various molecular line studies, and is located between the three continuum sources. The exciting stars of components A, B, & C have no significant relative motion with respect to each other, or with the molecular cloud. Using the 6

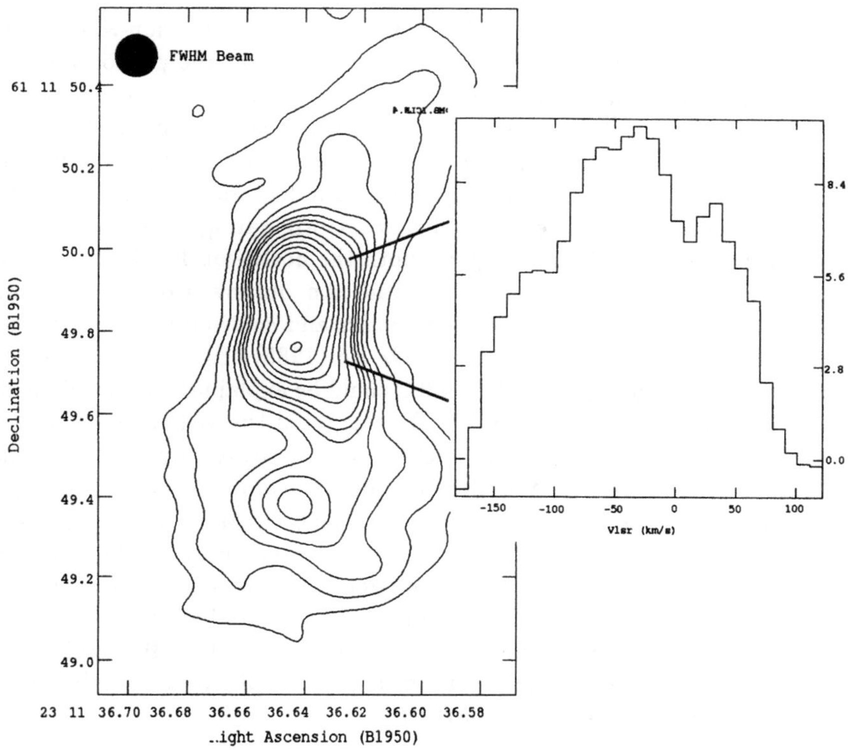

Fig. 3. Contour map of a 1.3 cm image of NGC 7538 IRS 1. Contours are at 2, 5, 10, 15, 20, 25, 30, 40...90% of the peak flux density of 0.044 Jy per beam. The insert shows a profile of the H66α radio recombination line integrated over the core of the continuum region. The broad recombination line may be tracing a wind photoevaporating from a neutral circumstellar disk.

and 2 cm observations of Wood and Churchwell (1989), the 3.6 cm observations of Van Buren *et al.* 1990, the 2 cm data of Garay Rodríguez, & van Gorkom, (1986), the 2 cm data from Gaume, Fey, & Claussen (1994), and the 1.3 cm data of Heaton, Little & Bishop (1989) we fit a spectral index, α, ($S \propto \nu^{\alpha}$), of 0.8±0.3 and 0.9±0.4 for components A & B respectively. The fitted spectral indices for components A & B are consistent with the canonical spectral index for a spherical symmetric, constant velocity stellar wind (i.e., $\alpha \approx 0.6$).

A possible qualitative model for the cometary H II region, component C, would also involve photoevaporation. Rather than photoevaporating material from a surrounding remnant accretion disk, as we have suggested for A & B, component C is the result of photoionized gas evaporated directly from the hot molecular core found to the east. Pressure exerted by the stellar wind of the central, exciting star would hinder the free expansion of the photoevaporated gas from the molecular core, increasing the density toward the ionized walls, resulting in a limb brightened morphology. The ionized gas that photoevaporates from the hot, dense core flows into a region of lower density to the west (the tail of the cometary H II region). If the H II region makes a non-zero angle to the line of sight, with the tail pointed toward the earth, the photoevaporating material will be blue-shifted with respect to the velocity of the molecular cloud, as is observed. The large ($\approx$ 35 km s^{-1}) north-south velocity gradient observed in the recombination line emission in the core of the H II region probably could not result from this photoevaporation model. However, an interaction between the cometary H II region and a wind originating from component B and perhaps A (see Fig. 1) may explain both the north-south velocity gradient, and the various structures seen in the tail. An outflow from the B and A components could greatly affect the kinematics and continuum structure of the cometary H II region. We note that Reid & Ho (1985) originally proposed that the continuum morphology of G34.3+0.2 C could be explained by a wind incident on the H II region. In this case Reid & Ho suggested the W44 supernova, 40 pc distant, as the origin of the wind, whereas we are suggesting that an outflow associated with the nearby A and B continuum components may affect the continuum structure of the cometary component C. Many of the properties explained by the moving star bow shock model may apply to the qualitative model we have presented here, since instead of the cometary's central star moving through neutral material, a wind(s) originating from the nearby continuum component(s) is moving past the central star of the cometary H II region.

Basic research in Radio Interferometry at the Naval Research Laboratory is supported by the Office of Naval Research.

References

Andersson M. & Garay, G. 1986, A&A, 167, L1
Carral, P., and Welch, W.J. 1992, ApJ, 385, 244
Fey, A. L., Claussen, M. J., Gaume, R. A., Nedoluha, G. E., & Johnston, K. J. 1992, AJ, 103, 234
Garay, G., Rodríguez, L. F. & van Gorkom, J. H. 1986, ApJ, 309, 553
Gaume, R.A., & Mutel, R.L. 1987, ApJS, 65, 193
Gaume, R.A., & Claussen, M.J. 1990, ApJ, 351 538
Gaume, R.A., Johnston, K.J. & Wilson, T.L. 1993 ApJ, in press (scheduled Nov 10 issue)
Gaume, R.A., Fey, A.L., & Claussen, M.J 1994 ApJ, submitted
Heaton, B.D., Little, L.T., & Bishop, I.S. 1989, A&A, 213, 148
Hollenbach, D., Johnstone, D. & Shu, F. 1993, ASP Conf. Series Vol. 35, p. 26
Kurtz, S., Churchwell, E., & Wood, D.O.S. 1993 ApJS, submitted
Mac Low, M-. M., Van Buren, D., Wood, D. O. S., & Churchwell, E. 1991, ApJ, 369, 395
Reid, M. J. & Ho, P. T. P. 1985, ApJ, 288, L17
Van Buren, D., Mac Low, M. M., Wood, D. O. S. & Churchwell, E. 1990, ApJ, 353, 570
Van Buren, D. & Mac Low, M. M. 1992, ApJ, 394, 534
Wood, D. O. S. & Churchwell, E. 1989, ApJS, 69, 831
Yorke, H. W., Tenario-Tagle, G., & Bodenheimer, P. 1983, A&A, 127, 313

From Cores to Stars

P. C. Myers

Harvard-Smithsonian Center for Astrophysics, MS 42, 60 Garden St., Cambridge MA 02138 USA

Abstract: Dense cores associated with low mass stars generally have mean density 3×10^4 cm^{-3} within 0.1 pc, and temperature 10 K, while cores of similar density forming more massive stars are generally larger and warmer, according to observations of molecular lines of NH_3, HC_3N, CS, and $C^{18}O$. These cores are near virial equilibrium, taking into account both the thermal and nonthermal parts of the line width. They are associated with the youngest stars, having the reddest broadband spectra and the greatest column density of circumstellar gas and dust. Their nonthermal motions probably arise from magnetohydrodynamic waves. Their nonthermal and thermal kinetic energy density are comparable when their associated star has mass $\sim 2\ M_\odot$; cores associated with less (more) massive stars are dominated by thermal (nonthermal) motions. Core maps are generally elongated, with aspect ratio $\sim$2, and a substantial fraction of cores with elongated maps are prolate. Core velocity gradients are of order one line width per map size, and correspond to rotational kinetic energy density less than 0.1 of the gravitational kinetic energy density. Core magnetic energy density is comparable to the nonthermal kinetic energy density, in the few cases where the Zeeman effect is detected. The nonthermal part of the line width in a core increases with the size of the map in the same line, as a power law with exponent 0.3-0.5. Equilibrium models of core-cloud systems, based on observed line widths and map sizes, indicate that cores in Orion A and B have star-forming mass infall rates up to an order of magnitude greater than in low-mass regions such as Taurus. Cores differ significantly in their typical size, velocity dispersion, and extinction from one molecular cloud complex to another. These differences correlate with the size and mass of the complex, and with the incidence of young stars in the complex.

1 Introduction

The last 25 years have seen the development of star formation as a relatively coherent area of astrophysical research, with contributions from radio, infrared, optical, and x-ray astronomy, and from theoretical studies of gas motions, chemistry, turbulence, magnetic fields, and radiative transfer. Radio observations of molecular lines at centimeter, millimeter, and submillimeter wavelengths indicate the size scales, densities, temperatures, and motions of circumstellar and interstellar gas. Far infrared continuum observations reveal the temperature, density, and distribution of interstellar dust. Near infrared, optical, and x-ray observations indicate the temperature and density of circumstellar dust, and of photospheric and coronal gas. Theoretical studies try to account for the observations in terms of physical processes, beginning with cold condensations of molecular gas and dust, and ending with newly formed stars. They are still trying!

Although this is a new field, and many surprises still lie ahead, tremendous progress has been made in the last 25 years. To illustrate some of the interesting results, consider the following surprises to past "conventional wisdom":

Interstellar molecules can't be polyatomic, because they are destroyed too rapidly by starlight (mid-1960s). But then radio astronomers found H_2O, NH_3, and H_2CO, and dozens of others, shielded by a few magnitudes of extinction.

Molecular lines are found only within a few arcminutes of an infrared source (early 1970s). But then CO maps revealed giant molecular clouds: molecular gas is not just "around" infrared sources; instead, infrared sources are young stars which formed in the molecular clouds where they are observed.

Stars form by purely inward motion (early 1980s). But then radio astronomers found bipolar outflows associated with most young stars. The outflow motions are believed to accompany infall onto a circumstellar disk.

Low-mass stars form in isolation (late 1980s). But then infrared array observations showed that many luminous IRAS sources are accompanied by tens to hundreds of point sources - low mass members of embedded clusters. Now most low-mass stars are thought to form in stellar groups.

In this contribution we summarize properties of "dense cores" revealed by spectral line observations and discuss their relation to properties of their associated young stars.

2 Core Properties

Properties of dense cores are known from observations of spectral lines from many molecules, including NH_3, CS, $C^{18}O$, HCO^+, and HC_3N. Comparisons of properties among some of these tracers are given by Swade (1989); Bachiller, Menten, & del Rio-Alvarez (1990); and Myers et al (1991). In this summary we concentrate on results based on observations of NH_3 lines, which have been observed in a large, and perhaps in the largest, number of cores. Most of the following discussion is based on papers by Benson & Myers (1989; BM) and Harju, Walmsley & Wouterlout (1993; HWW).

2.1 Basic Observational Properties

Searches for NH_3 emission have been made primarily toward regions of high visual extinction and toward IRAS sources. Presently available NH_3 surveys are not complete in terms of area coverage. Because of the integration time required to obtain useful results, few unbiased searches of large areas can be expected, until array receivers operating at 1 cm wavelength are put into operation. However for the nearest molecular clouds, the strong correlation between visual extinction and NH_3 line emission suggests that surveys of all suitably defined extinction peaks are fairly complete.

Table 1. Mean Core Properties in Taurus, Ophiuchus, and Orion.

Region	D(pc)	n(10^4 cm^{-3})	T_{kin}(K)	$\Delta v_{1/2}$(km s^{-1})
Tau, Oph				
(stars)	0.27	1	11	0.43
(no stars)	0.12	3	10	0.27
Ori A, B				
(stars)	0.17	4	16	0.8
(no stars)	0.17	3	15	0.7

Table 1 summarizes core diameter, density, temperature and line width, for 33 cores from the survey of BM, primarily in Taurus and Ophiuchus; and for 26 cores, in Orion A and B, from HWW.

Table 1 shows that NH_3 maps of cores have typical full width half maximum (FWHM)-size, D = 0.2 pc, density 3×10^4 cm^{-3}, kinetic temperature 13 K, and FWHM line width 0.6 km s^{-1} . Significant differences are also evident: in Taurus and Ophiuchus, cores with stars have size and line width larger than cores without stars, each by a factor of $\sim$ 2; and all cores in Orion A and B are warmer, by $\sim$ 5 K, and more turbulent, by a factor of $\sim$ 2. If only starless cores are considered, those in Orion exceed those in Taurus and Ophiuchus in size by a factor 1.4; in temperature by a factor 1.5; and in line width by a factor 2.6. The most significant difference is in the nonthermal component of the linewidth, computed by subtracting the thermal component in quadrature from the observed width (e.g., Myers 1983): starless cores in Orion have mean Δv_{NT} = 0.67 km s^{-1} , greater by a factor 3.1 than in Taurus and Ophiuchus, where Δv_{NT} = 0.21 km s^{-1} .

2.2 Virial Equilibrium

The linewidths, sizes, and densities of cores indicate consistency with virial equilibrium within a factor $\sim$ 2, a factor small enough to indicate rough balance between self-gravity and internal motions, but too large to make significant distinctions among stable equilibrium, unstable equilibrium, and gravitational infall (Myers et al 1991; HWW).

2.3 Rotation

Line velocities show systematic spatial patterns in many core maps, with a well-defined linear gradient over some or most of the map. These patterns in the maps of low-mass cores (BM) have been quantified by least-squares fitting a two-dimensional linear velocity field, yielding fits and estimates of their statistical significance (Goodman et al 1993). Most of the 43 analyzed maps have a gradient more significant than 3-σ. The typical gradient direction in the plane of the sky is uncorrelated with the principal axes of the core map, suggesting that the motion, either rotation or shear, is not a major factor in the core dynamics, on the 0.1 pc scale. This conclusion is also indicated by the deduced ratio of kinetic energy, assuming uniform rotation, to gravitational energy, typically < 0.1. The magnitude of the typical gradient is about 1 km $s^{-1}pc^{-1}$, or about one FWHM line width per FWHM map size. This indicates that the rotational broadening of an observed line is a negligible fraction of the observed width.

While the systematic motion is dynamically unimportant on the 0.1 pc scale, it is highly significant on the scale of the stellar systems which form in such cores. The specific angular momentum corresponding to the observed gradients lies in the range 3×10^{20} to 3×10^{22} cm^2 s^{-1}, while that for pre-main sequence binaries in Taurus is about 6×10^{19} to 2×10^{21} cm^2 s^{-1} (Simon et al 1992). These ranges overlap, but the typical core value exceeds the typical binary value by an order of magnitude. If multiple stellar systems derive their angular momentum from that observed on the core scale, they must lose close to 0.9 of their core angular momentum. Star formation from dense cores appear to have an "angular momentum problem" which is significant, but much smaller than the classical deficit between the Galactic shear and the typical stellar system.

2.4 Thermal and Nonthermal Motions

NH_3 line observations are useful for precise determination of both kinetic temperature and line width, When the observed width is corrected for broadening by saturation, hyperfine structure, and instrumental resolution, a significant excess over the thermal width remains. These "nonthermal" motions are evident in all spectral lines, even including lines narrower than those of NH_3, such as the J=4-3 line of HC_3N (Fuller & Myers 1993).

In most of the volume of a molecular cloud, as traced for example by lines of CO and ^{13}CO the nonthermal component of the line width exceeds the thermal component by an order of magnitude. Thus molecular cloud motions are supersonic. Furthermore, the supersonic motions vary with cloud size and density. For a typical ^{13}CO line width of 1.5 km s^{-1} and kinetic temperature 15 K in a massive molecular cloud, the ratio of observed to thermal line width is $\Delta v_{obs}/\Delta v_T = 9.7$. For dense cores in massive clouds this ratio is smaller: in the NH_3 line for starless cores in Orion, $\Delta v_{obs}/\Delta v_T = 3.5$, based on the data in Table 1. The supersonic line widths increase and the density decreases with increasing map size D, according to power laws $\Delta v_{obs} \sim D^p$, and $n \sim D^{-q}$, with $p = 0.3 - 0.5$ and $q = 1.0 - 1.5$ (Larson 1981). Trends of this type ("Larson's Laws") are seen from cloud to cloud in individual lines (e.g. Dame et al 1986), and from line to line in individual clouds (e.g. Snell 1981).

The supersonic motions probably arise from MHD waves, which can persist at the Alfven speed without shock dissipation, unlike supersonic hydrodynamic turbulence (Arons & Max 1975), which can propagate above cutoff due to nonlinear steepening in a wide range of cloud conditions (Zweibel & Josafatsson 1983), and which can account for the similarity of observed line widths and Alfven speeds (Myers & Goodman 1988).

The line width in a core is related to the mass of the star which tends to form in the core, according to the observations of 63 cores with associated IRAS sources by Wouterlout, Walmsley, & Henckel (1988) and by BM. Both the thermal and the nonthermal parts of the line width increase in the luminosity of the associated star, but the nonthermal part increases more rapidly, and the two trend lines cross at a stellar luminosity of about 10 $L_\odot$, corresponding to a stellar mass about 2 $M_\odot$ (Myers, Ladd & Fuller 1991). Thus in cores which form stars less massive than 2 $M_\odot$, thermal motions dominate nonthermal motions; and conversely.

2.5 Incidence of Young Stars

The earliest CO surveys revealed bright CO emission from signposts of the formation of massive OB stars, including compact H II regions, associated near-infrared sources, and OH and H_2O masers (e.g., Wilson et al 1974). Once the associated molecular gas was shown to be extended over many tens of pc (e.g., Kutner et al 1977) it was clear that massive stars form in "giant" molecular clouds.

Although line emission from massive star-forming regions is bright, it is difficult to interpret in terms of core-star genetics. The stellar winds and ionization from already formed stars are thought to alter initial conditions too much to trace the path from core to star. The "starless" cores associated with massive stars as in Table 1 were unknown until recently. Also, the brightest regions are generally more than 1 kpc distant, so the linear resolution of most radiotelescopes has generally been too coarse to resolve stellar masses of star-forming gas.

Cores associated with low-mass stars are fainter than their more massive counterparts, and it was not possible to study them systematically until the early 1980s, when sufficiently low-noise receivers at cm and mm wavelengths became available. Once detectable, more than 100 such cores were found in the nearest star-forming complexes, including Taurus and Ophiuchus (Myers, Linke & Benson 1983 [MLB]; Myers & Benson 1983). The IRAS satellite found that about half of the MLB cores are associated with an IRAS point source, and that half of these are previously unknown, optically invisible sources of luminosity 0.3 - 30 $L_\odot$ (Beichman et al 1986; Myers et al 1987). These findings suggest that low-mass stars form in the low-mass cores with properties in Table 1.

The size and density, temperature, and primarily thermal internal motions of low-mass cores match well the properties of the isothermal sphere (e.g. Chandrasekhar 1939). A similarity solution of the collapse of the "singular" isothermal sphere, the most unstable form of this system, was presented by Shu (1977), extending earlier models by Larson (1969). In this solution the onset of infall starts near the center, and propagates outward at the sound speed. This "inside-out" collapse has constant mass infall rate, $\frac{dm}{dt} = 0.975\frac{\sigma^3}{G}$, where σ is the isothermal sound speed and G is the gravitational constant. Modifications to this solution have led to a comprehensive model of a collapsing, rotating, magnetized core, which forms a protostar-disk system. The spectrum radiated by this system is in good agreement with observed spectra of the most obscured IRAS sources (see Shu, Adams & Lizano 1987 for a summary).

These models of low-mass star formation can be criticized because they do not account easily for the elongated shapes of many cores; for the lack of correlation between core map axes and the directions of core velocity gradients; for the formation of binaries and other multiple systems; and for the specific values of stellar masses. Nonetheless they are by far the most successful and comprehensive models of star formation now available.

2.6 Mass Infall Rates

The models of low-mass star formation described above are less successful for regions of more massive star formation, because massive cores are dominated by nonthermal motions, as Table 1 shows; and because these motions increase with map size, and thus with projected radius, as described in Section 2.4. Therefore massive cores are not isothermal. Furthermore the nonthermal motions in cores appear to increase with the mass of the star the core tends to form. This variation of core initial conditions with the mass of the associated star has given impetus to infall models which account for the nonthermal motions in the core (Myers & Fuller 1992, 1993; Caselli & Myers 1994).

These thermal and nonthermal ("TNT") models treat the nonthermal motions as a source of isotropic pressure. A spherically symmetric model of hydrostatic equilibrium is used to obtain the profile of density with radius needed for consistency with the observed variation of line width with map size. In this analysis, low-mass cores tend to have density about one-fifth that of more massive cores in Orion, at the same radius; and shallower density profiles (p E 1.2) than do the more massive cores (p E 1.6; Caselli & Myers 1994).

When a TNT model core collapses from the inside out, the wave of infall moves outward at the effective sound speed, which exceeds the isothermal sound speed by a factor which increases with radius. The mass infall rate is determined by the density structure in the hydrostatic equilibrium solution. In the typical massive core in Orion, the time to make a 10 $M_\odot$ star is 4×10^5 yr, the same as the time for the typical low-mass core to make a 1 $M_\odot$ star. Thus cores in regions of massive star formation are denser, and have mass infall rates up to 10 times greater than cores forming low-mass stars (Caselli & Myers 1994).

These TNT models account for an important difference between low-mass and massive star formation, but they do not address the tendency for massive stars to be accompanied by groups of lower-mass stars.

2.7 Shapes

Maps of cores in lines of NH_3, CS, and $C^{18}O$ have contours which are usually elongated, with aspect ratio close to 2 (BM; Lada, Bally & Stark 1991; Myers et al 1991). For cores in filamentary dark clouds, the long axis of the core map is usually aligned with the long axis of the more extended filament, which is 1-3 pc long, and has aspect ratio > 10. These dark lanes, so elongated in two dimensions, are statistically much more likely to be prolate than oblate: the symmetry axis of a cloud with high projected aspect ratio can have a much wider range of orientations in space if it is prolate than if it is oblate. Then the elongated core which is projected on a more elongated filament, and which is coaxial with the filament, is probably just the densest part of the filament, and both are very likely to be prolate (Myers et al 1991).

This departure from spherical symmetry implies that core structure cannot be explained by simple equilibrium between self-gravity and isotropic internal pressure. Magnetic fields and their waves have been suggested as a source of anisotropic pressure, but the long axes of filamentary dark clouds have no obvious relation to the magnetic field directions indicated by optical polarization (Goodman et al 1990).

Regardless of their origin, filamentary clouds appear common, and the collapse of a core in a filamentary cloud has been suggested as a way to make binaries and small stellar groups (Bonnell et al 1991).

2.8 Regional Variations

The differences in Table 1 between core properties in Taurus and Ophiuchus on the one hand, and Orion on the other, are one aspect of a group of properties which appear relatively similar within a star-forming complex, but which vary significantly from complex to complex. The typical column density of a core, the likelihood that the core will have an associated young stellar object, the ratio of nonthermal to thermal motions, the presence of concentrated stellar groups and massive stars, and the size, mass, and extinction of the complex, all appear to scale together. The smallest complexes, such as Chamaeleon 2, tend to have condensations with low extinction, weak lines in high-density tracers, low incidence of young stars, and no clusters. In contrast, Orion A is much larger, has many high-extinction cores, bright lines, many IRAS sources and T Tauri stars, and dense embedded star clusters. One can crudely rank complexes from Chamaeleon 3 to Chamaeleon 2 to Lupus to Taurus to Perseus and Ophiuchus to Orion B to Orion A in these properties (Myers 1991; Vilas-Boas, Myers & Fuller 1994; Ladd, Myers & Goodman 1994; Wiseman & Adams 1994).

I thank the organizers of this Ringberg Workshop for their invitation to participate, for their hospitality, and for their patience as this manuscript was being prepared.

References

Arons, J., & Max, C. E. 1975, ApJL, 196, 77.
Bachiller, Menten, & del Rio-Alvarez 1990, A&A , 236, 461.
Benson, P. J., & Myers, P. C. 1989, ApJS, 71,89 (BM).
Bonnell, I., Martel, H., Bastien, P., Arcoragi, J. P., & Benz, W. 1991, ApJ, 377, 553.
Caselli, P., & Myers, P. C. 1994, in preparation.
Chandrasekhar, S. 1939, An Introduction to the Study of Stellar Structure (New York: Dover).
Dame, T. M., Elmegreen, B. G., Cohen, R. S., & Thaddeus, P. 1986, ApJ, 305, 892.
Fuller, G. A., & Myers, P. C. 1993, ApJ 418, 273.
Goodman, A. A., Bastien, P., Myers, P. C., & Menard, F. 1990, ApJ 359, 363.
Goodman, A. A., Benson, P. J., Fuller, G. A., & Myers, P. C. 1993, ApJ,406, 528.
Harju, J., Walmsley, C. M., & Wouterlout, J. A. 1993, Astron. Astrophys. Suppl., 98, 51 (HWW).
Kutner, M. L., Tucker, K. D., Chin, G., & Thaddeus, P. 1977, ApJ, 215, 521.
Lada, E. A., Bally, J., & Stark, A. A. 1991, ApJ, 368, 432.
Ladd, E. F., Myers, P. C., & Goodman, A. A. 1994, ApJ, in press.
Larson, R. B. 1969, M. N. R. A. S., 145, 271.
Larson, R. B. 1981, M. N. R. A. S., 194, 809.
Myers, P. C. 1983, ApJ, 270, 105.
Myers, P. C. 1991, in Fragmentation of Molecular Clouds and Star Formation, eds. E. Falgarone, F. Boulanger, & G. Duvert (Dordrecht: Kluwer), p. 221.
Myers, P. C., & Benson, P. J. 1983, ApJ, 266, 309.
Myers, P. C., & Fuller, G. A. 1992, ApJ, 396, 631.
Myers, P. C., & Fuller, G. A. 1993, ApJ, 402, 635.
Myers, P. C., & Goodman, A. A. 1988, ApJL, 326, L27.
Myers, P. C., Fuller, G. A., Goodman, A. A., & Benson, P. J. 1991, ApJ, 376, 561.
Myers, P. C., Ladd, E. F., & Fuller, G. A. 1991, ApL, 372, L95.
Myers, P, C, Linke, R. A., & Benson, P. J. 1983, ApJ, 264, 517 [MLB].
Shu, F. H. 1977, ApJ, 214, 488.
Shu, F. H., Adams, F. C., & Lizano, S. 1987, ARAA, 25, 23.
Simon, M., Chen, W. P., Howell, R. R., Benson, J. A., & Slowik, D. 1992, ApJ, 384, 212
Snell, R. L. 1981, ApJS, 45, 121.
Swade, D. A. 1989, ApJS, 71, 219.
Vilas-Boas, J. W. S., Myers, P. C., & Fuller, G. A. 1994, ApJ, submitted.
Wilson, W. J., Schwartz, P. R., Epstein, E. E., Johnson, W. A., Etcheverry, R. D., Mori, T. T., Berry, G. G., & Dyson, H. B. 1974, ApJ, 191, 357.
Wiseman, J. A., & Adams, F. C. 1994, ApJ, submitted.
Wouterlout,, J. A., Walmsley, C. M., & Henkel, C. 1988, A&A , 203, 367.
Zweibel, E. G., & Josafatsson, K. 1983, ApJ, 270, 511.

Surveying Giant Molecular Clouds for Low Mass Stars: NIR Imaging of the DR22 and S184 Regions

S. T. Megeath

Max–Planck–Institut für Radioastronomie, Auf dem Hügel 69, 53121 Bonn

Abstract: Due to recent advances in infrared detector array technology, near–infrared imaging is emerging as a powerful technique for surveying the embedded stellar population of giant molecular clouds. We discuss the strengths and limitations of this technique through the analysis of two regions, DR22 and S184. Near–infrared images of these regions show dense clusters of low–intermediate mass stars. We analyze the properties of the stars and the initial mass functions of the regions using photometry in the near–infrared bands.

1 Introduction

Most of this meeting has been devoted to the gas and dust which comprise giant molecular clouds (GMCs), but little has been said on the stars which form in and eventually determine the evolution of molecular clouds. The lack of discussion on this topic is the result of a lack of data on the stellar content of GMCs. The huge extinction through molecular clouds ($A_V > 10$) obscures stars from optical UBVRI band imaging with CCD cameras. Radio continuum observations with telescope arrays can map the gas ionized by massive stars with subarcsecond resolution, but these observations trace only the most massive component of the stellar population. Far–infrared data, especially IRAS data, is invaluable in measuring the total luminosity of stars in a molecular cloud core, but the luminosity is again dominated by the most massive stars. In the past few years, this situation has been radically altered by the development of large format near–infrared (NIR) arrays. With these arrays and 2 meter class telescopes, surveys of the stellar content can be made which can detect stars with masses close to the substellar limit (0.08 $M_\odot$) at distances of 2 kpc.

The first infrared surveys with arrays concentrated on the Orion region. Surveys of the Trapezium region and the Orion B cloud showed clusters of low mass stars surrounding sites of high mass star formation (Lada et al. 1991 and Zinnecker et al. 1993). The result from the Orion B cloud survey was that 58% to 96% of the embedded sources were found in dense clusters; this suggests that most stars form in clusters and not in isolated dense cores (Lada et al. 1991).

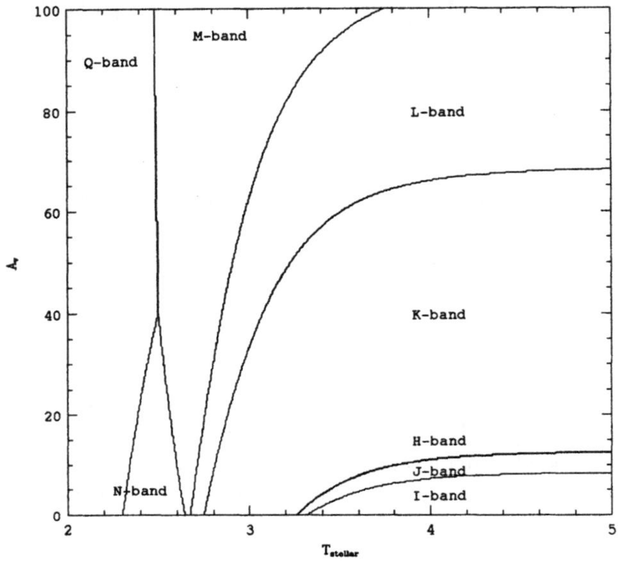

Fig. 1. Optimal Bands for Molecular Cloud Surveys

This figure displays the optimal wavelength band for detecting an embedded star as a function of reddening and photospheric temperature. The curves trace the region of parameter space where the signal to noise ratio for the two adjacent bands are equal. In the regions between the curves, the signal to noise ratio is dominated by the designated band. Note that the optimal H-band region lies in a narrow strip, between the J and K-band regions, which is not resolved in our plot.

Following the Orion surveys, a growing list of regions have been surveyed in the 1-2.5 μm regime. Given the sensitivity of the NIR arrays, regions out to 3 kpc can be efficiently surveyed for young (10^6 years), low-mass stars (< 1 $M_\odot$). In almost all surveys of regions showing the signposts of high mass star formation, dense clusters of low mass stars are detected. For a recent review of these regions, see Zinnecker et al. 1993.

In the following contribution, we discuss why the 2.2 μm regime is important for surveying the stellar content, and we present the detection and analysis of two young clusters associated with regions which have recently formed massive stars. The analysis illustrates the methods used to interpert NIR photometry of young stars and clusters. We will discuss the ambiguities inherent in such an approach, and some possible observations which may circumvent these problems.

2 Why a 2.2 μm Survey?

The rapid pace of infrared detector array development has revolutionized infrared astronomy. In the past few years, relatively large format infrared arrays have been developed with capabilities close to that of an ideal detector. Currently, such arrays can cover all the infrared windows available to ground based telescopes. Given this newfound flexibility, we must ascertain which infrared bands are optimal for molecular clouds surveys.

In Fig. 1 we have plotted the optimal wavelength for detecting a star of temperature $T_\star$ and extinction A_V. We have calculated the signal to noise ratio for a star in 8 different wavelength bands I (0.9 μm), J (1.25 μm), H (1.65 μm), K (2.2 μm), L (3.4 μm), M (4 μm), N (10 μm) and Q (20 μm). These bands are the windows in the infrared available for ground based telescopes. We have assumed that the noise in each band is dominated by the sky background, a combination of OH skyglow and thermal emission (see Gillett 1987). We have also assumed that the detector characteristics, telescope and camera transmission, telescope size, and seeing are the same for all bands.

The result is that for a wide range of reddenings and temperatures, the K–band gives the highest signal to noise ratio. Only for very cool objects ($T < 500$ K), very deeply embedded stars ($A_V > 60$), and for objects with low reddening ($A_V < 10$) do other bands dominate. Thus, K–band surveys will give the most complete census of pre–main sequence stars embedded in molecular cloud cores.

We should note that this plot only shows the *relative* signal to noise ratio in each band. Due to the increase in background at mid-infrared wavelengths, only very luminous objects may be detected at 10 and 20 μm . Although the 10 and 20 μm bands may be the most sensitive regime to detect very low temperature stars, only the most luminous (and consequently, the most massive) stars will be detectable. For example, with a 1 minute integration on a 2 meter class telescope, we can detect a 1 $L_\odot$ star with a photospheric temperature of 1000 K to a distance of 3 kpc in the K–band (the optimal band for 1000 K blackbodies). For a 300 K object with an equivalent distance, we would be limited to detecting 20 $L_\odot$ objects in the N–band (the optimal band for detection of 300 K blackbodies) given identical telescopes and integration times. For this reason, surveys of GMCs at wavelengths longer than 2 μm will be limited to the detection of massive or nearby YSOs until the launch of a large aperture (> 1 meter for a spatial resolution of 1″ or better) space based telescopes.

3 The Detection of Young Clusters Near DR22 and S184

We have recently surveyed several molecular cloud cores using infrared cameras designed for the 1–2.5 μm regime. Using a camera with a 256 × 256 NICMOS3 array on the University of Hawaii 2.2 m telescope and and a camera with a 126 × 126 NICMOS2 array on the Palomar 5 meter [1], we obtained data on two high mass star forming regions:

DR22 is a compact HII region embedded in a ridge of dense molecular gas. A diffuse ionization front has been detected adjacent to the southern edge of

[1] The 5m Palomar Observatory is operated jointly under a collaborative agreement between the California Institute of Technology and Cornell University

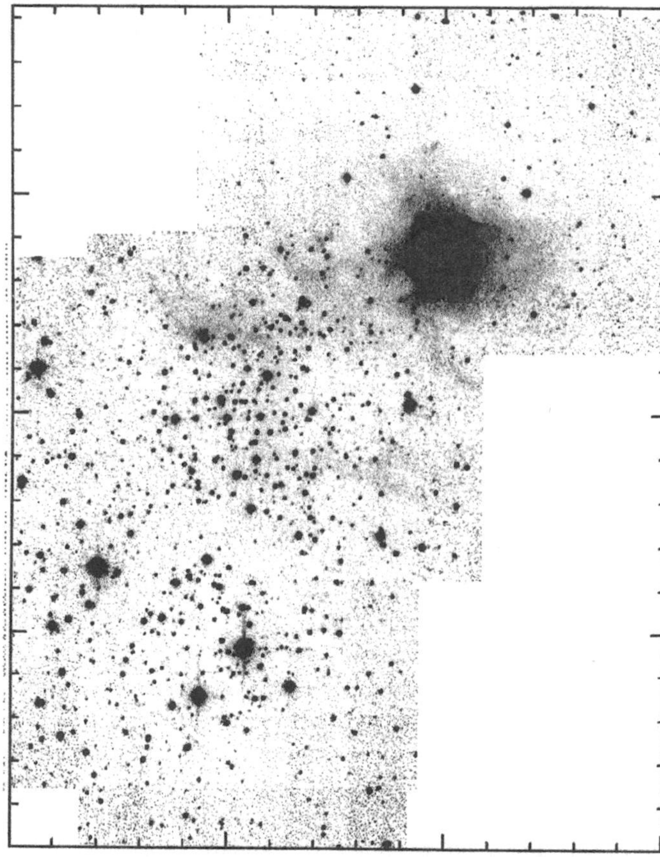

Fig. 2. DR22 Image

The 2.2 μm mosaic of the DR22 region, taken with a NICMOS3 camera on the University of Hawaii 2.2 m telescope. The maximum dimensions of the mosaic are $4'.9 \times 6'.2$ or 4.28×5.41 pc.

the ridge, suggesting that the ridge is being ionized by, and may have been swept up by, a previous epoch of massive star formations (Odenwald 1986, Megeath, unpublished results). The distance is approximately 3 kpc.

S184, or NGC 281, is a giant HII region in the Perseus arm with a distance of 2.2 kpc. A study of Elmegreen and Lada (1978) revealed a H_2O maser embedded in a molecular cloud associated with the HII region. Recent mapping of the this cloud in isotopomers of CO and CS and the 20 cm continuum shows that the maser is in a dense ridge of molecular gas at the HII /molecular gas interface (Carpenter et al. 1993, Carpenter et al. 1990, Megeath, unpublished results), making this region a prime candidate for induced star formations.

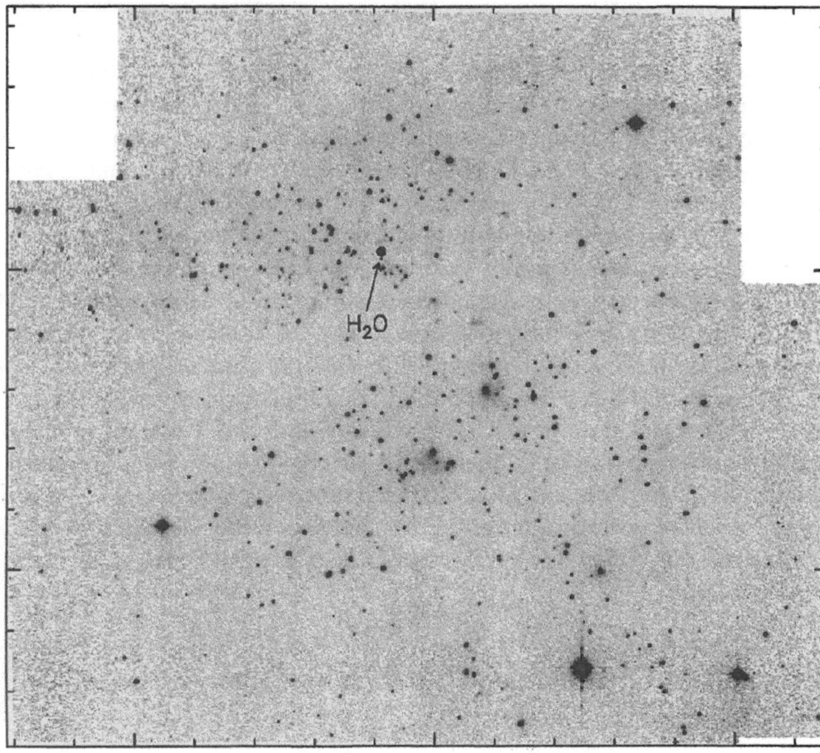

Fig. 3. S184 Image

The 2.2 μm mosaic of the S184 region, also taken from the U. of H. 2.2 m telescope. The dimensions of the mosaic are $4\overset{'}{.}0 \times 4\overset{'}{.}5$ or 2.56×2.88 pc.

Inspection of the resulting 2.2 μm mosaics in Figs. 2 and 3 immediately shows an element missing from previous studies: stars.

The DR22 mosaic (Fig. 2) shows several interesting features. First, the compact HII region detected by the VLA is located in the northwestern corner of the image. A cluster is detected to the southeast of the HII region. The cluster stars are typically faint, and the cluster shows a centrally condensed core. A faint nebulosity is apparent on the northern edge of the cluster, this nebulosity delineates the ionization front detected with the VLA (Megeath, unpublished results).

The image of S184 (Fig. 3) also shows a dense cluster of faint stars associated with the molecular gas ridge (see Carpenter et al. 1993 for first detection of this cluster). The most striking feature of the cluster is the dust lane, running northwest to southeast, which bisects the 2.2 μm mosaic. This dust lane is coincident with a 210 $M_\odot$ dense core detected in CS by Carpenter et al. (1993). A faint nebulosity is seen surrounding the two brightest stars in the southern part of the cluster, further evidence that these stars are associated with the dense gas.

We completed our study of these regions by taking deep J, H, and K–band images of the central 1.′5 × 1.′5 of the cluster. The goal was to obtain a deep census of the population from K–band imaging and obtain color information on the embedded stars. We also obtained observations of reference positions, taken 5′ west of the cluster positions. These images are used to constrain the properties of the field stars. By assuming that the number of field stars is the same for the cluster and reference positions, we can correct our data of young clusters for field star contamination. This assumption is clearly not completely valid, extinction from molecular gas associated with the cluster will reduce the number of background stars; however, with this problem in mind we can still place limits on the field star contamination. To a limiting magnitude of 18 in the K–band, we find 175 stars in the DR22 cluster and 177–216 stars in the S184 cluster. The surface densities of the central cores (i.e. inner $\sim$ 0.25 pc) are 169 pc^{-2} and 377–426 pc^{-2} for DR22 and S184 respectively. The range in values for S184 is the result of uncertainties in the field star contamination due to the large extinction observed through the S184 cluster (but not the DR22 cluster, see the following section).

4 Constraining the Masses and Evolutionary States of Young Cluster Members

The study of galactic clusters has relied primarily on optical photometry. Using CCD cameras and sophisticated software for finding stars and performing digital photometry, photometric magnitudes can easily be obtained for thousands of cluster stars. Using photometry in several bands, optical astronomers can construct color vs. magnitude and color vs. color diagrams (typically m_V vs. $B-V$ and $U-V$ vs. $B-V$) which are used to place stars in an H–R diagram. A comparison of the H–R diagram with evolutionary stellar models gives a mass and age for each cluster star.

Using NIR arrays and the photometry software developed for optical CCD cameras, photometry can be efficiently obtained for hundreds of near infrared sources. Unfortunately, the conversion of the NIR data into a H–R diagram for the cluster is problematic. The NIR luminosity of a Young Stellar Object, or YSO, depends on a mixture of photospheric emission, disk emission, extinction from a dust envelope, and extinction from dust in the surrounding molecular cloud. Furthermore, the variation of color with spectral type is much smaller in the NIR than in the optical. Consequently, the classification of YSOs on the basis of NIR photometry is difficult and fraught with ambiguities.

To understand the near infrared properties of young stars, it is important to first examine the characteristics of nearby YSOs. We can classify these as:

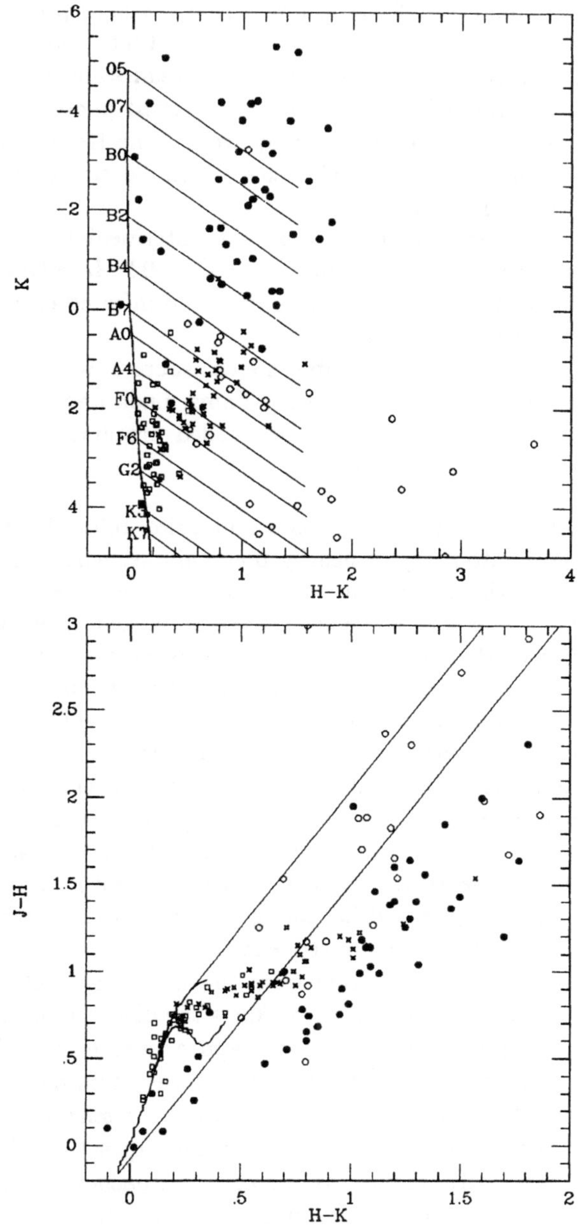

Fig. 4. Color–Magnitude and Color–Color Diagrams of Nearby YSOs

The color–magnitude (top) and color–color (bottom) diagrams for nearby young stellar objects. The open squares are weak line T Tauri stars from Strom et al. (1989), the stars are classical T Tauri stars from Strom et al., the filled circles are Herbig Ae/Be stars from Hillenbrand et al. (1992), and the open circles are the dense core sources of Myers et al. (1987) which include several Class I objects. In the color–magnitude diagram, we have added the main sequence and the extinction tracks corresponding to an A_V of 20 mag for each listed spectral type. In the color–color diagram, we have drawn the main sequence, the giant branch, and the two extinction tracks which bracket the possible colors for main–sequence and giant stars (Koornneef 1983 and Draine 1989).

T Tauri stars are young ($< 10^7$ yrs), low mass (< 3 $M_\odot$) pre–main sequence stars (Bertout 1989). T Tauri stars can be divided into two separate groups: Classical T Tauri stars which show strong Hα emission and bright excess emission in the infrared (compared to the expected photospheric emission) and Weak Line T Tauri stars which show weak or no Hα emission and little excess emission in the infrared. The excess infrared emission is thought to arise in circumstellar disks.

Herbig Ae/Be stars are thought to be the intermediate mass (>3 $M_\odot$) equivalents of T Tauri stars (Hillenbrand et al. 1992). Like Classical T Tauri stars they often exhibit bright emission lines and infrared excesses, and may have circumstellar disks as well.

Class I Objects are optically invisible stellar sources which show a rising spectrum with increasing wavelength (Adams, Lada and Shu 1987). These objects are thought to be the dust enshrouded precursors of T Tauri stars (and possibly Herbig Ae/Be stars).

We have constructed color–magnitude and color–color diagrams (Fig. 4) from the published photometry of nearby pre–main sequence stars and deeply embedded source. In the color–magnitude diagram we have plotted K–band magnitude vs. $H - K$ color. We have chosen these bands since many stars detected in the K and H bands are too reddened to detect in the J–band. The main caveat is that the K–band can be dominated by disk emission, which adds an extra dimension of uncertainty. Nevertheless, we must rely on the K–band to detect the most deeply embedded and faintest sources.

For comparison, we have included the main sequence (based on the photometry of Koornneef 1983) in the color–magnitude diagram and for each listed spectral type, we have drawn the reddening track corresponding to an A_V of 20 mag (using the reddening law in Draine 1989). A reddened main sequence star of a given type will fall along the appropriate reddening vector; consequently, for an embedded main sequence star we can obtain the mass for a star in the m_K vs. $H - K$ diagram from the intersection of the reddening track and the main sequence (assuming we know the distance to the star). If the stars are pre–main sequence stars, they will appear above the reddening track for an equivalent mass main sequence star. If we falsely assume that a pre–main sequence star is on the main sequence and project the star along a reddening track onto the main sequence, we will obtain a mass higher than the actual mass of the star. This mass will decrease as the star approaches the main sequence and will finally converge to the correct mass. Consequently, for a reddened star in the m_K vs. $H - K$ diagram which does not have an a priori determination of the age, we obtain a range of possible masses, the upper limit of which is given by the intersection of the reddening track and the main sequence.

It is evident from the color–magnitude diagram that the intermediate and low mass YSOs can be distinguished on the basis of their K–band luminosities. Most Herbig Ae/Be stars in the diagram show a 2.2 μm luminosity in excess of that for a reddened B4 star, conversely most T Tauri stars show a luminosity below that of a reddened B4 star. Thus, with some overlap, the reddening track of the

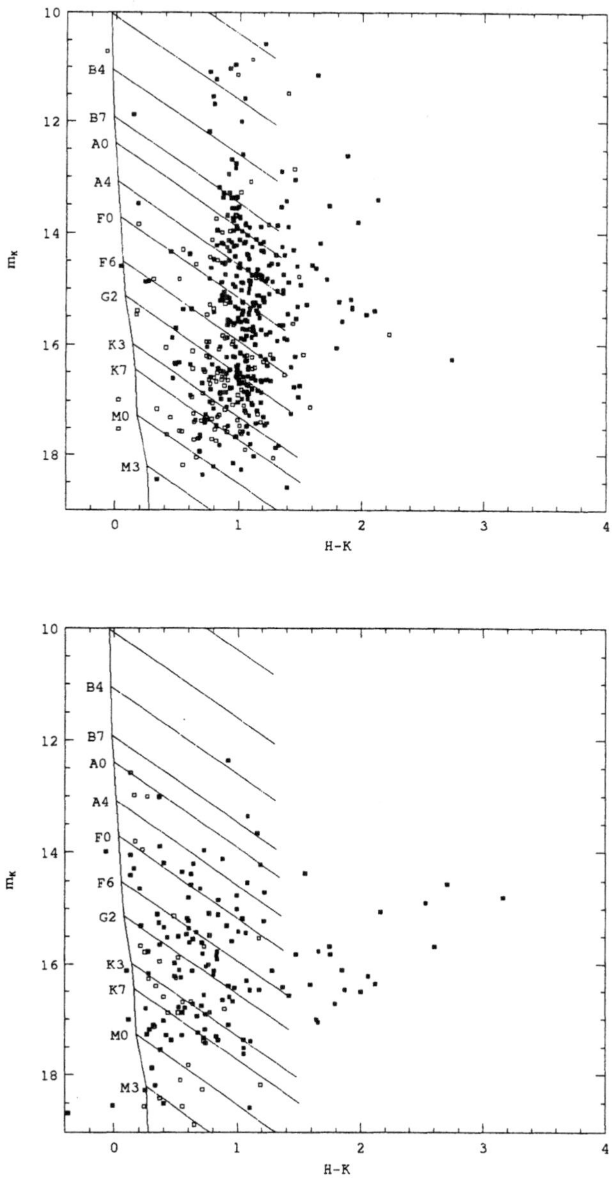

Fig. 5. Color–Magnitude Diagrams of DR22 and S184

The color-magnitude diagrams of DR22 (top) and S184 (bottom). The filled squares are stars in the cluster field and the open squares are from a reference field 5′ west of the cluster. The curves are the main sequences at a distance of 3 and 2.2 kpc for DR22 and S184 respectively. The lines represent the extinction tracks for an A_V of 20 mag.

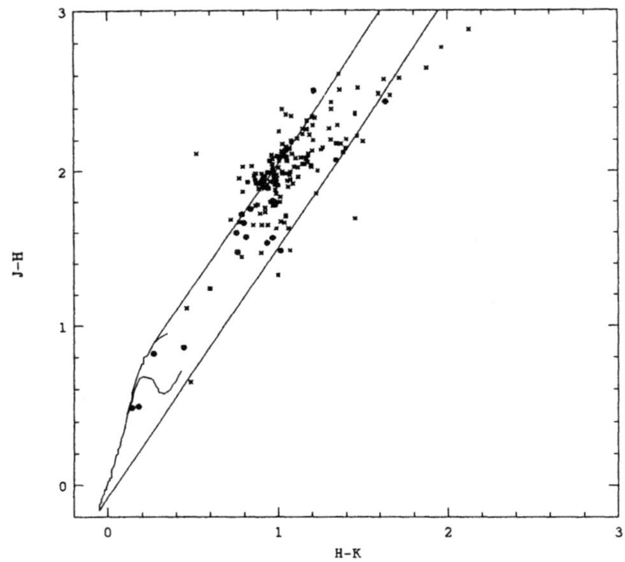

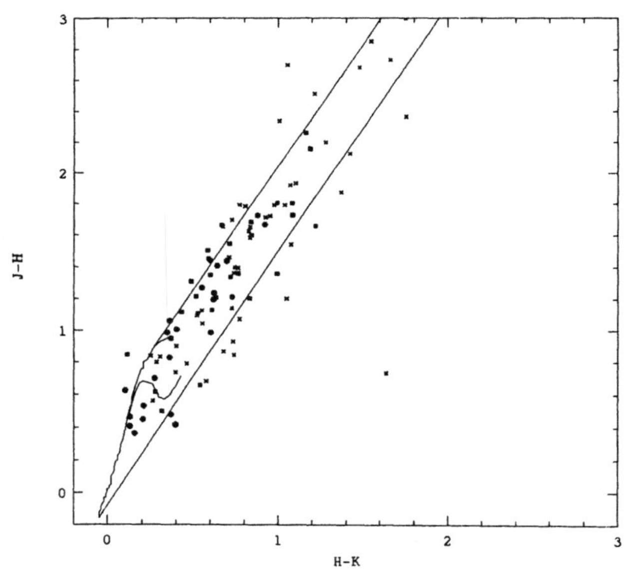

Fig. 6. Color–Color Diagram of DR22 and S184

Above are the color–color diagrams of DR22 (top) and S184 (bottom). For DR22, the circles are stars with $m_J < 16$, the squares are stars with $16 < m_J < 17$, and the stars are stars with $m_J > 17$. The J–band image is complete down to 19th magnitude. For S184, the circles are stars with $m_J < 17$, the squares are stars with $17 < m_J < 18$, and the stars are stars with $m_J > 18$. The J–band image is complete down to 20th magnitude.

B4 star is a convenient division between low mass and intermediate mass pre-main sequence stars. The class I objects of Myers et al. (1988) fill the reddened half ($H - K > 1.5$) of the color–magnitude diagram and are indistinguishable from reddened T Tauri stars. In others words, we cannot distinguish between a pre–main sequence star (i.e. T Tauri or Herbig Ae/Be star) deeply embedded in a GMC core and a Class I object enshrouded in an accreting envelope solely on the basis of J, H, and K–band colors (see also Lada and Adams 1992).

To compare the colors of YSOs with the colors of more evolved stars, we have plotted the main sequence and the giant branch in the color–color diagram with two reddening vectors which bracket the possible location of reddened main sequence and giant branch stars (from the photometry of Koornneef 1983 and the extinction law of Draine 1989). It is apparent from this diagram that pre–main sequence stars often fall below the lower reddening track, i.e. their colors are not consistent with the colors of main sequence or giant branch stars. This is due to an excess emission which can dominate over the photospheric emission for wavelengths greater than 2 μm . Such an excess is thought to be the result of luminous disks surrounding young stars (Lada and Adams 1992). It is apparent from the color–color diagram that Ae/Be stars typically have higher excesses than T Tauri stars, and can be distinguished on the color–color diagram. Also note that weak line T Tauri stars typically have colors similar to main sequence G and K stars.

Using excess 2 μm emission as an indicator of the existence of a circumstellar disk, Skrutskie et al. 1990 and Strom et al. 1989 have examined the frequency of occurrence for disks around pre–main sequence stars in the Taurus–Auriga region as a function of age. The result is that approximately half of the stars with ages less than 3 Myrs have disks. If all solar type stars initially have disks, then 50% of all disks have a lifetime less than 3 Myrs. Disks are observed around stars up to an age limit of 10 Myrs, at which the frequency of occurrence for disks drops dramatically. If the stars in the Taurus-Auriga region are representative of stars in the S184 and DR22 clusters, we can use the existence of disks as an age indicator. In particular, if $> 50\%$ of the stars do not show excesses, the cluster age should be in excess of 3 Myrs.

An analysis of the DR22 color–magnitude diagram (Fig. 5) reveals that 68% of the stars detected in the H and K–bands DR22 have a maximum mass of 3 $M_{\odot}$, and that 87% fall below the B4 line and are consistent with T Tauri stars. These numbers indicate that the DR22 cluster is dominated by low mass stars. Further examination of the the DR22 color–magnitude diagram shows a remarkable feature. The cluster fall into a narrow line in $H - K$ space. This implies that the cluster has a substantial amount of foreground reddening, but little reddening is seen through the cluster. We can infer that this cluster is not embedded in a molecular core.

The color–color diagram (Fig. 6) also shows that the stars in the DR22 cluster occupy a very narrow region in $J - H$ vs. $H - K$ space, indicative of little internal reddening. Furthermore, the 2 μm excesses detected toward Herbig Ae/Be stars and classical T Tauri stars are apparent only for a small percentage of stars. In fact, we find more than 50% of the sources lack excesses and show colors consistent with weak line T Tauri stars or main sequence stars. If the same processes govern the evolution in disks in this cluster as the Taurus-Auriga region, we expect this cluster to have an age > 3 Myrs.

From the color–magnitude diagram, we have inferred that at least 13% of the stars in DR22 are intermediate mass stars, yet these stars do not have the large excesses commonly observed for Herbig Ae/Be stars. Recent results of Hillenbrand et al. (1993) suggest that the disappearance of disks around intermediate stars occurs in less than 1 Myr. Accordingly, we expect excesses towards intermediate mass stars to be rare in all but the youngest clusters.

In contrast, the S184 cluster shows a wide range in reddening, indicating that this cluster *is* embedded in a molecular cloud (Fig. 5). A comparison of the cluster color–magnitude diagram shows that 98% of the stars detected in the H and K–bands fall below the B4 reddening track, the dividing line between T Tauri and Herbig Ae/Be stars. We also find that 91% of the stars have an upper mass limit of 3 $M_{\odot}$. Thus, the S184 cluster is populated mainly by low mass stars.

The color–color diagram shows that the stars in S184 are shifted downwards relative to the stars in the DR22 region (Fig. 6). This could be explained by infrared excesses, in this case we expect more than 50% of these sources have excesses. For this reason, we expect the age of this cluster to be less than 3 Myrs.

In conclusion, we find that the clusters are composed primarily of a mix of intermediate to low mass stars, with the low mass stars dominating the mix. We find that the DR22 cluster members do not show excesses which are indicative of circumstellar disks while most stars in the S184 cluster seem to show excesses. If we compare these statistics with the disk lifetimes calculated from stars in the Taurus-Auriga region, we find an age of less than 3 Myrs for S184 and more than 3 Myrs for DR22. This age difference may also explain why the S184 cluster is embedded in a molecular cloud and the DR22 cluster is not.

5 The K–Band Luminosity Function

The ultimate goal of NIR surveys of molecular clouds is the determination of the initial mass function for individual star forming regions. There are two fundamental problems we have to overcome to determine the mass function. First, there is no unique conversion from J, H, and K–band magnitudes to mass for an embedded star at a known distance. Second, we typically have an incomplete sample of stars, and many of the sources detected in the K–band are not detected in the J or H–bands.

Despite the ambiguities in the individual properties of stars comprising embedded clusters, we can learn a great deal by examining the gross properties of the cluster. In particular, the K–band luminosity function, the number of stars per K–band magnitude interval, provides a valuable diagnostic for studying the stellar population. Since the K–band is the most sensitive wavelength band for detecting embedded pre–main sequence stars, the K–band luminosity function, or KLF, is derived from the most complete sample of the embedded population.

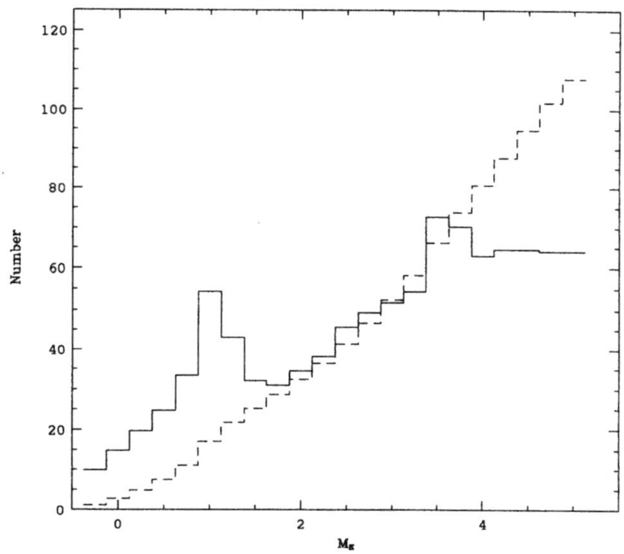

Fig. 7. Model KLF for Naked and Embedded Cluster

The solid line shows the theoretical KLF for a 0.3 Myr coeval cluster presented in Zinnecker et al. 1993. The dashed line shows the same cluster embedded in a molecular cloud. We have assumed that the stars are evenly distributed throughout a homogeneous cloud with extinctions ranging from $A_K = 0 - 5$.

What do we expect the KLF to look like for a young cluster? Lada et al. (1993) write the KLF as the following:

$$\frac{dN(m_K)}{dm_K} \propto 10^{\gamma m_K/2.5\beta} \tag{33}$$

where $N(m_K)$ is the number of stars brighter than m_K, γ is the power law index of the initial mass function (1.35 for a Salpeter IMF), and β is the power law index for the mass–luminosity relationship, which can vary from 3.1 for solar mass main sequence stars to ~ 1 for 10^6 old year protostars (Lada et al. 1993). A more detailed analysis of coeval clusters was performed by Zinnecker et al. 1993. In that work, the theoretical pre–main sequence evolutionary tracks were used in conjunction with a non–power law Miller-Scalo IMF (Miller and Scalo, 1979). The K–band magnitudes were calculated by assuming a blackbody photospheric spectrum. The resulting KLFs are not power laws, instead they show peaks due to deuterium burning in the stars. As the cluster ages, progressively lower mass stars go through their deuterium burning phase, and the K–band magnitude of the peak increases (i.e. becomes fainter) as the cluster ages. Thus, if a peak is apparent in a cluster KLF, the magnitude of the peak can be used as an age indicator.

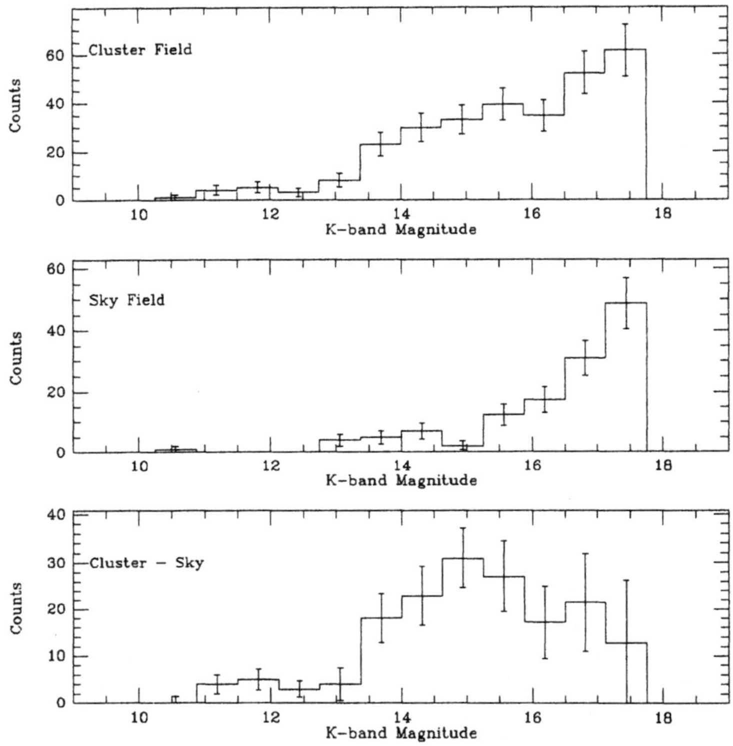

Fig. 8. K–band Luminosity Function for DR22

Above are the KLF for the DR22 cluster (top), the sky counts from the nearby reference field (middle), and the field star subtracted cluster KLF (bottom). The KLF is corrected for incompleteness to 17th mag.

The effect of extinction may alter the shape of the cluster KLF. If the intrinsic KLF is a power law, the observed KLF for an embedded cluster will have the same power law index (Megeath et al. 1994). If, however, there are peaks in the KLF, extinction from the surrounding molecular cloud can smear the peak, and can even alter a peaked KLF into a monotonic KLF with a roughly power law form. To illustrate this, in Fig. 7 we have plotted the Zinnecker et al. KLF for a 0.3 Myr unobscured cluster and a 0.3 Myr embedded cluster with visual extinctions ranging from 0–50.

What do KLFs for real clusters look like? In Fig. 8, we display the KLF for the DR22 cluster. We have also measured the K–band field star counts from a field 5′ west of the source and subtracted the resulting distribution from the cluster KLF. The field star subtracted KLF is peaked around 15th magnitude.

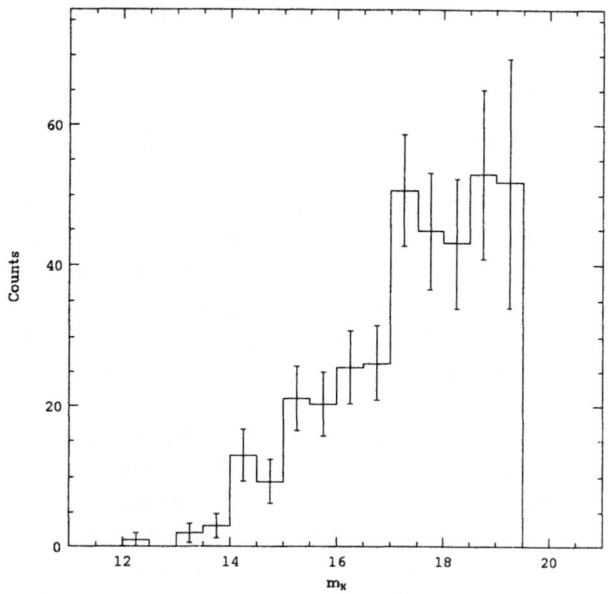

Fig. 9. S184 Cluster K–band Luminosity Function

The KLF for the S184 cluster corrected for incompleteness to 18th magnitude. Since there is a large amount of extinction through the cluster, we have not subtracted the field star contribution. Given that S184 is an outer galaxy region with a relatively high galactic latitude (galactic coordinates = 123.06-6.31), the field star contamination has a minimal effect on the KLF. The cutoff at 19.5 mag is the sensitivity limit of our data.

Given a distance of 3 kpc and measuring the extinction towards the DR22 cluster from the $H-K$ and $J-H$ colors, we estimate the absolute magnitude of the peak is $M_K \sim 1.2$. Comparing this value to the absolute magnitude of the peaks in the 0.3 Myr, 0.7 Myr, 1.0 Myr and 2.0 Myr cluster models presented in Zinnecker et al. 1993, we get an age for the DR22 region of 0.3 Myr. The young age of the cluster could be the result of an overestimate of its distance, indeed the kinematic distance is uncertain: 3.4 ± 1.8 (Reifenstein et al. 1970 - however, Dickel et al. 1978 derive a *minimum distance* of 3 kpc based on the extinction towards the region). If the cluster is at 1.6 kpc, the peak is at $M_K \sim 2.6$ and the resulting age is < 1 Myr. In either case, we find an age of < 1 Myr, much shorter than the age we estimated from expected disk lifetimes. It is not clear why this discrepancy arises. We can suggest four possible solutions:

- The stars in our reference frame do not accurately account for the field star contamination.

- The peak is the result of dynamical evolution. In particular, the lower mass stars may be preferentially ejected from the cluster due to energy equipartion among the cluster members.
- The timescale for disk dissipation is < 1 Myrs in young clusters due to dynamical interactions or disruption of disks by ionization fronts and stellar winds.
- The peak results from a turnover in the IMF of DR22 at a mass $> 0.6\ M_{\odot}$.

Further observations are necessary to understand which of the four solutions is correct.

The S184 KLF shows a distinctly different form (Fig. 9). Only at the faintest magnitudes, $m_K \sim 18$, is there any hint of a turnover in the KLF. The region between 13.5th and 18th magnitude can be well fit by a power law, $10^{\alpha m_K}$ (note that a power law function of luminosity is an exponential function of magnitude), with $\alpha = 0.20 \pm 0.04$. The S184 KLF is consistent with the peaked 0.3 Myr KLF calculated by Zinnecker et al. (1993) smeared by several magnitudes of extinction.

In both S184 and DR22, the KLFs can be fit models using current theoretical pre–main sequence tracks and a Miller–Scalo IMF. The reader should note that the slope of the KLF, particularly in the case of S184, is dependent on the flattening (compared to a power law IMF) of the Miller–Scalo IMF around 0.5 $M_{\odot}$. For this reason, we can rule out a Salpeter power–law IMF; however, we cannot rule out the field star IMF derived by Miller and Scalo. An independent estimator of age is needed to test whether the cluster IMF is equivalent to the field star IMF.

6 What Next?

Although 2.2 μm imaging is a powerful method for *detecting* young clusters, characterizing the clusters is problematic. In particular, there is no unambiguous indicator of the age. We have applied two techniques to constrain the ages of the clusters, an examination of IR excesses and an examination of the K–band luminosity function. For one cluster, DR22, we get two inconsistent results: > 3 Myr and < 1 Myr. Clearly, another approach is needed.

The most promising technique is spectroscopy in the J, H and K–bands. Spectroscopy should add detailed information on the photospheric and disk properties of young stars. From this information, we hope to obtain accurate photospheric temperatures for young stars. Combined with J, H, and K photometry, this information should give an unambiguous value for the age and mass of a given cluster member. However, it is not clear that we can accurately unveil the underlying photospheric emission from the line and continuum emission due to disks and winds. Observational efforts in the next few years should show whether spectroscopy is an effective tool for the study of young clusters.

This contribution is derived from work done at Cornell University in partial fulfillment of the requirements for the degree of Doctor of Philosophy in Astronomy. I would like to thank Terry Herter for guiding me through my thesis work and John Rayner for his generous assistance in obtaining, reducing, and understanding my thesis data. Finally, I would also like to acknowledge Andrea Moneti, Malcom Walmsley, and Tom Wilson for their insightful comments on this contribution.

References

Adams, F. C., Lada, C. J., and Shu, F. H.: *Astrophys, J.* **312** (1987) 788

Bertout, C.: *Annu. Rev. Astron, Astrophys.* **27** (1989) 351

Carpenter, J. M., Snell, R. L. and Schloerb, F. P. *Astrophys. J.* **362** (1990) 147

Carpenter, J. M., Snell, R. L., Schloerb, F. P. and Skrutskie, M. F.: *Astrophys. J.* **407** (1993) 657

Dickel, H. R., Wendker, H. J. *Astron. Astrophys.* **66** (1978) 289

Draine, B. T.: in *Proc. 22nd Eslab Symposium on Infrared Spectroscopy in Astronomy*, ESA SP-290, (1989), 93

Elmegreen, B. G. and Lada, C. J.: *Astrophys. J.* **219** (1978) 467

Gillett, F. C.: in *Infrared Astronomy with Arrays*, eds C. G. Wynn-Williams and E. E. Becklin (University of Hawaii) (1987) 3

Hillenbrand, L. A., Strom, S. E., Vrba, F. J., and Keene, J.: *Astrophys. J.* **397** (1992) 613

Hillenbrand, L. A., Strom, S. E., Merrill, K. M., and Gatley, I.: in *Massive Stars: Their Lives in the Interstellar Medium*, eds. J. P. Cassinelli and E. B. Churchwell (ASP Conference Series, Vol. 35) (1993) 141

Koornneef, J.: *Astron. Astrophys.* **128** (1983) 84

Lada, C. J. and Adams, F. C.: *Astrophys. J.* **393** (1992) 278

Lada, C. J., Young, E. T., and Greene, T. P.: *Astrophys. J.* **408** (1993) 471

Lada, E. A., DePoy, D. L., Evans, N. J., II, Gatley, I.: *Astrophys. J.* **371** (1991) 171

Megeath, S. T.: in preparation (1994)

Miller, G. E., and Scalo, J. M.: *Astrophys. J. Supp.* **41** (1979) 513

Myers, P. C., Fuller, G. A., Mathieu, R. D., Beichman, C. A., Benson, P. J., Schild, R. E., and Emerson, J. P.: *Astrophys. J.* **319** (1987) 340

Odenwald, S., Shivanandan, K., Campbell, M., Fazio, G., Schwartz, P. and Moseley, H.: *Astrophys. J.* **306** (1986) 122

Reifenstein, E. C., III, Wilson, T. L., Burke, B. F., Mezger, P. G., and Altenhoff, W. J.: *Astron. Astrophys.* **4** (1970) 357

Skrutskie, M. F, Dutkevitch, D., Strom, S. E., Edwards, S., Strom, K. M., and Shure, M. A: *Astron. J.* **99** (1990) 1187

Strom, K. M., Strom, S. E., Edwards, S., Cabrit, S. and Skrutskie, M. F.: *Astron. J.* **97** (1989) 1451

Zinnecker, H., McCaughrean M. J., Wilking, B. A.: In. *Protostars and Planets III*, eds E. H. Levy and J. I. Lunine (Tuscon: Univ. of Arizona Press) (1993) 429

The Search for Protostars - With Dust Emission as a Tracer

P.G. Mezger

Max-Planck-Institut für Radioastronomie, Auf dem Hügel 69, 53121 Bonn, FRG

Abstract: A critical discussion of the determination of molecular cloud parameters using spectral line data is presented. A series of specific cases where H_2 column densities and densities from millimeter and sub-millimeter dust continuum data are given.

1 Historical Introduction

This meeting commemorates 25 years of molecular spectroscopy. Summarizing here recent observations of dust emission from protostellar condensations - and especially arriving at the conclusion that there is a density/temperature regime $n_H \simeq 10^6 cm^{-3}$, $T \simeq 20K$, where molecular transitions cease to be reliable tracers of H_2 column densities - may appear like a stroll into the Lion's den. But then - history tends to repeat itself: After the discovery of atomic hydrogen by Ewen and Purcell in 1951 the leading astonomes thought that all prbelmes connected with interstellar matter (ISM), such as spiral structure and star formation, could be solved by observations of this line. Only fifteen years later, it dawned on the astronomical community that they actually saw only one half and –in my opinion the more boring half– of all of the ISM, namely, the diffuse, warm intercloud gas.

Already in the early sixties there was a general suspicion that at higher densities H may turn into H_2 which - due to its symmetry - has no transitions in the radio and mm-regime. But only after the detection of the NH_3 and H_2O lines at 1cm wavelength by Townes and his collaboraters did the ice break: Giant Molecular Clouds (GMCs) with hydrogen masses as high as $10^6 M_\odot$ were discovered. Their association with compact HII regions and Far Infrared sources signalled that finally the birth sites of star formation had been discovered.

After 25 years of relentless observations more than 85 molecules with several hundred transitions have been identified in interstellar space and many U(nidentified) transitions still wait for their identification. Nevertheless, we are approaching a situation which has some similarities with the situation in the mid-sixties mentioned above: Spectroscopists begin to realize that - while easily tracing a total of some $10^9 M_\odot$ of H_2 in the Milky Way traced by spectral line radiation from molecules such as CO and CS–they miss a frraction of $<10^{-4}$ of this mass of H_2. This fraction is the gas contained in high density, low kinetic temperature regions which are recognized as being the earliest stages of star formation. The lack of tracer molecules can be explained by depletion, that is, the formation of molecular ice mantles around refractory cores and by a drastic change in IS chmeistry at low temperatures.

Henk van de Hulst, the Dutch astronomer who as a graduate student had predicted the possibility to detect the 21cm line, had also shown in his 1946 published thesis a way out of his dilemma: Dust grains are immersed in an Interstellar Radiation Field (ISRF) which in the solar vicinity has a mean radiation density corresponding to that of a 3K black body radiation field, but with a spectral energy distribution corresponding to that of a (diluted) 10^4K black body. The grains, absorbing with their geometrical cross section at optical and ultraviolet (UV) wavelengths, must acquire temperatures as high as 17K, in order to reemit tha t absorbed energy, since their absorption cross section decreases with λ^{-m}, where m<2. Only in the vicintiy of luminous objects with greatly increased radiaiton densities can dust temperatures attain values as large as 100K.

At first, van de Hulst's prediction made it only into astronomical text books. Only after Frank Low's invention of the Ga-doped Germanium bolometer in the sixties did it become possible to actually observe the extended FIR dust emission. The IRAS Infrared Astronomical Satellite, conceived during the seventies and put into an orbit in 1983, surveyed the sky in four bands centered on 100, 60, 25 and 12 μm. It detected more than 250 000 IR sources–but up to now, not a single protostar. Fig. 1 shows why: The quiescent GMS's without embedded luminous sources are heated by the interstellar radiation field (ISRF) only from the outside where warm dust, detectable by IRAS, forms a this layer up to extinction $A_v \sim 1$. The dust inside this layer is heated by the far infrared emission from this outer layer, plus the mid IR part of the primary ISRF (see Mathis et al. 1983). This dust will be too cold to be observable even in the 100μm IRAS band.

The search for protostars and what we learned about the morphology and physical state of the earliest evolutionary stages of stars from dust observations at mm/submm-wavelengths will be the topic of my talk.

2 Star Formation: Earlier Results

The present-day star formation rate, as derived from the production rate of Lyman continuum photons for the case of bimodal star formation, is $\Phi \sim 4$ $M_\odot yr^{-1}$. The mass spectrum in the range $0.1 < M/M_\odot < 100$ can be approximated by Salpeter's (1955) IMF, $\Phi(M) \sim M^{-2.5}$. Bimodal star formation (Güsten and Mezger, 1983) postulated that one deals, in the interarm region, with the spontaneous formation of stars of all masses. In the main spiral arms, this is supplemented by induced formation of only massive stars $M > 3M_\odot$.

Stars form in molecular clouds. Fig. 2 shows the morphology of the Orion A cloud as a typical example of a massive star forming interarm GMC. The total extent of the cloud with an average density of $< n_{H_2} >$ a few 100cm^{-3} is best

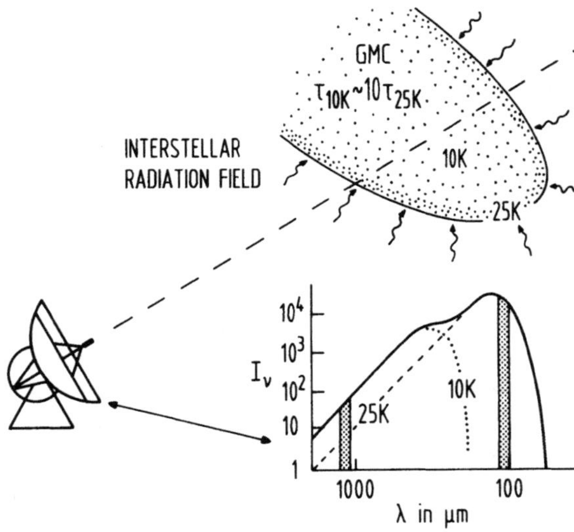

Fig. 1. The spectrum of a quiescent cloud which is heated from outside by the Interstellar Radiation Field.

seen in CO emission (Maddalena et al. (1986)). Gas with densities of $10^4 cm^{-3}$ outlined by CS emission forms a NS-ridge (Fig. 2a) which containes $10^4 M_\odot$. The two massive star-forming cores OMC1 and 2 (Fig. 2b) have densities of 10^5 to 10^6 cm^{-3} and each contain $\sim 1.5 \times 10^3 M_\odot$of gas. These cores also contain a dense ridge surrounded by a less dense envelope. Massvie stars seem to have formed close to the center of this ridge.

If thermal pressure pressure were the only force to counteract gravity most of the extended CS ridge seen in Fig. 2b should be in the process of collapse and star formation. The fact that the actual star-forming cores (Fig. 2c, d) have hundred times higher densities indicates that other forces–e.g. magnetic fields and turbulence–are at work.

3 The Case of NGC2024: Molecular Spectroscopy Meets Its Limits

In 1988 Mezger et al., in the cloud core associated with the compact HII region NGC2024, detected (what they considered to be) the first isothermal protostars. The cloud core shows again a core/envelope structure and has a total mass of $M_H = 500 M_\odot$, of which the central ridge structure accounts for $140 M_\odot$ and the protostellar condensations for another $70 M_\odot$. This discovery was met with substantial scepticism. This is because, among other arguments, Schulz et al. (1991) had suggested that molecules were depleted and therefore were no longer reliable tracers of H_2 column densities.

In two subsequent papers Mezger et al. (1992a) and Mauersberger et al.

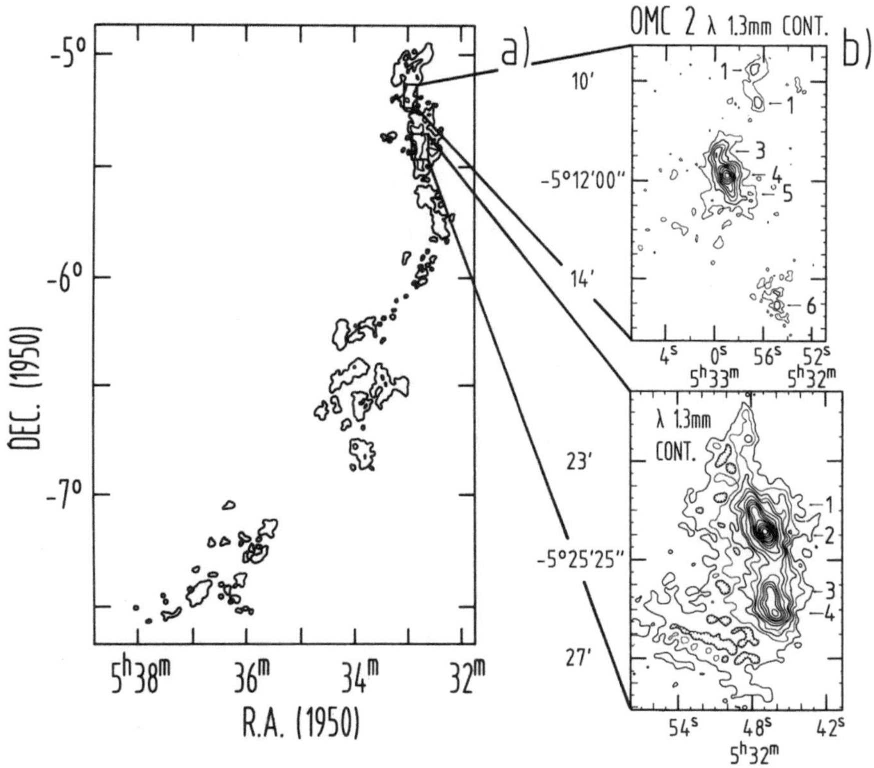

Fig. 2. Morphology of the massive star forming Orion GMC: a) The GMC complex (which consists of the Orion A and B clouds) as outlined by CO mapping (Maddalena et al., 1986). Each cloud has a mass of $10^5 M_\odot$. b) The dense ridge in the Orion A cloud as outlined in CS(1-0) emission (Tatemutsu and Umemoto, 1992). It has a mass of $10^4 M_\odot$. c) The dense cloud cores OMC1/2 outlined in λ=1300μm dust continuum (Mezger et al.1990b) with masses of 1.7×10^3 and 1.5×10^3 $M_\odot$.

(1992) presented new observations: An 8″, 870μm dust emission MRT map which reveals a seventh condensation; and $C^{34}S$ and $C^{18}O$ MRT maps with angular resolutions of 16" and 12". These definitely show that molecular and dust emission peaks do not coincide. Fig. 3a shows the 870μm map. The column densities derived from dust emission and from $C^{34}S$ and $C^{18}O$ transitions, respectively, along the fidge line indicated by the dashed curve, are shown in Fig. 4b. Note that in general, only $C^{18}O$ is considered to be a reliable tracer of H_2, while $C^{34}S$ traces higher density regions, but its abundance relative to H_2 is rater uncertain. It is obvious in the case of the core of NGC2024 that $C^{18}O$ traces only the low density envelope, while $C^{34}S$ traces, at least roughly, the medium density ridge. The protostellar condensations, however, are visible only in dust emission. Maps made the Very Large Array (angular resolution 3″)

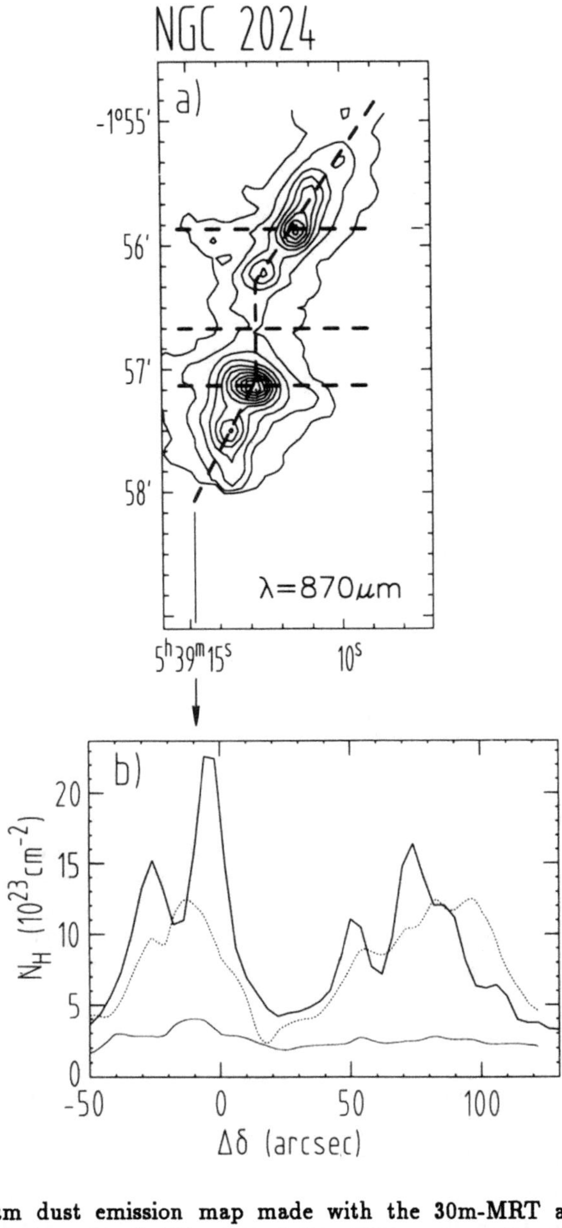

Fig. 3. A 870μm dust emission map made with the 30m-MRT and smoothed to an angular resolution of 11″. The dashed line follows approximately the elongated axis of the ridge. Along this ridge H_2 column densities have been obtained from the 870μm dust surface brightness (heavy solid curves), the $C^{34}S$ (dashed curves) and the $C^{18}O$ luminosities (thin solid curves; see text). Column densities along the ridge line are shown as Fig. 3b. Offset coordinates refer to the position $\alpha = 05^{\circ}39^{m}12^{s}$, $\delta = -01^{\circ}57'04''$ (1950), from Mezger et al. 1992.

indicate that NH_3 clumps are found close to the dust maxima. However, it is also found that NH_3 must be depleted, since the relative abundace is more that two orders of magnitude lower that the value of 10^{-8} found for low density dust clouds (Mauersberger et al. 1992). Considering these facts the authors

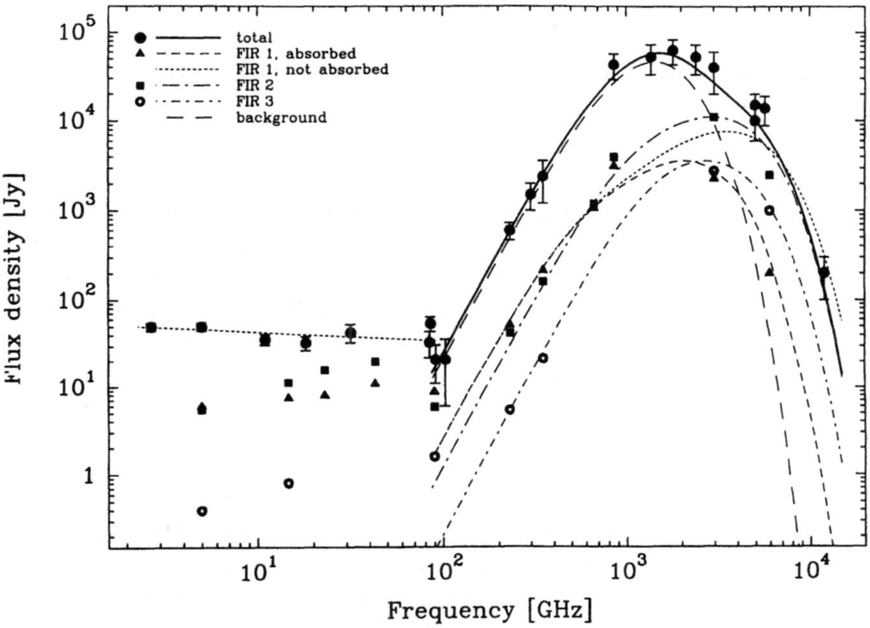

Fig. 4. Filled circles and a solid line indicate the continuum spectrum of Sgr B2. Only flux densities attributable to dust emission alone are used. Filled triangles mark the spectrum of the compact warm dust component FIR1; filled squares, FIR2; and open circles, FIR3. The various curves relate to our spectral fits. The legend at the upper left indentifies the individual components. Note that for FIR1 both the 'unabsorbed' and the 'absorbed' (i.e., the observed) spectra are shown (see text).

arrived at the following conclusions: Molecular transitions in NGC2024 only trace the central ridge of the cloud core with a density of $n_H = 3 \times 10^6 cm^{-3}$, but not the protostellar condensations with average densities of $2.5 \times 10^8 cm^{-3}$. Gas and dust in the condensations FIR1,2,3 have about the same temperature, $T_{dust} \sim T_{gas} \sim 20K$, while in the condensations FIR4, 5 (and perhaps in FIR6) the gas temperatures derived from ammonia are $\simeq 40K$, but the dust temperatures appear to be lower. The two latter condensations also show molecular outflows and may indicate that the stellar cores have already formed.

One problem remaines, however: The narrow widths of the NH_3 lines were not compatible with a virial mass as large as the mass derived from dust emission. To reconcile the narrow NH_3 lines with the large protostellar masses derived from dust emission Mauersberger et al. (1992) suggested that ammonia is present in only a small part of the dust condensations, and would have to be far from most of the mass in the dust condensation and not distributed symmetrically. Ongoing measurements with the Plateau de Bure array of IRAM in the continuum and $C^{34}S$ line toward FIR5 and 6 show that the $C^{34}S$ emission is offset relative to the dust peaks. These data will further constrain the parameters of any model of this cloud.

4 Massive Star Formation in Main Spiral Arms and in the Galactic Center

In a series of papers entitled 'Dust emission from star forming regions' and 'Anatomy of the Sagittarius complex' we have investigated the morphology of some of the very compact GMCs which are located in main spiral arms and in the inner 300pc of our Galaxy: W49/51 (Sievers et al., 1991); M-0.13-0.08 (Zylka et al. 1990) and Sgr B2 (Gordon et al., 1993) In both cases the GMCs have masses comparable with that of GMCs in the interarm region (i.e. some $10^5 M_\odot$) but are much more compact ($< n_H >= 10^{4-5} cm^{-3}$). The massive star formation activity as measured by the luminosity to mass ratio, L_{IR}/M_H, is ~40 to 80 in W49/51 ~5 in SgrB2 and <1 in M-0.13-0.08.

Typically, the spectrum can be decomposed into components representing warm (50-100K) and cold (≃20K) dust. Fig. 4 shows as an example the composite spectrum of Sgr B2 (Gordon et al., 1993). This HII/GMC complex consists of three compact (≃0.4pc) warm dust components with young and massive stars (revealed by their ultra-compact HII regions) embedded in an extended (4x11pc) cold dust envelope. These warm dust components contribute more than half of the total luminosity ($L_{IR} \sim 4\ 10^6 L_\odot$) but only ~5% of the total mass, $M_H \sim 8 \times 10^5 M_\odot$. The compact warm dust component FIR1 is so deeply embedded in the cold dust envelope that its emission is being absorbed at $\lambda < 100\mu m$. The envelope becomes opaque at $\lambda < 200\mu m$. FIR1 must be embedded at ~1/4 of the total optical depth of the envelope. If it were located on the far side of the envelope, it would have been hard to discriminate FIR1 from the cold dust emission.

It is of special interest that IRAS does not reveal the presence of Sgr B2 although the envelope fills the IRAS beams. The reason is that the dust in the envelope is to cold to be seen against the background emission provided by the cloud skin which is heated by directly absorbing the ISRF, whose mean intensity in the central region is ≃50 times that in the solar vicinity (see also Fig. 1).

More recently these observations have been extended. In the spiral arm region of the Milky Way (R≃3kpc) we have surveyed the dust emission from very massive cores of GMCs, which have been found in CS surveys (Solomon et al, 1987). The LIR/MH ratio shows that the parental GMCs are active star forming clouds but the lack of extended HII region appears to indicate that part of these cores are at much earlier evolutionary stages than W49/51 (Mooney et al., in prep.).

We have continued to survey the dust emission from compact GMCs with high angular resolution (Zylka et al., in prep.). Although the Lyc-photon production rate in the inner 300pc amounts to 10% of the total production rate in the Milky Way, we do not find many massive stars in early evolutionary stages. This may indicate that the star formation rate in the central region varies with time.

5 Morphology of Low and Medium-Mass Star Forming Clouds: The Ophiuchus Cloud

The nearby (D$\simeq$160pc) Ophiuchus cloud complex extends for tens of parsecs. It is comprised of filamentary clouds of low density molecular gas (Fig. 5a). In its westernmost part, however, this cloud containes a massive cloud fragment of size 2pc x 2pc, mass M_H =500 to 600$M_\odot$and column densities of a few 10^{23} H-atoms cm^{-2} averaged over a 90"-beam. In this fragment $C^{18}O$ observations (Wilking and Lada, 1983; se also the review by Wilking, 1991) reveal the presence of a ridge of size 1 x 2pc which extends from SE to NW (Fig. 5b). In the following, I refer to this fragment as the ρ Oph cloud

This cloud appears to form low-mass stars over its whole extent. Wilking et al. (1989 and references therein) within an area of 4.3pc^2, have identified a total of 78 Young Stellar Objects (YSO) as members of a young cluster of low-mass stars (see Fig. 5b). About one half of these YSOs appear to be dust-embedded stars and the other half T Tauri stars.

We have mapped with SEST and MRT the 1300μm dust emission within the dense ridge outlined in $C^{18}O$ (Mezger et al., 1992b). The detection limit in the different fields ranges from 0.2-0.04mJy per beam-area. We have detected altogether four cloud cores with masses ranging from M_H=1-15$M_\odot$ and typical densities of $n_H = 10^6 cm^{-3}$. Three of these cores were structureless (see e.g. Oph B2 shown in Fig. 5d); only the core SM1 (Fig. 5c) shows four protostellar condensations with masses M_H=0.4-3$M_\odot$, and typical densities of n_H $\sim$10cm^{-3}. The sum-mm spectrum of the source SM1 is compatable with $T_{dust}$$\sim$15K to 20K (Ward-Thompson et al. 1989; André et al. 1993). The southernmost of the condensations, labelled FIR4 in Fig. 5c, is associated with the ultra-compact HII region IRAS1623. The star in the source appears to be the driver for one of the youngest and best collimated CO bipolar outflow found so far.

The active star-forming core SM1 may represent the low-to-medium mass protostellar pendant of the high-mass star-forming core NGC2024. Its ridge-structure containes a total mass of 15$M_\odot$, of which one third has collapsed into isothermal protostars. As in the case of NGC2024 the cold high-density condensations FIR1-4 in SM1 are barely visible in molecular isotopic transitions (Mauersberger, priv.comm.).

Of the 80 YSOs identified in the Ophiuchus cloud as members of a young stellar cluster, one half are identified as T-Tauri stars with estimated lifetimes of 4×10^5yr (Wilking and Lada, 1983). If the Star Formation Rate in this cloud was approximately constant over the past million years, and if statistics can be applied to the small number of four isothermal protostars, we derive a typical time scale for isothermal protostellar evolution of $\tau_{iso} \sim 4\times10^4$yr. This time

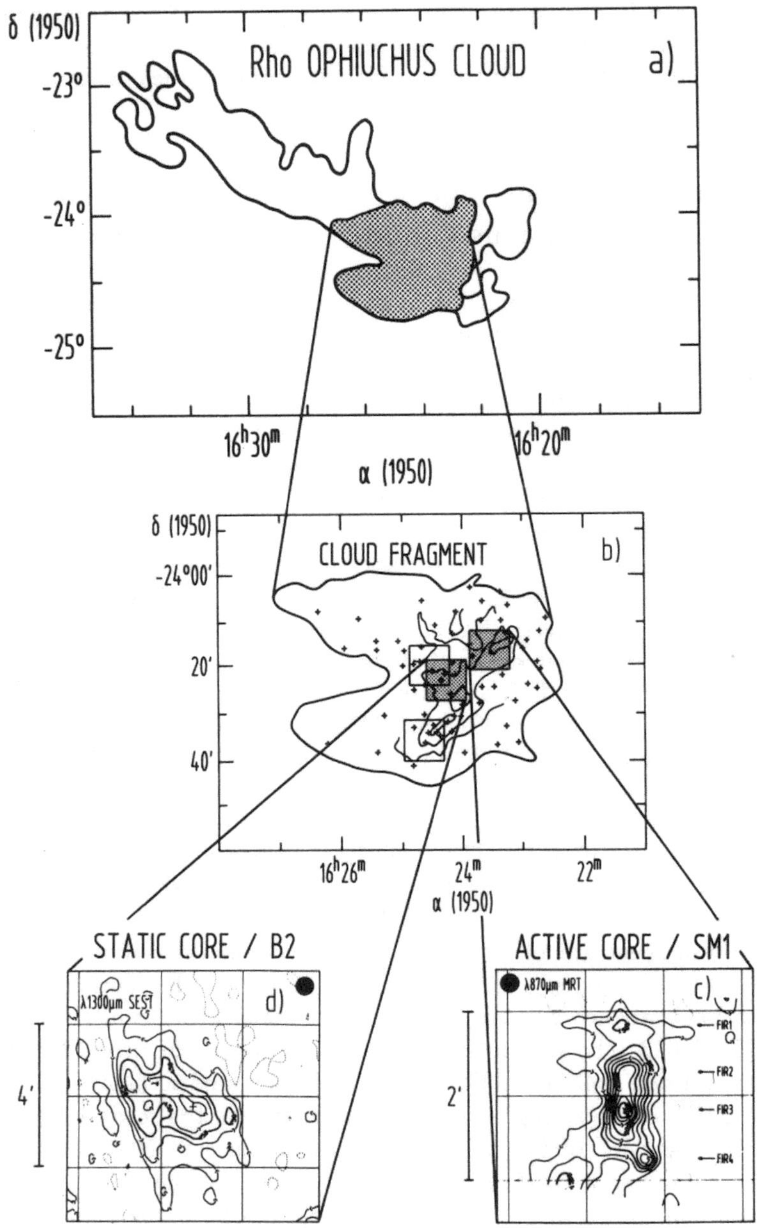

Fig. 5. Morphology of the low- and medium-mass star forming Ophiuchus cloud. a) The low-density envelope (of mass $M_H = 3000 M_\odot$) is shown as outlined in CO-emission. The densest part (shown stipled) with $M_H = 500 M_\odot$ and $< n_H >= 10^4 cm^{-3}$ is shown in b) together with the positions of 78 YSOs, observed at 2.2μm by Wilking et al. (1989) and identified as members of a young star cluster. c) The active star forming cloud core SM1 ($M_H \sim 15 M_\odot$) and d) the static core B2 ($M_H \sim 15 M_\odot$) are shown as observed in λ=870μm in (c) and λ=1300 μm (d) dust emission (adapted from Mezger et al. 1992b). The angular resolution is 8″ in (c) and 22″ in (d).

scale accounts for only ~5% of the total pre-MS evolution time of medium mass stars and thus can explain the scarceness of isothermal protostars. It is also of interest that 10^4 yr corresponds to the free-fall time of a protostar which begins contracting at a mean density of $\sim 10^6 cm^{-3}$, as is presently found for the static cores B1, B2 and F1.

Lada, Bally and Stark (1991; hereafter LBS) have surveyed in CS(2-1) with an angular resolution of 1.8′ of the densest part of the Orion B GMC. Forty-two cloud cores were found; LBS determined that for most of these CS cores, masses were $M_H < 100 M_\odot$. We have followed up their work by surveying 25 of these LBS cores with SEST (θ=23″) and subsequently with the IRAM MRT (θ=12″) and detected about half of it (Launhard et al., in prep.). Only in the case of the well investigated cloud core containing NGC2024 (see Section 3) did the LBS mass estimate agree with our results based on λ=1300μm dust emission. in all other cases, the core masses estiamted on the basis of dust emission are ~10% of the values given by LBS. It is possible that this difference is caused by the different angular resolutions used.

6 Summary and Conclusions

1) In the regime $n_H \simeq 3 \times 10^6 cm^{-3}$, Td~Tg≃20K molecular transitions cease to be reliable tracers of hydrogen column densities. At these densities dust and gas also begin to be thermally coupled.

2) Although GMCs in the central ~300pc of the Milky Way are as massive and compact as GMCs in the main spiral arms their present rate of massive star formation as measured by the luminosity-to-mass ratio LIR/MH appears to be lower by factors of 10.

3) Free-fall collapse of isothermal protostars begins at densities of $n_H = 10^6 cm^{-3}$.

4) The lifetime of isothermal protostars without a stellar core is of order 4×10^4yr.

References

André P., Ward-Thompson D., Barsony M. 1993 ApJ (in press)

Gordon, M.A., Berkermann, U., Mezger, P.G., Zylka, R., Haslam, G., Kreysa, E., Sievers, A. 1993, A&A (in press)

Güsten R., Mezger P.G., 1983, Vistas in Astronomy 26, 159

Lada E.A., Bally J., Stark A.A. 1991 ApJ 368, 432

Maddalena R.J., Morris M., Moscowitz J., Thaddeus P. 1986 ApJ 303, 375

Mathis, J.S., Mezger, P.G., Panagia, N., 1983 A&A 128, 212

Mauersberger R., Wilson T.L., Mezger P.G., Gaume R., Johnston K.J.: 1992 A&A 256, 640

Mezger, P.G., Chini, R., Kreysa, E., Wink, J.E. 1988 A&A 191, 44

Mezger P.G., Wink J.E., Zylka R. 1990 A&A 228, 95
Mezger P.G., Sievers A., Zylka R., Haslam CGT. Kreysa E., Lemke R. Mauersberger R. Wilson T.L. 1992a A&A 256, 631
Mezger, P.G., Sievers A., Zylka R., Haslam C.G.T., Kreysa E., Lemke R. 1992b A&A 265, 743
Salpeter E.E., 1955 ApJ 121, 161
Solomon P.M., Rivolo A.R., Barrett J., Yahil A., 1987 ApJ 319, 730
Schulz A., Güsten R., Zylka R., Serabyn E. 1991 A&A 246, 570
Sievers A.W., Mezger P.G., Gordon M.A., Kreysa E., Haslam C.G.T., Lemke R. 1991 A&A 251, 231
Tatematsu K., Umemoto T. 1991 'Molecular cloud cores in the Orion A Cloud', Proc. of workshop on 'Young Star Clusters and Early Stellar Evolution' held in Vulcano, Italy
Ward-Thompson D., Robson E.I., Whittet D.C.B., Gordon M.A., Duncan W.D. 1989, MNRAS 241, 119
Wilking B.A., Lada C.J. 1983 ApJ 274, 69 8
Wilking B.A., Lada C.J., Young T.E. 1989 ApJ 340, 823
Wilking B.A. 1991 ESO Scientific Report No.11, (B. Reipurth ed.) p. 159
Zylka R., Mezger P.G., Wink J.E. 1990 A&A 234, 133

Molecular Emission from Disks Around Young Stars

Steven V. W. Beckwith[1] *and Anneila I. Sargent*[2]

[1]Max-Planck-Institut für Astronomie, Heidelberg, Germany
[2]California Institute of Technology, Pasadena, CA, USA

Abstract: Results and a discussion of disks around stars are presented, with examples taken from the millimeter and submm wavelength range.

1 Introduction

A wide variety of observations indicate that low-mass pre-main sequence stars are surrounded by disks (e.g.. Beckwith & Sargent 1992, and references therein). An important clue to the presence of a disk is the detection of small-scale molecular line emission from, for example, CO and its isotopes. The emission originates in the outermost parts of the disks, but is nevertheless quite compact by the standards of molecular line astronomy, typically subtending angles less than 10″ and requiring interferometric observations. Corresponding physical scales are several thousand AU, considerably larger than the Solar system and larger than the minimum disk sizes implied by infrared and millimeter-wave continuum emission. In a few cases, the emission regions are elliptical, reminiscent of inclined circular disks.

Molecular emission, mainly in CO and its isotopes, has now been observed from more than a dozen young stars. Velocity-resolved maps reveal the dynamical state of the gas near the stars, some of which appear to be dominated by stellar gravity. Exploration of the chemistry and physical state of this gas has just begun. This article is a brief review, in a rapidly changing field, of recent observational results on molecular emission from disks.

Several other reviews of circumstellar disks that provide a broader background for the interested reader include Shu, Adams, and Lizano (1987), Bertout (1989), Beckwith and Sargent (1992), Sargent & Beckwith (1993), and Beckwith (1993). A brief summary of the main results is given below. The remainder of the article concentrates on the relationship between molecular clouds and disks and a few of the current problems in disk research that have emerged as a result of molecular line observations.

2 Disk Properties

2.1 Molecular emission

The high spatial resolution afforded by millimeter-wave interferometers was a necessary prerequisite to the study of circumstellar disks. As a result, most of our information has come within the last six or seven years. To date, the total number of objects observed in molecular lines is relatively small. Even though extant maps are noisy, the general features of maps made with different interferometers are quite reproducible.

HL Tauri was first solar-mass pre-maing sequence star detected in small-scale molecular emission (Beckwith et al. 1986). It is one of the strongest emitters and, therefore, continues to be the prototype star/disk system. The molecular emission is extended along a northwest-southeast line with an axial ratio of order six, similar to a flattened disk viewed at high inclination (Sargent and Beckwith 1987, 1991; Guilloteau 1993; Hayashi et al. 1993). The major axis of this structure is almost exactly perpendicular to the jet of emission seen by Mundt et al. (1988) along PA=45°, and so the morphology alone strongly suggests that we are viewing a disk highly inclined to the line-of-sight. Figure 1a is a velocity-integrated map of the ^{13}CO $J = 1 - 0$ emission near HL Tauri. The dynamical state of the gas is controversial. Sargent & Beckwith (1987, 1991)

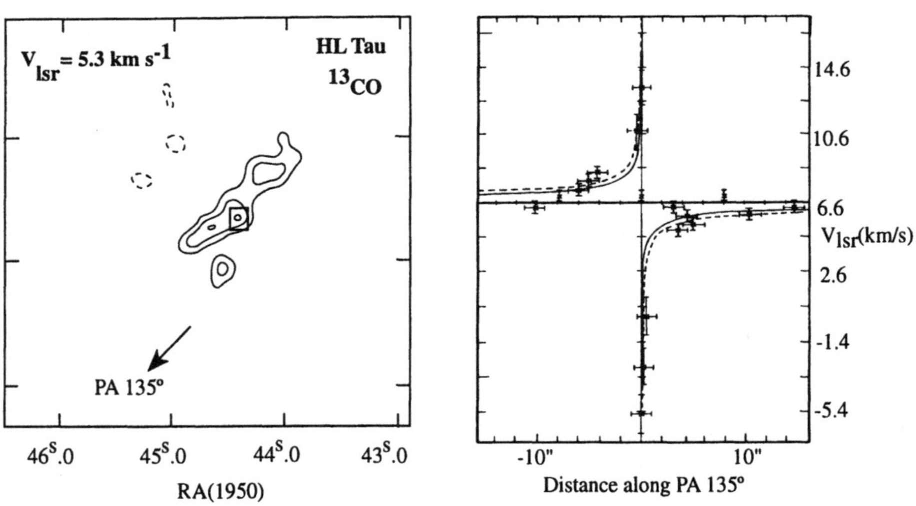

Fig. 1. a) A 1 MHz channel map of the ^{13}CO emission (4.0–6.6 km s^{-1}) near the star HL Tau from Sargent & Beckwith (1991). The box marks the optical position of the star. The peak contour is 150 mJy beam^{-1}. b) The apparent rotation curve derived from individual velocity channel maps. The solid and dashed lines plot the Keplerian law $v \propto (GM/r)^{\frac{1}{2}}$ for 0.55 and 1 $M_\odot$, respectively Sargent & Beckwith 1991.

interpreted the maps at different velocities as being consistent with Keplerian orbits around a $\sim 1\,M_\odot$ star. Figure 1b shows their derived rotation curve for the gas. More recently, Guilloteau (1993) pointed out that emission to the south of the star confuses this interpretation, and there may be substantial contamination by emission from the outflowing wind. By contrast, Hayashi et al. (1994) fitted a model of *infalling* gas in the plane of the disk to their own interferometer maps of HL Tauri. The principle features of the infall model are the presence of emission along the entire major axis near the systemic velocity, not expected in the Keplerian model, and slight offsets along the minor axis of the highest velocity gas with equal but opposite displacements for velocities on opposing sides of the systemic velocity. Figure 2 shows their position-velocity maps for the major and minor axes. Unfortunately, the data do not clearly point to one or the other of these models.

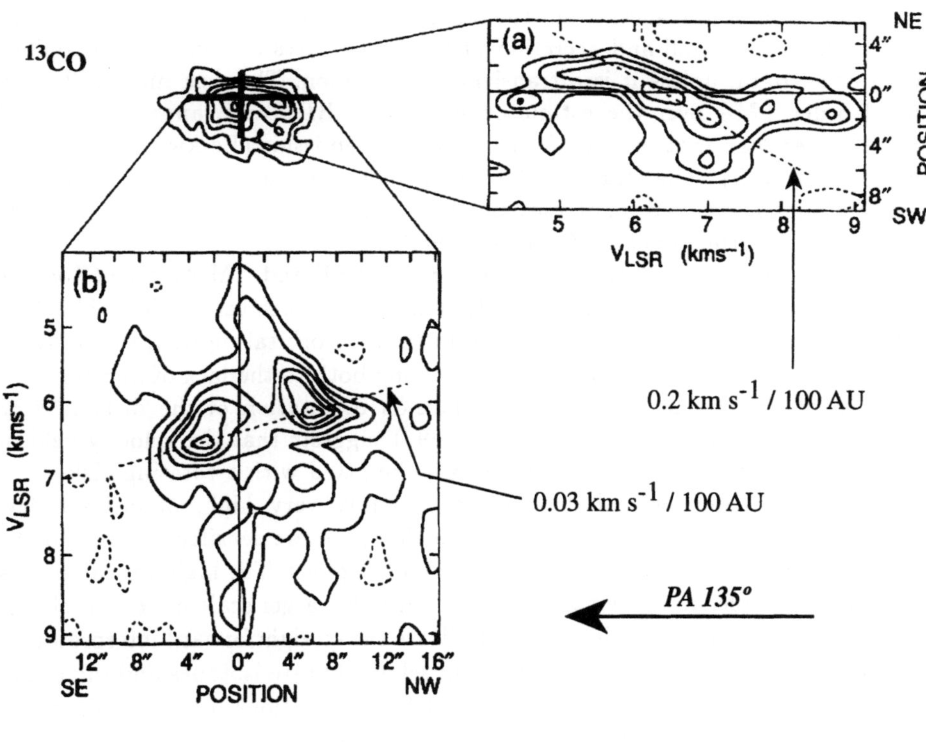

Fig. 2. a) The position-velocity diagram of ^{13}CO emission from HL Tauri along the major axis of the disk at PA=135° from Hayashi et al. (1994). Offsets are from the stellar position. b) The position-velocity diagram of HL Tauri along the minor axis. The relatively larger velocity gradient along the minor axis of the disk cause Hayashi et al. to favor infall as the dominant component at large radii.

One can also combine rotation and infall to produce a better match to the small-scale features at the expense of introducing additional free parameters into the model. Although it is too soon to identify the dominant dynamical component, the number of new observations of good quality make it likely that the velocity field near HL Tauri will be reconstructed accurately. There are at least two different new interpretations, in addition to the orbital interpretation, which are stimulating progress in this field. We think it likely that an unambiguous interpretation will be possible for the next generation of maps of HL Tauri. As discussed below, observations of other molecules may better delineate the pure disk emission and make it possible to determine the dominant factors in the velocity field of the the gas.

Either mass infall or mass loss will explain the minor axis displacements in the channel maps. In both cases, the rate of mass loss or accretion is rather large, nearly $10^{-5}\,M_\odot\,yr^{-1}$. This rate would normally be considered unacceptably high for mass loss from a T Tauri star; one expects to see more swept up molecular material at large distances from the stars. Hayashi et al. therefore favor mass infall. But mass infall and accretion at such high rates also come into conflict with other data, unless it is a transient phenomenon. The disk mass will grow to of order $1\,M_\odot$ over the estimated lifetime of HL Tauri, approximately 10 times larger than that inferred from the particle emission (Beckwith et al. 1990, hereafter BSCG). The star may be younger or the particle mass larger than conventionally assumed, or it may be that mass is periodically released in FU Orionis type outbursts (*e.g.*, Lin et al. 1994).

Two other stars, T Tauri (Weintraub et al. 1989) and GM Aurigae (Koerner et al. 1993a) also have velocity fields indicative of orbital motion. The GM Aurigae maps are the more persuasive in favor of orbital motion; the T Tauri emission is difficult to interpret uniquely owing both to the lack of line emission models for the source and the nearly face-on orientation of the disk. A very recent announcement by Dutrey et al. (1993) suggests that the velocity field of molecular emission from GG Tauri can also be fitted with a rotating disk.

The line center optical depths are probably very high, of order a few hundred (Beckwith and Sargent 1993), so the maps are sensitive mostly to the temperature of the gas except in the very outer regions. The high optical depths suggest that molecules in winds, envelopes, and the larger scale molecular clouds can contribute to the emission and complicate the straightforward interpretation of the maps. It also means that the use of relatively rare isotopes and molecules excited only at very high densities should be a propitious means of observing these emission regions. The disk interpretation rests on economy of hypothesis: it is the simplest framework in which the data can be reconciled within the uncertainties.

2.2 General properties of disks

Most of the information about disks is inferred indirectly from spectral energy distributions, the shapes of spectral lines, and characteristics of scattered light near the stars. There are very few images of disks in molecular lines, and those provide information only about the outer parts. Of the indirect studies of disks, the concentration has thus far been on low-mass pre-main sequence stars, the T Tauri stars in their different subcategories: classical, weak-line, and naked (Strom et al. 1989, hereafter SSECS; Cohen et al. 1989; André et al. 1990, André and Montemerle 1993. Observations of intermediate mass stars, the Herbig Ae/Be stars have been made only recently (e.g.. Hillenbrand et al. 1992; Natta et al. 1992), and the disk interpretation for these stars is not yet secure.

High mass stars present a more difficult case (see discussion in Sargent & Welch 1993). They are more distant than the nearest low-mass stars, and they are usually found in close proximity to many other stars within dense molecular clouds. The principle difference betweeen the high and low mass stars is that the former evolve quickly to the main sequence before completely shedding the spherical halo which is a remnant of cloud collapse – in fact, the high mass stars may continue to accrete material from the cloud throughout its pre-main sequence lifetime (Stahler et al. 1980). The halo or envelope radiates strongly in the same spectral domains as the disk, making it virtually impossible to separate the two without exceedingly high spatial resolution (see, e.g., the discussions in Tereby et al. 1993). In many low-mass stars, the contribution from the envelope can be shown to be negligible compared to that from the disk (Beckwith et al. 1990, hereafter BSCG), and it is possible to derive properties of the disks from spectral information alone.

For the low mass stars, the following general properties emerge from observations of various spectral properties (taken from Beckwith 1993):

1. In a typical sample of young stellar objects, roughly half of the sample is observed to have disks using one or more of the criteria in the last section, plus or minus a factor of two (Strom et al. 1989; BSCG).
2. Disk masses are of order 0.1 $M_\odot$; they rarely, if ever, exceed the mass of the central star (BSCG; André et al. 1990; Adams, Emerson, and Fuller 1990; Henning et al. 1993; Terebey, Chandler, and André 1993).
3. The disk radii are almost always greater than a few tens of AU, although these are only lower limits (Beckwith and Sargent 1991). They are similar in size to the present day Solar System.
4. The disk luminosity is usually less than the *apparent* stellar luminosity, L_*, but it often exceeds $\frac{1}{2}L_*$ and occasionally exceeds L_* by substantial amounts (Cohen et al. 1989; Strom et al. 1989).
5. The disk *photospheric* temperatures are usually well characterised by power-laws in radius: $T(r) \sim r^{-q}$ (Adams, Lada, and Shu 1988). The power-law index, q, is typically between 0.5 and 0.75 (BSCG). Surface temperatures at a characteristic radius of 1 AU are between about 70 and 400 K; colder disks would probably escape detection by present observing techniques.
6. Disk lifetimes are of order a few million years (Skrutskie et al. 1990, BSCG).
7. Accretion rates estimated from the disk luminosities are typically between 10^{-8} and 10^{-6} $M_\odot$ yr^{-1} (Cabrit et al. 1990).
8. Most disks have associated winds driving molecular outflows and optical jets (Edwards et al. 1992).

9. There is apparently a strong correlation between the rotation period of the stars and the presence of an inner disk (Bouvier et al. 1993; Edwards et al. 1993). The sense is that stars with disks have longer rotational periods than stars without disks.

These properties are readily derived for disks which are no more than 100 AU in radius, subtending $\lesssim 1''$ at the 150 pc distance to the nearest dark clouds. The best resolution of millimeter-wave interferometers is currently $\sim 1''$. Therefore, molecular emission studied to date does not originate from the same region as the continuum radiation and may not share the same properties. Figure 2 plots the local temperature versus radial distance for the region surrounding a typical young star indicating the origins of the various emission components.

The physical properties of the disk giving rise to molecular emission are not yet well determined. It is quite likely that the gas is cold, at most a few tens of degrees and probably ~ 15 K, owing to its large distance from the central stars. The densities can only be guessed based on an extrapolation from the particle emission at much smaller radii: $n(r) \sim 10^{?} r^{-2.5}\,\mathrm{cm}^{-3}$ (Beckwith & Sargent 1991, 1993). But there might be substantial depletion of the molecules onto grains in these high-density environments. As noted below, molecules such as CS are depleted by large factors, making it even more difficult to measure densities by comparing molecules with different critical densities. The actual measurement of gas physical conditions remains an important outstanding problem in the study of these molecular disks.

3 Directions for Molecular Line Studies

Molecular line observations of circumstellar disks are relatively new, and there are many obvious directions for research. The rapid improvements in interferometers will open up this field dramatically in the next few years. These directions can be broadly divided into two categories: 1) providing compelling examples of molecular disks that are gravitationally bound to the central stars, and 2) expanding the molecular observations to include more stellar types, different molecules, and to address specific questions about the effects of the environment on disk formation and properties.

Providing compelling examples of molecular disks is important to ensure that the interpretation of the existing maps is correct. There are three clear avenues to improve our existing knowledge:

1. Go to even higher angular resolution with higher sensitivity. As the resolution gets better, interference by the larger-scale ambient cloud emission is suppressed, the Keplerian velocities are higher (thus further eliminating cloud emission), and the small scale structure is dominated much more by the gravity of the star.
2. Observe stars where the background and wind emission are small compared to that from the disk.

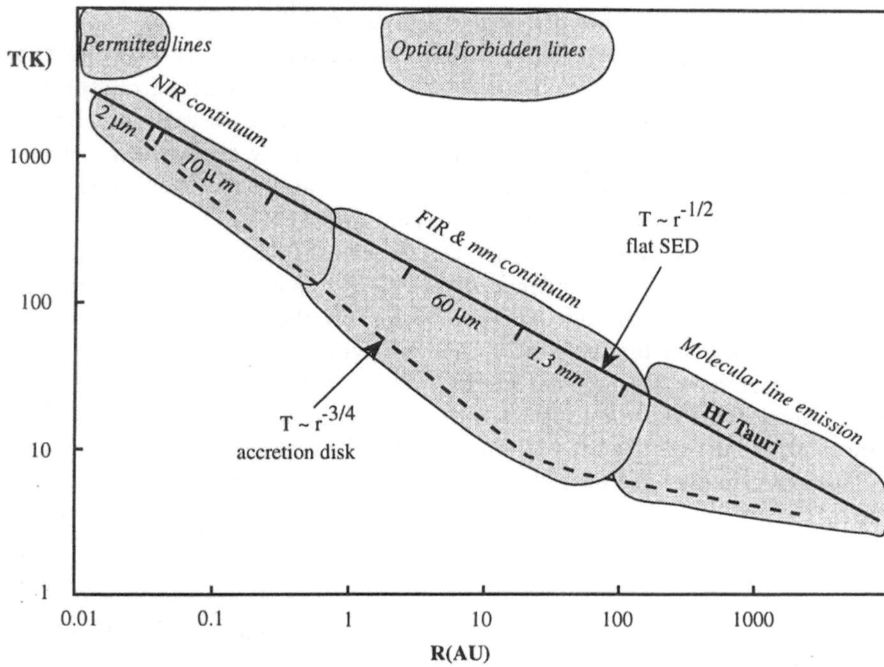

Fig. 3. The temperature as a function of distance within a disk for HL Tauri (solid line) and a theoretical accretion disk (dashed line). Also shown are the regions primarily emitting the 2, 10, and 60 μm continuum radiation as well as the locus of the molecular lines from disks. The optical permitted and forbidden lines arise in regions above and below the disk plane.

3. Observe molecules and their isotopes which are preferentially confined to the disks and which are weak emitters from the cloud and wind.

Continuing improvements in maps of HL Tau at better and better resolution at first made the case for disk emission more secure (Sargent & Beckwith 1991) and subsequently began to complicate the interpretation as weak features deviated from the simplest models (Hayashi et al. 1994). Millimeter-wavelength resolution has not yet reached the point where the molecular maps overlap the spatial regions responsible for the infrared and millimeter continuum radiation, and, therefore, do not actually test the conclusions reached from spectral energy distributions. To do so will require angular resolution of at least $0''.5$, about three times better than the best extant maps.

Ambient cloud emission and molecular emission from winds complicate the interpretation of maps. One can search for stars in regions of low ambient emission to minimize the interference. Furthermore, calculations of the line-center optical depth show that even ^{13}CO has very high optical depths within disks (Beckwith & Sargent 1993), so observations of rarer isotopes should show disk emission while simultaneously minimizing the emission from other components. Both approaches are being used successfully.

GM Auriga is relatively isolated, without strong mass loss, and is somewhat older than most classical T Tauri stars. Maps of the ^{13}CO emission from this star show an inclined molecular disk whose velocity field matches well with bound, orbiting gas (Koerner et al. 1993a). There is a particularly good match between the morphology of emission at different velocities and the calculations for a somewhat inclined disk by Beckwith & Sargent (1993).

New maps of GG Tau in ^{13}CO by Dutrey et al. (1993) are also consistent with Keplerian motion in an inclined disk. GG Tau is a quadruple system (see below). The continuum and line emission both show absence of gas very close to the star as if the inner binary cleared the disk dynamically.

The authors have recently detected $C^{17}O$ from HL Tauri using the OVRO interferometer. This very rare species will only be seen when the optical depths of the more abundant species are extremely high. The emission demonstrates very high concentrations of gas near the star and supports the disk interpretation.

3.1 Stellar type

No surveys of large samples have been carried out. There are a few examples of limited surveys and a variety of observations of individual stars.

Skrutskie et al. (1991) used the Massachusetts 15m to search for CO emission from 8 stars in the Ursa Major stream, a nearby (25 pc) open cluster 3×10^8 yr old. They established only upper limits typically of $3 \times 10^{-5}\,M_\odot$ for any associated disks, showing that the molecular phase of disk evolution cannot last more than a few million years. The dust evolves on similar time scales (Skrutskie et al. 1990) in the inner regions of the disks.

On the youngest end of the scale, the star L1551 IRS 5 is embedded in dense molecular gas with a variety of different components. Observations of this gas do not show an unambiguous molecular disk, even though the thermal continuum is best explained by a $\sim$ 50 AU radius particle disk (Keene and Masson 1990; Koerner et al. 1994). The $C^{18}O$ emission (Sargent et al. 1988) is extended perpendicular to the direction of the larger scale molecular outflow seen in CO (Snell et al. 1980), consistent with a disk; the velocity field is not sufficiently resolved to make a meaningful comparison to orbital motion.

FU Orionis stars and their precursors are thought to have disks to explain their infrequent outbursts among other properties (Hartmann et al. 1992). The star V1331 Cygni is probably an FUor precursor. Recent interferometer maps of CO and its isotopes reveal a rich circumstellar environment with a disk component, red and blue-shifted wind components, and more extended spheroidal emission that is likely associated with the star (McMuldroch et al. 1993). The disk mass of $0.5 \pm 0.15\,M_\odot$ is somewhat greater than that of a typical T Tauri star disk.

3.2 Molecular species

The densities in circumstellar disks should be large, orders of magnitude greater than the $\sim 10^6\,\mathrm{cm}^{-3}$ seen in the densest molecular clouds. Molecular emission usually associated with high densities – CS & HCN, for example – ought to be good tracers of circumstellar disks.

It is, therefore, surprising that such molecules are not strong emitters in classical T Tauri star disks (Blake et al. 1992; van Langevelde et al. 1993), even though they are present in the larger circumstellar clouds (*e.g.,* Menten & Walmsley 1985; Ohashi et al. 1993). Depletion factors between 25 and 50 for CS are derived by Blake et al. for the HL Tau disk.

Ohashi et al. 1993 surveyed 6 T Tauri stars and 11 embedded young objects in CS(J=2-1) with the Nobeyama interferometer. Infrared and millimeter-wave continuum emission from these objects implies the presence of disks in almost all cases. They detected two of the T Tauri stars, HL and DG Tau, and nine of the embedded objects including L1551 IRS 5; the CS emission is *not* well-centered on the T Tauri stars, so it is possible that none of the T Tauri disks has been detected in CS. This relatively weak emission in the T Tauri disks might result from depletion of CS relative to other species or small, sharply bounded disks; Ohashi et al. favor the former interpretation.

Molecular processing in the dense disks probably alters the relative abundance of molecular species. As the disks evolve, molecules will also freeze out on grain surfaces causing a general depletion of the molecular gas. This process is expected prior to the formation of larger, planetary bodies, so the time scales are of some interest to construct time histories of the disk evolution.

3.3 Companion stars

Binary and multiple star systems may even be more common among low-mass pre-main sequence stars than among evolved field stars (Leinert et al. 1993; Ghez et al. 1993; Simon et al. 1992). The binaries T Tau (projected $\delta = 100\,\mathrm{AU}$), IRAS 16293-2422 ($\delta = 1000\,\mathrm{AU}$) and the quadruple system GG Tau (projected separations between 40 and 1500 AU) have been mapped in CO and its isotopes.

IRAS 16293-2422 shows two compact radio continuum sources which are most probably individual stars in a binary system. Mundy et al. (1990, 1992) observed $C^{18}O$ and NH_3 from this system resolving the emission in both molecules surrounding the binary. The Figure 4 shows the maps of the source IRAS 16293-2422 together with a model of the source.

The T Tau emission is consistent with a circumbinary disk (Weintraub et al. 1989) with an orbital velocity field, although the rather low inclination of a few degrees makes the velocity interpretation problematic as discussed above.

GG Tauri consists of two binaries separated by peaks on both does not obviously destroy the molecular disks (Simon & Guilloteau 1992; Skrutskie et al. 1993; Kawabe et al. 1993; Koerner et al. 1993b). The Dutrey et al. (1993) results discussed above demonstrate that molecular emission may be seen even when the inner parts are cleared away. Their resolution of the inner hole in the

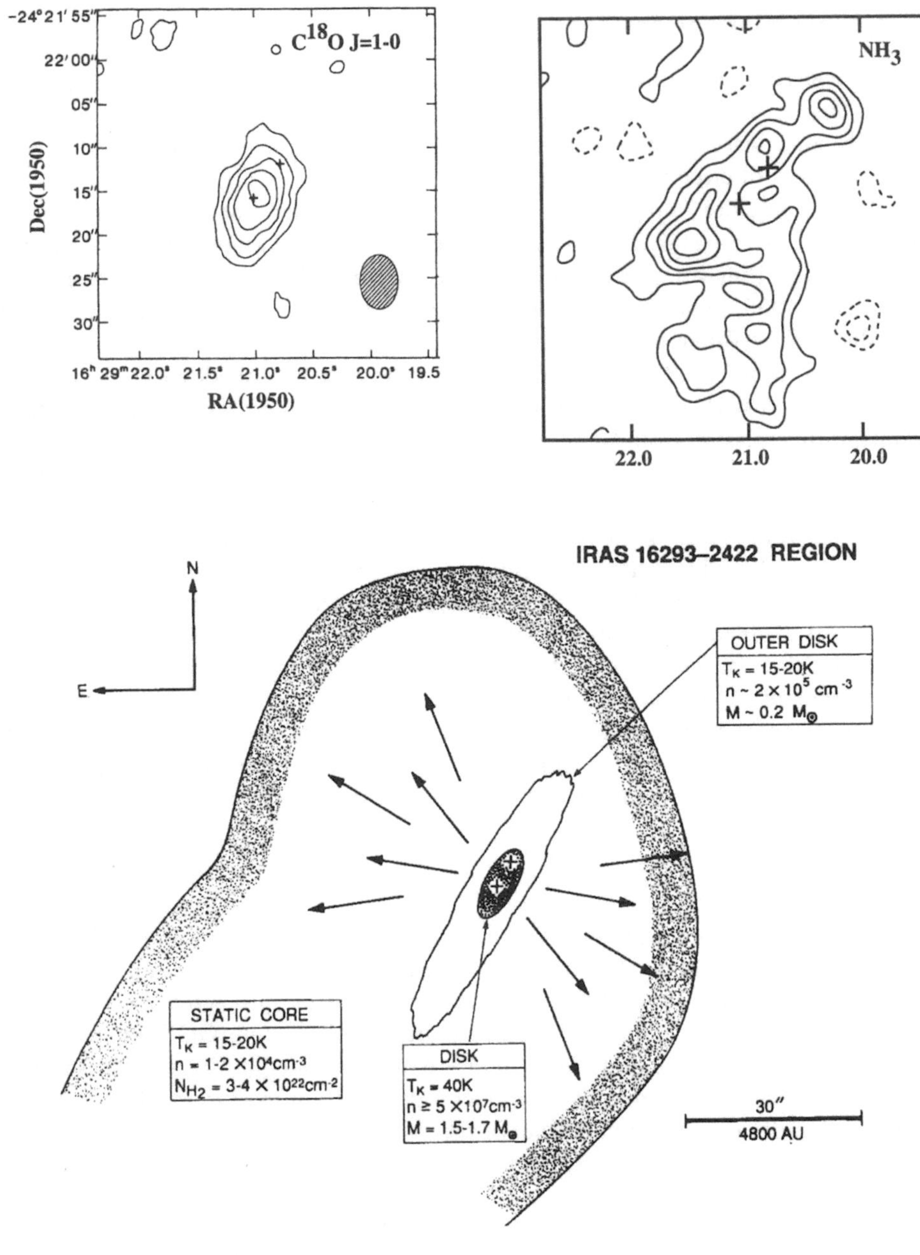

Fig. 4. The upper left part of the figure is the $C^{18}O$ map of IRAS 16293-2422 from Mundy et al. 1990. The upper right shows the NH_3 map from the same authors. The bottom is their schematic diagram of the circumstellar disk and environs. The two crosses mark the position of radio continuum sources which are thought to be stars separated by 6″ (1000 AU). If the continuum sources are, indeed, stars, this is a circumbinary disk.

northern GG Tauri star pair shows that the inner holes may be quite large, of order hundreds of AU, without decreasing the molecular emission from the outer parts.

L1551 IRS 5 also has two compact radio continuum sources within the center of the molecular and infrared emission (Rodriguez et al. 1986). It is not yet known if these sources indicate two separate stars or are produced in some other way; however, this object may, too, be a binary.

4 Summary

The evidence for disks from molecular observations alone is now quite strong. Small-scale emission from CO and its isotopes is commonly seen toward young stars. Its relatively small size, clear association with pre-main sequence stars, depletion of common molecules, highly flattened morphology, orientation perpendicular to well-collimated jets, and fairly regular velocity fields differentiate the emission from that seen in molecular clouds and molecular winds and suggest high volume densities characteristic of circumstellar disks. The disk hypothesis has the virtue of accounting for a wide variety of different observations, line and continuum spanning 4 orders of magnitude in wavelength, with few adjustable parameters.

There is as yet no firm agreement about the dynamics of the molecular gas. Keplerian rotation curves, mass infall, and interactions between outflowing winds and molecular cloud material can be used to fit the observations of different stars. The resolution of the maps is not yet sufficient to guarantee that stellar gravity should dominate the molecular line shapes, at least in the commonly used model of a circumstellar disk. There is relatively little information about the physics of the molecular emission, the volume densities and local temperatures in the outer regions of the disks. Molecular abundances are certainly different in these disks than in molecular clouds, and it will take somewhat more effort to identify those species which give reliable measures of the local conditions.

With the buildup of the interferometers at Owens Valley (OVRO), Hat Creek (BIMA), Plateau de Bure (IRAM), and Nobeyama, we can expect a steady flow of new results in the next few years. The quality and ease of these observations is increasing, owing to more telescopes in each array, improved receivers, and correlators with larger bandwidths. Some of the current ambiguities might soon be resolved with better quality maps in more strategically useful lines. Large samples of stars will be especially important to determine just how strongly winds and background cloud emission contribute to the present maps.

References

Adams, F. C., Emerson, J. P., and Fuller, G. A. 1990, *Ap. J.*, **357**, 606.

Adams, F. C., Lada, C. J., and Shu, F. H. 1988, *Ap. J.*, **326**, 865 (ALS).

André, P., Montmerle, T., Feigelson, E. D., and Steppe, H. 1991, *Astron. Ap.*, **240**, 321.

André, P., & Montemerle, T. 1993, *Ap. J.*, in press.

Beckwith, S. V. W., Sargent, A. I., Chini, R., and Güsten, R. 1990, *Astron. J.*, **99**, 924 (BSCG).

Beckwith, S. V. W. and Sargent, A. I. 1991, *Ap. J.*, **381**, 250.

Beckwith, S. V. W. and Sargent, A. I. 1992, in *Protostars and Planets III*, ed. G. Levy and J. Lunine (Tucson:U. Arizona Press), p. 521.

Beckwith, S. V. W. and Sargent, A. I. 1993, *Ap. J.*, **402**, 280.

Bertout, C. 1989, *Ann. Rev. Astron. Ap.*, **27**, 351.

Beckwith, S. V. W. 1993, in *Theory of Accretion Disks – 2*, a NATO Advanced Workshop, ed. W. Duschl, F. Meyer and W. Tscharnuter)

Beckwith, S., Sargent, A. I., Scoville, N. Z., Masson, C. R., Zuckerman, B., and Phillips, T. G. 1986, *Ap. J.*, **309**, 755.

Blake, G. A., van Dishoeck, E. F., and Sargent, A. I. 1992, *Ap. J. Lett.*, **391**, L99.

Bouvier, J., Cabrit, S., Fenandez, M., Martin, E. L., and Matthews, J. M. 1993, *Astron. Ap.*, in press.

Cabrit, S., Edwards, S., Strom, S. E., and Strom, K. M. 1990, *Ap. J.*, **354**, 687.

Cohen, M., Emerson. J. P., and Beichman, C. A. 1989. *Ap. J.*, **339**, 455.

Dutrey, A., Guilloteau, S., and Simon, M. 1993, *IRAM Newsletter*, No. 12, November 8, 1993, p. 5.

Edwards, S., Strom, S. E., Herbst, W., Attridge, J., Merrill, K. M., Probst, R., and Gatley, I. 1993, *Astron. J.*, in press.

Edwards, S., Ray, T. P., and Mundt, R. 1992, in *Protostars and Planets III*, ed. G. Levy and J. Lunine (Tucson:U. Arizona Press), p. 567.

Genzel, R. & Stutski, J. 1988, *Ann. Rev. Astron. Ap.*, **27**, 41.

Ghez, A. M., Neugebauer, G., and Matthews, K. 1993, *Astron. J.*, November 1993.

Guilloteau, S. 1993, in *The Cold Universe*, ed. T. Montemerle, in press.

Hayashi, M., Ohashi, N., and Miyama, S. 1994, *Ap. J. Lett.*, submitted.

Hartmann, L., Jones, B. F., Stauffer, J. R., & Kenyon, S. J. 1991, *Astron. J.*, **101**, 1050.

Hartmann, L., Kenyon, S., and Hartigan, P. 1992, in *Protostars & Planets III*, ed. E. H. Levy, J. I. Lunine, M. S. Matthews, Tucson:U. of Arizona Press.

Henning, Th., Pfau, W., Zinnecker, H., and Prusti, T. 1993, preprint Jena astrophysics.

Hillenbrand, L. A., Strom, S. E., Vrba, F. J., and Keene, J. 1992, *Ap. J.*, **397**, 613.

Kawabe, R., Ishiguro, M., Omodaka, T., Kitamura, Y., Miyama, S. M. 1993, *Ap. J. Lett.*, **404**, L63.

Keene, J., & Masson, C. R. 1990, *Ap. J.*, **355**, 635.

Koerner, D. W., Sargent, A. I., & Beckwith, S. V. W. 1993a, *Icarus*, in press.

Koerner, D. W., Sargent, A. I., and Beckwith, S. V. W. 1993b, *Ap. J. Lett.*, in press.

Koerner, D. W., Sargent, A. I., Padin, S., and Scoville, N. Z. 1994, in preparation.

Leinert, Ch., Haas, M., Richichi, A., Zinnecker, H., & Mundt, R. 1991, *Astron. Ap.*, **250**, 407.

Leinert, Ch., Zinnecker, H., Weitzel, N., Christou, J., Ridgway, S. T., Jameson, R., Haas, M., and Lenzen, R. 1993, *Astron. Ap.*, in press.

Lin, D. N. C., Hayashi, M., Bell, K. R., and Ohashi, N. 1994, in preparation.

McMuldroch, S., Sargent, A. I, & Blake, G. A. 1993, preprint.

Menten, K. M., and Walmsley, C. M. 1985, *Astr. Ap.*, **146**, 369.

Mundy, L. G., McMullin, J. P., Grossman, A. W., and Sandell, G. 1993, *Icarus*, in press.

Mundy, L. G., Wooten, H. A., and Wilking, B. A. 1990, *Ap. J.*, **352**, 159.

Natta, A., Palla, F., Butner, H. M., Evans, N. J., and Harvey, P. M. 1992, *Ap. J.*, **391**, 805.

Ohashi, N., Kawabe, R., Hayashi, M., & Ishiguro, M. 1993, to appear in *Planetary Systems: Formation, Evolution, and Detection*, conference held at Caltech, Dec. 1992.

Rodríguez, L. R., Cantó, J., Torrelles, J. M., and Ho, P. T. P. 1986, *Ap. J. Lett.*, **301**. L25.

Sargent, A. I. 1989, in *The Formation and Evolution of Planetary Systems*, eds. H. A. Weaver and L. Danly (Cambridge:Cambridge Univ. Press), p. 111.

Sargent, A. I. and Beckwith, S. V. W. 1987, *Ap. J*, **323**, 294.

Sargent, A. I. and Beckwith, S. V. W. 1991, *Ap. J. Lett.*, **382**, L31.

Sargent, A. I., Beckwith, S., Keene, J., and Masson, C. 1988, *Ap. J.* **333**, 936.

Sargent, A. I., & Welch, W. J. 1993, *Ann. Rev. Astron. Ap.*, in press.

Shu, F. H., Adams, F. C., and Lizano, S. 1987, *Ann. Rev. Astron. Ap.*, **25**, 23.

Skrutskie, M. F., Dutkevitch, D., Strom, S. E., Edwards, S., and Strom, K. M. 1990. *Astron. J.*, **99**, 1187.

Skrutskie, M. F., Snell, R., Dutkevitch, D., Strom, S. E., Schloerb, F. P., and Dickman, R. L. 1991, *Astron. J.*, **102**, 1749.

Skrutskie, M. F., Snell, R. L., Strom, K. M., Strom, S. E., Edwards, S. et al. 1993, *Ap. J.*, in press.

Simon, M., Chen, W. P., Howell, R. R., Benson, J. A., Slowik, D. 1992, *Ap. J.*, **382**, 212.

Simon, M., & Guilloteau, S. 1992, *Ap. J. Lett.*, **397**, L47.

Stahler, S. W., Shu, F. H., and Taam, R. E. 1980, *Ap. J.*, **241**, 637.

Strom, K. M., Strom, S. E., Edwards, S., Cabrit, S., and Skrutskie, M. F. 1989, *Astron. J.*, **97**, 1451.

Terebey, S., Chandler, C. J., and André, P. 1993, *Ap. J.*, in press (Sept. 1993).

van Langevelde, H., van Dishoeck, E. F., & Blake, G. A. 1993, in *IAU Colloq. No. 140, Astronomy with Millimeter and Submillimeter Wave Interferometry*, A.S.P. Conference Ser. San Francisco:BookCrafters, in press.

Weintraub, D. A., Masson, C. R., and Zuckerman, B. 1989, *Ap. J.*, **344**, 915.

Late Type Stars Associated with Maser Emission

K. J. Johnston

United States Naval Observatory, Washington, D. C.

Abstract: An account of the development of the study of stellar envelopes by means of maser lines is given.

1 Introduction

Masers are found in the circumstellar envelopes of evolved stars which are in the process of mass loss or have gone through this phase. These stars are post main sequence stars, red giants, supergiants, protoplanetary nebulae, and young ionization bounded planetary nebula. The masers trace out the kinematic as well as the physical structure of the circumstellar envelope and also allow for the study of the maser phenomenon in this environment. Further these masers may be used as celestial distance indicators for examle determining the distance to the center of the galaxy. In this article we shall trace the developments in understanding the the circumstellar envelopes of late type stars since the discovery of these masers in 1968.

2 Historical Perspective

In the late sixties, a new class of infrared stars was discovered. Wilson and Barrett (1968) found that these stars were associated with strong 1612 OH line maser emission. This emission was double peaked in its spectrum, which led to the conclusion that the masers are located in an expanding in circumstellar shell. This was followed shortly by the discovery of water vapor maser emission from these stars.

In the seventies this was followed by high angular resolution measurements to determine the brightness temperature of the emission to confirm that this was indeed maser emission. These measurements detected only the strongest maser features and did confirm that the emission was indeed due to a maser process as the brightness temperatures exceeded 10^7 K. Models were developed for maser emission in the circumstellar envelope of a variable, oxygen rich late M type star undergoing mass loss, creating an expanding envelope of gas and dust which emits significant radiation at IR wavelengths. Again the OH 1612 maser emission confirmed that this model was valid for a "thin" spherical shell of emission. SiO masers were discovered which also gave qualitative support to these models. A scheme for pumping the 1612 OH maser emission was developed. It was also found that the optical and maser emission variations were correlated.

In the eighties the technology of mapping celestial sources advanced with the fabrication of large synthesis arrays such as the MERLIN array at Jodrell Bank and the Very Large Array (VLA) of the National Radio Astronomy Observatory. Detaied maps showed that the emission arises from an expanding shell of gas and dust. The double peak velocities come from the front and back of the shell while the emission at intermediate velocities comes from slices through the shell. The emission appears to be irregular or clumpy, the effects of turbulence and excitation of molecules on velocity coherence are very evident. It was also found that the shell size was correlated with mass loss rate. As a follow on to the correlation of optical and maser emission variations, the lag time between variations in the peaks of the 1612 OH maser emission could be used to determine distances to these stars given the apparent shell size from the aperture synthesis observations. .

At the present time the research on these stars is directed at understanding the mass loss mechanism, dynamics of the shell including asymmetric features due to rotation, magnetic fields and turbulence, detailed excitation mechanisms for the masers in the shell and interaction with the interstellar medium or local radiation field.

Table 1 25 Years of Research on OH-IR Stars

1968 Discovery of OH 1612 Emission from IR stars Wilson and Barrett

1969 Discovery of H_2O Emission from late type stars Knowles, Mayer, Cheung, Rank and Townes

1970 VLB measurements of 1612 OH features of NML Cyg show sizes of masers 0.08", masers coincident with star to 5", all masers within a radius of 1" Wilson, Barrett nad Moran

VLB measurements of H_2O masers of VY C Ma Burke, Papa, Papadopoulos, Schwartz, Knowles, Sullivan, Meeks and Moran

1971 Photometric Determination of Mass Loss Rate Gehrz and Woolf

1972 Correlation of OH Luminosity with IR Excess Wilson, Schwartz, Neugebauer, Harvey and Becklin

Correlation of Maser Intensity with IR Variations Harvey, Bechis, Wilson and Ball

1974 Discovery of SiO Masers oCeti, VY CMa, R leo Buhl, Snyder, Lovas and Johnson W Hya Thaddeus, Mather, Davis and Blair

Correlation of H_2O Maser with Optical/IR Variability Schwartz, Harvey and Barrett

1975 Effect of Radiation Pressure/ Simple Models Kwok

1976 Thermal SiO Stellar Velocity Determination Reid and Dickenson

Model for Dynamics fo Mass Loss, Simple Spherical Model Goldreich and Scoville

1612 IR Pumping Scheme Elitzer, Goldreich and Scoville

1977 Confirmation of Thin Spherical Spherical Shell Model for 1612 OH Masers Reid, Muhleman, Moran, Johnston, Schwartz

1978 VLB "Spot" Size Measurements of VY CMa H_2O Masers Rosen, Moran, Reid, Walker, Burke, Johnston and Spencer

1979 VLB "Spot" Size Measurements of Mira stars Spencer, Johnston, Moran, Reid and Walker

1980 1612 OH Maser Time Lag/Shell Size Correlation Jewell, Webber and Snyder

1981 MERLIN Map of OH Masers of OH127.8 Booth, Kus, Norris and Porter
VLA Map of OH Masers of VX Sgr Baud, Habing, Mathews and Winnberg
Correlation of Shell Size with Mass Loss Rate Bowers, Johnston and Spencer

1985 VLA Maps of H_2O Maser Associated with Mira Stars Johnston, Spencer and Bowers
Distances of OH/IR Sources Diamond, Norris, Rowland, Booth and Nyman Herman, Baud, Habing and Winnberg

1986 Composite Study of VX Sgr Chapman and Cohen

1987 Optical/Radio Positions of Stars/Masers deVegt, Kleine, Johnston, Bowers and Spencer Baudry, Mazurier, Perie, Requieme and Rousseau (1990)

1988 u Orion Not Represented by a Simple Spherical Outflow Bowers and Johnston

1989 Large Sample of H_2O Maser Emission from Miras Lane et al.

1990 Continuum/H_2O Maser Map of W Hya Reid and Menten

1991 Proper Motion of OH Masers in u Orion Chapman, Cohen and Saikia
Complex Velocity Fields/ 3D Models Bowers

1992 SiO/H_2OOH Maser Emission Located with Common Velocities Bowers
VLB of SiO Masers Colomer, Graham, Kirchbaum, etc.

1993 Additional Evidence for Complex Flows

3 Description of Phenomenon

The circumstellar OH, H_2O and SiO masers have allowed the development of a model for the circumstellar envelopes of oxygen rich late type stars. These shells can be modeled to first order as predominately spherical shells undergoing radial expansion at a nearly constant velocity. The masers have a clumpy distribution in the shell. The OH masers appear to be located in a "thin" shell about 50 to 100 stellar radii, the H_2O masers 5 to 30 stellar radii and the SiO masers 3 to 5 stellar radii. The mass loss rates are between 10^{-7} to 10^{-4} Solar masses/yr. The spectral profiles depend on the coherence of the maser region which is related to velocity gradients along the line of sight, excitation, molecular density and pumping mechanisms. The density of the shell can be modeled as depending r^{-2} if there is a uniform mass loss from the star. This is probably correct to first order. The star pulsates with a period of about 300 days with the intensity of the maser and IR emission following this variation.

There are various pumping schemes for the masers. The 1612 OH masers associated with OH/IR sources is caused by radiative excitation of the OH rotational states by IR photons at 35 and 53 microns followed by radiative cascade to the ground state. The H_2O masers are most likely collisionally pumped. A detailed model has been put foreward by Cooke and Elitzer (1985). The physics giving rise to the SiO masers is the vibrational transitions become optically thick, then vibrational deexcitation rates decrease for higher J values which leads to overpopulation of the higher J levels. Model are not yet available to account in detail the high photon rates observed for these masers.

4 Scientific Issues

There are problems with this simple picture. Using the standard theory of radiative transfer in a thin spherical shell, and accounting for saturation and beaming, the emission at the emission at the stellar velocity if far weaker than generally observed (Alcock and Ross, 1986). The observed emission at this velocity appears to come from a much thicker shell than a thin annulus as predicted by the thin shell model.

The masers may be used a three dimensional probes of the circumstellar environment. The velocity field of the star VX Sgr increases more slowly than previously beleived. Models must account for the clumpyness of the maser emission, molecular abundance, turbulence, excitation, pumping mechanisms, etc.

5 Summary

The discovery of maser emission from the envelopes of larte oxygen rich stars has allowed an unprecedented look in the envelope of thser stars as well as using them a distance indicators.

Mapping the Molecular Emission of Circumstellar Envelopes with the Plateau de Bure Interferometer

M. Guélin

IRAM, 300 rue de la piscine, F-38406 St. Martin d'Hères, France

Abstract: Millimeter wave interferometers allow to study with arc second resolutions the molecular emission from opaque circumstellar envelopes. We present here recent data obtained with the IRAM Plateau de Bure interferometer.

1 Introduction

The dusty envelopes of late type stars are fascinating objects on their own; they are also interesting for what they teach us about IS chemistry. From their velocity field and density profile, we can study the mass loss during a crucial phase of stellar evolution: since the gas expands at nearly constant velocity, the velocity maps yield a 3-D view of the molecule spatial distribution. The spatial distribution of the different molecular species shows how photochemical, molecule-molecule and grain-surface reactions proceed with time in a well behaved environment.

The high mass loss phase ($M \geq 10^{-5} M_{\odot} yr^{-1}$) is short lived ($10^4$ to $10^5 yr$) and observable only in medium or high mass stars ($M \geq 1.5 M_{\odot}$). It is thus not surprising that the closest massive envelopes lie few hundred parsecs away and have small angular sizes. These envelopes are too opaque to be studied optically and must be observed at IR and mm wavelengths. The construction of large millimeter-wave interferometers, in particular the IRAM Plateau de Bure interferometer (Guilloteau et al. 1992), has provided a major breakthrough in their investigation.

2 Molecular emission in IRC+10216

The most remarkable and probably closest massive envelope surrounds the bright IR object IRC+10216 (CW Leo – Becklin et al. 1969). Its outer radius, observed in the mm lines of ^{12}CO, the most abundant molecule and the best shielded from photodissociation after H_2, is $R = 3'$ (Guélin & Cernicharo, *in preparation*). Its warm 'core', where about 50 different molecules are detected (see e.g. Fig. 1), has a radius of $\simeq 15''$. The envelope expands at a fairly constant velocity, $v_{exp} = 14.5$

kms^{-1}; the mass loss rate is estimated to be presently $1 - 2\ 10^{-5} M_{\odot} yr^{-1}$ and was probably several times larger a few thousand years ago (Sahai & Wannier 1985). The envelope mass is $\geq 0.5 M_{\odot}$.

The expansion motion most probably results from acceleration of the dust by stellar radiation in the region $R = 2 - 10\ R*$ (see e.g. Winters et al. 1994). High resolution infrared observations (e.g. Bloemhof et al. 1985) show that this region, which is also the region where the dust is formed, has an outer diameter of 0.4″. The distance of IRC+10216 is poorly known; from luminosity considerations, it is estimated to lie between 150 and 250 pc. For the sake of simplicity, we adopt here the commonly used distance of 200 pc, for which 1″ corresponds $3\ 10^{15}$ cm; the radius of the central star ($\sim 1000 R_{\odot}$) is then 0.02″.

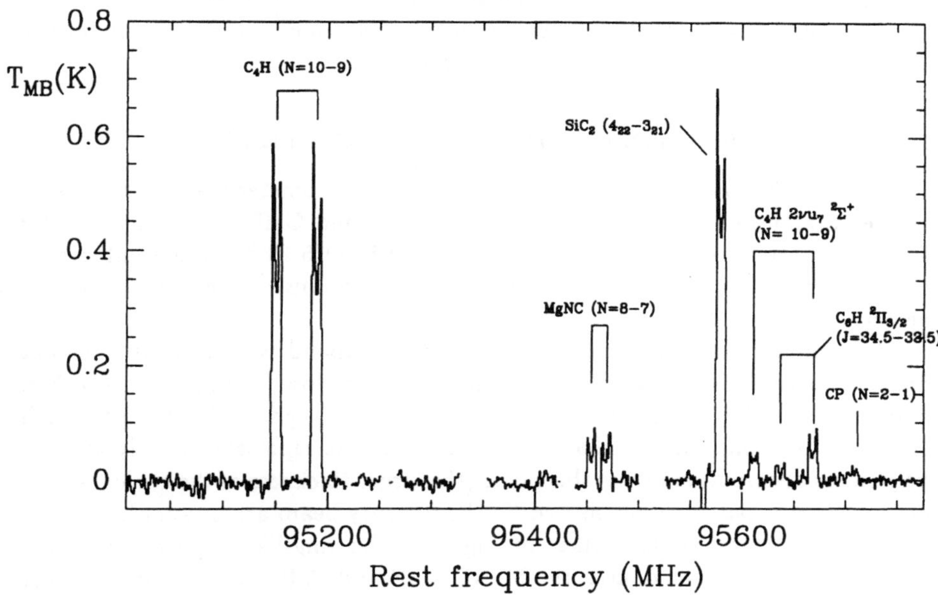

Fig. 1. The spectrum of IRC+10216 between 95 and 96 GHz, observed with the IRAM 30-m telescope (Cernicharo, Guélin, Kahane *in preparation*). The ordinate scale is T_{MB} (K). The portions of the spectrum contaminated by lines lying in the receiver upper sideband have been deleted.

The IRAM Plateau de Bure interferometer is ideally suited to study the dense core of IRC+10216. Its primary field has a FWHP of 55″ at 90 GHz and its synthetized beam can be as small as 2″. The large effective area of the instrument (460 m^2 since April 1993) makes it sensitive enough to map with a few arcsec resolution tens of molecular lines like those of Fig. 1. The possibility to observe simultaneously two 500 MHz-wide IF bands with a good frequency resolution, allows to study 5–10 lines in a single observing session.

A programme of mapping the λ 3 mm lines of the chemically most significant molecules is under way at Bure (Lucas et al. 1992, Guélin et al. 1993, Lucas 1994, Lucas et al. *in preparation*). 14 molecules have been observed to date: CO, CS,

C_3S, SiO, SiS, SiC_2, HC_5N, the carbon-chain radicals C_nH, $n = 2$–6, and the metal compounds NaCl and MgNC. The synthetized beam FWHP range from 3″ to 10″ (for CCH and C_5H, two of the very first maps observed with the Bure interferometer). In addition, vibrationally excited lines of C_4H (Lucas 1994) and HCN (Lucas & Guilloteau 1992) have been observed.

Complementary programmes, with larger fields of view, but lower resolution and poorer sensitivity, are also in progress with the BIMA (Bieging & Tafalla 1993, Bieging 1994 and references herein) and the NMA (Takano et al. 1992) interferometers; they include the mapping of different 3 mm lines of C_4H, SiS, SiC_2 and HCN, as well as that of lines of CN, C_3N, HNC and HC_3N. Finally, low resolution maps (HPBW= 12–17″) of higher energy 2 mm and 1.3 mm lines of HC_3N, C_4H and SiS were observed with the IRAM 30-m telescope (Audinos et al. 1994, Kahane et al. *in preparation*). These maps set constrains on the excitation conditions throughout the envelope.

2.1 Molecule distribution in the 'core' of IRC+10216

Figure 2 and 3 shows four examples of the Bure interferometer maps: the emission of the 95 and 98 GHz GHz lines of MgNC, C_3H and C_4H, and of the 91 GHz line of SiS, observed with a spatial resolution of 5″ (1.5 10^{16} cm) and integrated over a 13 kms^{-1}-wide band centred at $v_{sys} = -26.5$ kms^{-1}, the source systemic velocity.

The outer IRC+10216 envelope is know to expand with a fairly constant radial velocity (-14.5 kms^{-1}), so that the velocities close to v_{sys} plotted in Fig. 2 arise from a thin conical sector, axed on the line of sight and cutting through the envelope along the meridian plane. The contours represent roughly the line brightness distribution across this plane (the larger thickness of the conical sector at the edges is partly compensated by the primary beam attenuation).

The first thing which strikes in Fig. 2 is the ring-like appearance of the brightness contours: obviously, the MgNC, C_3H and C_4H emissions arise mostly from a few arcsec thick shell with a radius of $\simeq 15''$. The second is that, except for a factor of 5 of difference in intensity (the peak intensities in Figs. 2 & 3 are normalized to 1; note the higher noise on the MgNC map), the three molecules have very similar brightness distributions. The maps have the same clumpy appearance (the bright clumps coincide almost on both maps) and show the same intensity minima to the N and the SE. The central 'hole' seems more pronounced in C_4H than in MgNC (see the dotted negative contours), but this largely comes from the observing procedure. Whereas the MgNC map of Fig. 2 results from a combination of the data observed at Bure with a single-dish map (observed with the IRAM 30-m telescope), the C_4H and C_3H maps are the combination of the interferometer map with a ***single*** 30 m telescope spectrum and lacks spacings < 24 m (see Guélin et al. 1993).

The ring, or hollow shell appearance of the brightness contours of Fig. 2 is not common to all molecules. One third of the species mapped so far show centrally peaked distributions. Fig. 3 shows the brightness distribution of the J=5–4 line of SiS; the angular scale and velocity interval are the same as in Fig. 2. The SiS

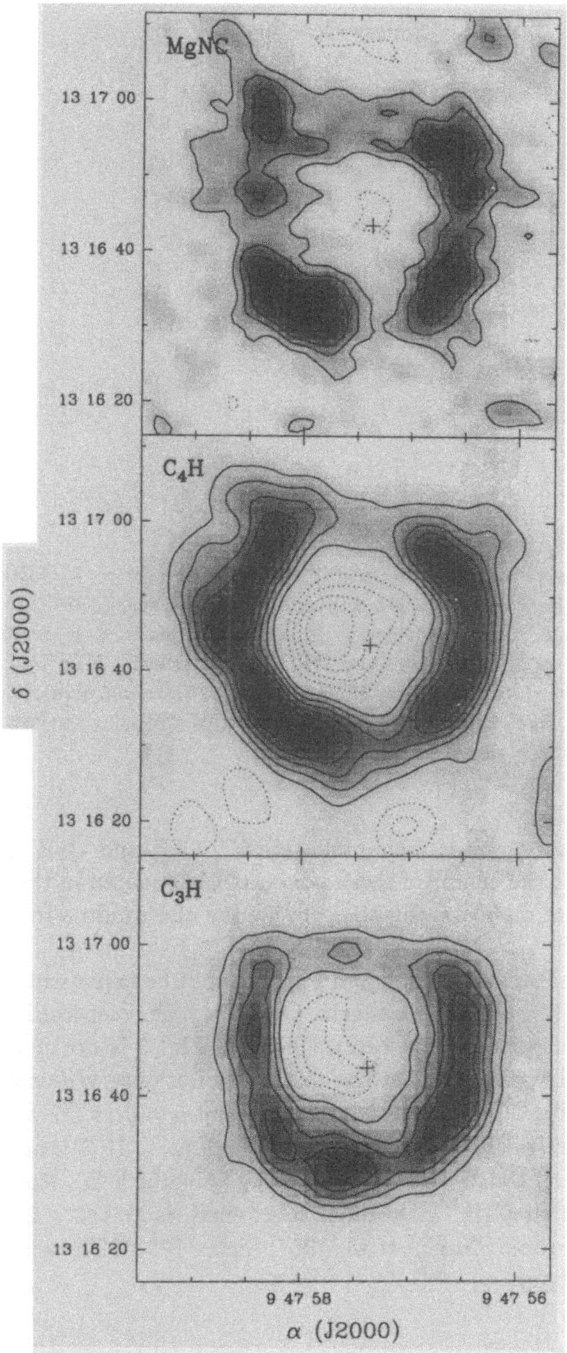

Fig. 2. The intensity distributions of the MgNC, N=8–7, C_4H, N= 10–9, and C_3H, $^2\Pi_{1/2}$J= 9/2–7/2 lines ($\nu = 95$ GHz, 95 GHz and 98 GHz), observed in IRC+10216 with the Plateau de Bure interferometer (Guélin et al. 1993). The intensities are integrated over the velocity interval -33 kms^{-1} $< v < -21$ kms^{-1} and normalized to 1. The angular resolution is $5'' \times 5''$. A point-like continuum source of 66 mJy, corresponding to the thermal emission of hot dust in the inner envelope, has been removed from the maps; its position is marked by a cross; note that the cross does not lie at the center of the shell-like intensity contours, but 2-3″ more to the west. The maps are not corrected

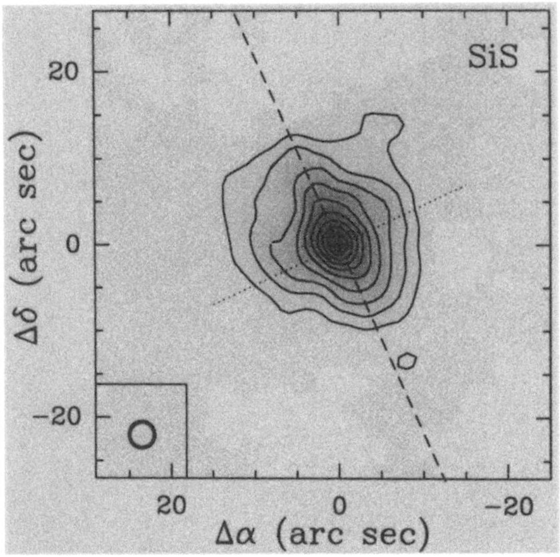

Fig. 3. The intensity contours of the SiS, J= 5–4 line (ν = 91 GHz), observed in IRC+10216 with the Plateau de Bure interferometer (Lucas, 1994). As in Fig. 2, the intensity is integrated over the velocity interval -32 kms^{-1} $< v < -21$ kms^{-1}. The angular resolution is $3'' \times 3''$ (see the HPB in the insert). The inner contours are elliptical with a major axis at $PA = 25°$ and a major to minor axis ratio of $\simeq 1.5$. The major axis (dashed line) could denote the direction of a bipolar outflow and the minor axis (dotted line) the plane of an equatorial torus

source is much more compact than the MgNC, C_3H and C_4H sources and fits inside the 'hole' at the center of these latter. It is elongated in the NS or NE-SW direction, i.e. more or less along the line joining the minima in the MgNC and C_4H 'rings'.

The differences between Fig. 2 and Fig. 3 cannot be explained by the envelope structure and/or line excitation effects. Obviously, the 'central hole' in Fig. 2 is not due to lack of gas, and the central peak of Fig. 3 is not just an excitation artifact: a centrally peaked brightness distribution is observed as well in the lines of the easy to excite CO molecule (dipole moment $\mu = 0.11$ D), as in those of the hard to excite species SiS ($\mu = 1.7$ D) and NaCl ($\mu = 10$ D), and not observed in C_4H (μ= 0.7 D), HC_5N ($\mu = 4$ D) and MgNC ($\mu = 5$ D). Rotational lines of vibrationally excited C_4H, with upper level energies $> 150 - 200K$, show also a ring-like distribution (Guélin et al. 1987, Lucas 1992). These differences must result from chemistry.

2.2 Chemistry in IRC+10216

The temperature, density and radiation in IRC+10216 are highly contrasted, laying the ground for a large variety of chemical processes. The atmosphere of the star ($R* < R \leq 2R*$) is dense and warm enough ($T_{phot} \simeq 2300$ K) for exo- and endothermic 3-body reactions to proceed efficiently. It must be close to thermodynamical chemical equilibrium; the most abundant C- and Si-bearing molecules are there CO, HCN, H_2C_2, SiO and SiS (Tsuji 1973, McCabe et al. 1979, Lafont et al. 1982). During the high mass loss episode, these 'parent' molecules are expelled with the upper atmosphere layers into the envelope. The gas temperature drops quickly in this process and most most of the 'heavy' atoms ($A \geq 12$ a.m.u.) and molecules condense onto grains as soons as $R >$ few$R*$, (Winters et al. 1994).

Further out, in the outer envelope ($R > 500R_*$), the ambiant interstellar UV dissociates the 'parent' molecules (CO, H_2C_2, HCN,..) still present in the gas phase. The radicals and ions produced by photodissociation fuel a rich gas phase chemistry which can form fairly complex molecules, such as the long cyanopolyynes HC_nCn, n= 2–6 (Glassgold et al. 1986, Nejad & Millar 1987, Cherchneff et al. 1993). The grains travelling from the inner to the outer envelope are covered with mantles formed of adsorbed molecules. These mantles may partly evaporate in the outer envelope, when exposed to UV radiation (photodesorption) or to shocks, releasing new molecules and radicals in the gas. At the outer edge of the envelope, the UV photons dissociate even the most resistant species, CO and H_2.

In Cherchneff et al. (1993a,b) model, SiS is a 'parent' molecule coming from the stellar atmosphere. Its distribution, should peak on the star, as is observed in Fig. 3. C_4H is formed from CCH by radical-radical reactions, in the outer envelope. It should thus have a shell-like distribution, as is observed in Fig. 2. For a mass loss rate of 2 $10^{-5} M_{\odot} yr^{-1}$ and a 'normal' interstellar UV field, Cherchneff et al. predict a radius of the C_4H shell of 4–5 10^{16} cm, in perfect agreement with what is observed.

The observed maps are however much too detailed to be fully explained by simple chemical models. The C_4H, C_3H and MgNC rings in Fig. 2 show bright patches which, remarkably, have nearly exactly the same positions (the small differences can be caused by the relatively high noise in the map of the weak MgNC emission).

C_3H and C_4H, in the model of Cherchneff et al., are formed by different types of reactions and their abundances should peak at different distances from the central star. There is no prediction for MgNC in this model, but it is likely that its formation and destruction (hence its peak abundance radius) have little to do with those of C_3H and C_4H. The great similarity of the C_3H, C_4H and MgNC maps, in particular as concerns the extent of the emission, implies either that the shell observed in Fig. 2 delineates the edge of the dense envelope, or that the three molecules are formed and destroyed together at a much faster rate than calculated by Cherchneff et al. (1993).

In the first case, the lower gas density and enhanced UV radiation to the outside of the shell would combine to reduce drastically the molecular column densities. Judging from Fig. 2, the edge should be sharp: the mass loss rate should have briskly increased some 10^3 yr ago. In the second case, one should consider processes faster than two-body gas phase reactions. Desorption of molecules from grain mantles could offer an interesting alternative; desorption could result from irradiation by interstellar UV photons or, if the molecules are weakly adsorbed, from shocks. In both cases, the bright spots on Fig. 2, which are observed for molecules with very different excitation conditions, must correspond to high density clumps.

2.3 Bipolar structure

The bright rings on Fig. 2 appear dimmer to the N and the SW. A similar pattern is observed for about all the molecules showing a shell-like brightness distribution. Very probably, these dark areas correspond to density holes in the envelope. The holes define a N-SW axis very roughly aligned with the major axis of the SiS source (Fig. 3).

The *inner* SiS brightness intensity contours are elliptical with a major axis oriented at $PA \simeq +25°$ and a major-to-minor axis ratio of 1.5 (after correcting for the 3″ wide beam). Within the uncertainties, this orientation is the same as that of the 0.4″-wide 2.2 μm source ($PA \simeq 20°$, Dyck et al. 1987), and of the diffuse optical nebulosity (Becklin et al. 1979–see however Le Bertre 1988). Its constancy over such a large range of angular sizes suggests the outflow from the central star may not be as spherically symmetrical as indicated by the CO outer envelope and the fairly regular velocity field. It may be bipolar with a preferential axis roughly perpendicular to the line of sight: IRC+10216 could be at the verge of becoming a protoplanetary nebula.

It is interesting to note in this respect that *i)* a bipolar structure (polar outflow and equatorial torus) with a symmetry axis at $PA = 20°$, inclined by about 25° on the plane of the sky was suggested by Serkowski and by Dyck et al. from polarization measurements (see Dyck et al. 1987); the diamond-shaped SiS *outer* contours on Fig. 3 may trace the outflow and the torus; and *ii)* the eastward 2-3″ shift of the center of the C_4H, C_3H and MgNC shells with respect to the IR/radio continuum source (and of the SiS source), is easily explained if the central star is a binary system (Guélin et al. 1993); the orbit of the binary system may fix the equatorial plane and polar axis of the envelope.

3 The very fast outflow in CRL 618

Whereas bipolar outflow and shock chemistry are only suggested in IRC+10216, they are flagrant in the more evolved stellar envelope CRL 618. CRL 618, which is thought to be a 'proto planetary nebula' at the end of the high mass loss phase, has already developped a compact HII region at the center of its thick dusty envelope (Kwok & Bignell 1984). The central star, hidden by dust, has reached the temperature of a B0 star; it is surrounded by two optical nebulosities showing high expansion velocities in the Hα and [NII] lines (Carsenty & Solf 1982). The distance to CRL 618, although very uncertain, is thought to be about 1.7 kpc.

Molecular line observations show, like in IRC+10216, an extended molecular envelope ($R \simeq 40''$) with a compact dense core ($R = 3''$), expanding at fairly constant velocity (20 kms^{-1}– Bachiller et al. 1988, Phillips et al. 1992). Inside the core, one observes in the lines of half a dozen of molecules a bipolar outflow with a very large velocity: 150–200 kms^{-1} (Cernicharo et al. 1989).

The envelope core and the fast molecular outflow have been mapped in several molecular lines, including the 2–1 (Cernicharo et al. 1989) and 3–2 (Neri et al. *in preparation*) lines of CO, using the IRAM 30-m telescope (resolution 12″ and 9″). The HCN 1–0 line has been mapped with a 3″ resolution with the IRAM Plateau de Bure interferometer (Neri et al. 1992). The high velocity HCN emission arises from a barely resolved source ($2'' \times < 0.5''$) elongated in the EW direction and about aligned with the optical lobes. Like in Hα, the HCN blue lobe is to the E and the red lobe to the W and the outflow axis is probably inclined by 45° with respect to the line of sight (Fig. 4).

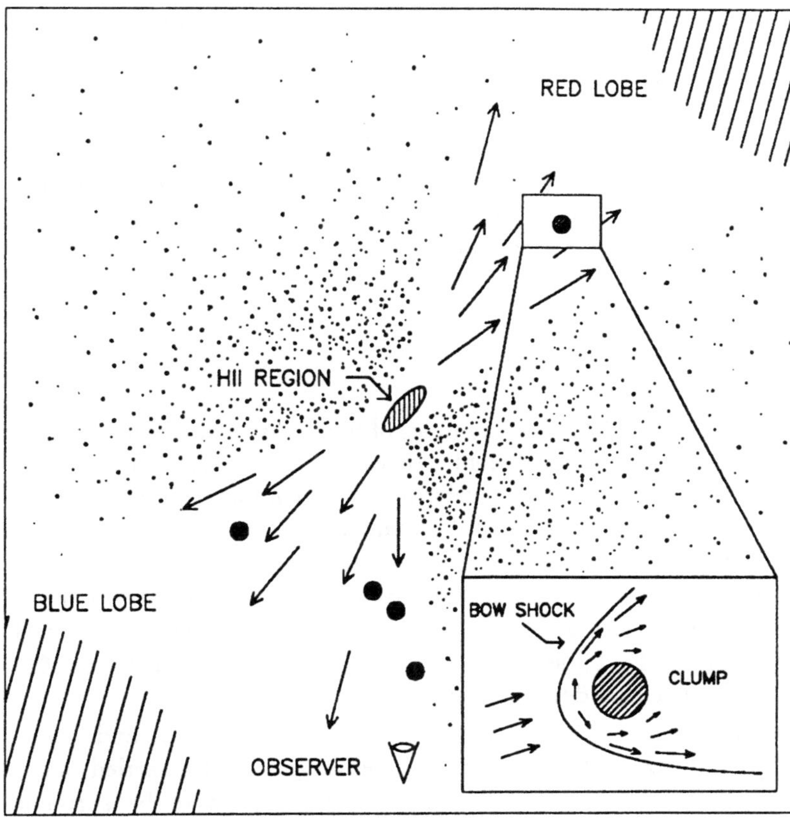

Fig. 4. Morphology of the envelope core and high velocity outflow in the proto planetary nebula CRL 618, as derived from HCN J=1–0 interferometric data and from Hα measurements. The HII region lies at the center of a compact and rather spherical core, itself surrounded by a low density halo (extending beyond the edges of the figure). A highly energetic flow collimates stellar material along an axis inclined by about 45° on the line of sight. Impacts of the flow on dense envelope clumps induce strong shocks. The high velocity wings of the HCN line arise from turbulent flows behind these shocks (from Neri et al. 1992).

Observing molecules with such large velocities with respect to the surrounding gas (200 kms^{-1} versus 20 kms^{-1}) is surprising. It is known that almost all molecules are dissociated in a shock with $\Delta v > 50$ kms^{-1} (Hollenbach & McKee 1980). Unless the molecular gas is progressively accelerated to 200 kms^{-1}, which is unlikely in view of the short distances and time scales involved, the gas must have been shocked at some point and its molecules dissociated; the observed molecules must have reformed after the shock. The outflow of CRL 618, whose geometry and time scales are roughly known, offers thus a text-book case of chemistry in post shock regions.

Molecule reformation after a fast shock should occur fairly rapidly after the gas has recombined and cooled down below 4000 K. Neufeld & Dalgarno (1989) have modelled the chemistry in post type C shocks (Δv up to 100 kms^{-1}) and found H_2, CO and HCN to be the first molecules to reform after the shock; the time scale for the HCN abundance to build up is a few years for a density of 10^5 cm^{-1} (and larger for lower densities). According to the interferometric data, the redshifted high-velocity HCN is confined into a clump located at 3 10^{16} cm from the star. From this, it can be inferred that the HCN reformation time is < 50 yr. The relative abundances of HCN and CO in the high-velocity gas are consistent with those calculated by Neufeld & Dalgarno (1989) for carbon-rich gas.

It will be interesting to reobserve the HCN line in several years to check whether the outflow geometry and molecular abundances have changed.

4 Other envelopes

CRL 618 and IRC+10216 are examples of high mass loss carbon stars (possibly symbiotic stars) at two different stages of their evolution. To follow the metamorphosis of the cocoon-envelopes and to compare the fate of C-rich and O-rich stars, it is necessary to study many more objects.

Other high resolution maps of molecular emission have been made with the Bure interferometer on CRL 2688, another 'proto planetary' C-rich nebula (HC_5N, SiS, see Lucas 1994) and on NGC 7027, a planetary nebula (HCO^+, see Cox et al. 1994). These maps, like those of IRC+10216 and CRL 618, show clumpy structures and deviations from spherical symmetry.

An ambitious programme aimed at measuring the sizes and asymmetries of a large number of thick circumstellar envelopes is also carried out at Bure and Pico Veleta. Fifty stars of various types, known to be surrounded by conspicuous envelopes, have been observed in the mm lines of ^{12}CO (Neri et al. *in preparation*); a subset of this sample has also been measured in SiO (Lucas et al. 1992). The Bure observations have been made in the 'snapshot' mode. For most objects, the sampling of the UV plane is too poor to yield maps of the molecular emission as detailed as those of Figs. 2 & 3; they allow however to derive for each envelope the radial brightness distributions in the different velocity-channels, as well as to measure the degree of ellipticity and the position angles of the envelopes.

The SiO data in most envelopes (but to the notable exception of IRC+10216) show unusual line brigthness distributions. As noted above, when one observes a spherically symmetric envelope expanding at a constant radial velocity, each velocity-channel sees the emission from a conical sector axed on the line of sight. The extreme velocities arise from two narrow cones on the near and rear sides of the envelope, whereas the median velocities arise from the vicinity of the meridian plane. The source size in the different velocity-channels should normally decrease when the velocity departs from the median velocity. Such a decrease is not observed for most envelopes, which shows the expansion velocity may not be constant. This could mean that the gas is accelerated throughout the whole region where SiO is observed – 100–300 $R*$ or so! Alternately, the envelopes may have formed from successive outflows, each having it proper velocity (Sahai and Bieging 1993).

This article was largely written when the author was at the Max Planck Insitut für Radioastronomie. It is a pleasure to acknowledge the MPIfR for its hospitality and the Alexander von Humboldt Stiftung for financial support during this stay.

References

Audinos, P., Kahane, C., Lucas, R. 1994, *Astron. Astrophys. in press.*

Bachiller, R., Gomez-Gonzalez, J., Bujarrabal, V., Martin-Pintado, J. 1988 *Astron. Astrophys.* **196**, L5.

Becklin, E.E., Frogel, J.A., Hyland, A.R., Kristian, J., Neugebauer, G. 1969 *Astrophys. J.* **158**, L133.

Bloemhof, E.E., Danchi, W.C., Townes, C.H. 1985 *Astrophys. J.* **299**, L37.

Bieging, J.H., Tafalla, M. 1993 *Astronomical J.* **105**, 576.

Bieging, J.H. 1994, in IAU Coll. 140, *Astronomy with Millimeter & Submillimeter Wave Interferometry*, Hakone, Japan Oct. 1992.

Carsenty, U., Solf, J. 1982, *Astron. Astrophys.* **106**, 307.

Cernicharo, J., Guélin, M., Martin-Pintado, J., Penalver, J., Mauersberger, R. 1989, *Astron. Astrophys.* **222**, L1.

Cherchneff, I., Glassgold, A.E., Mamon, G.A. 1993 *Astrophys. J.* **410**, 188.

Cherchneff, I., Glassgold, A.E. 1994 *Astrophys. J. in press.*

Cox, P., Bachiller, R., Guilloteau, S., Forveille, T., Huggins, P. 1994 in *37th Herstmonceux Conference Proceedings*, Cambridge, UK, July 1993.

Dyck, H.M., Zuckerman, B., Howell, R.R., Beckwith, S. 1987 *Publ. Astr. Soc. Pacific*, **99**, 99.

Glassgold, A.E., Lucas, R., Omont, A. 1986 *Astron. Astrophys.* **157**, 35.

Guélin, M., Cernicharo, J., Navarro, S., Woodward, C.R., Gottlieb, C.A., Thaddeus, P. 1987 *Astron. Astrophys.* **182**, L37.

Guilloteau et al. 1992 *Astron. Astrophys.* **262**, 624.

Hollenbach, D., McKee, C.F. 1980 *Astrophys. J.* **241**, L47.

Keene, J., Young, K., Phillips, T.G., Buttgenbach, H., Carlstrom, J.E. 1993 *Astrophys. J. in press.*

Kwan, J., Linke, R.A. 1982 *Astrophys. J.* **254**, 587.

Lafont, S., Lucas, R., Omont, A. 1982 *Astron. Astrophys.* **106**, 201.

Le Bertre, T. 1988 *Astron. Astrophys.* **203**, 85.
Lucas, R. 1992 in *Astrochemistry of Cosmic Phenomena*, Ed. P.D. Singh, (publ. Kluwer: Dordrecht), p. 389.
Lucas, R., Guilloteau, S. 1992 *Astron. Astrophys.* **259**, L23.
Lucas, R. *et al.* 1992 *Astron. Astrophys.* **262**, 49.
Lucas, R. 1994, in IAU Coll. 140, *Astronomy with Millimeter & Submillimeter Wave Interferometry*, Hakone, Japan Oct. 1992.
McCabe, E.M., Smith, R.C., Clegg, R.E.S. 1979 *Nature*, **281**, 263.
Nejad, L.A.M, Millar, T.J. 1987 *Astron. Astrophys.* **183**, 279.
Neri, R., Garcia-Burillo, S., Guélin, M., Guilloteau, S., Lucas, R. 1992 *Astron. Astrophys.* **262**, 544.
Neufeld, D.A., Dalgarno, A. 1989 *Astron. Astrophys.* **340**, 869.
Phillips, J.P., Williams, P.G., Mampaso, A., Ukita, N. 1992 *Astron. Astrophys.*
Sahai, R., Wannier, P.G. 1985 *Astrophys. J.* **299**, 424.
Sahai, R., Bieging, J.H. 1993 *Astrophys. J.*
Takano, S., Saito, S., Tsuji, T. 1992 *Publ. Astr. Soc. Japan*, **44**, 469.
Truong-Bach, Morris, D., Nguyen Q. Rieu 1991 *Astron. Astrophys.* **249**, 435.
Tsuji, T. 1973 *Astron. Astrophys.* **23**, 411.
Winters, J.M., Dominik, C., Sedlmayr, E. 1994 *Astron. Astrophys. in press.*

High Resolution Studies of SiO Masers

Alain Baudry

Observatoire de Bordeaux, BP 89, 33270 Floirac, France

Abstract: The discovery of SiO maser emission from Orion and late type stars is briefly reviewed and the main spectroscopic properties of the SiO molecule are given. Several rotational transitions in various vibrational states of ^{28}SiO, ^{29}SiO or ^{30}SiO have been observed in the direction of about 225 evolved stars. The SiO molecule is strongly masing and provides a new tool to explore the dynamics and structure of regions extending from the stellar photosphere to the circumstellar dust shell. In this review we discuss recent observations made with sensitive radio interferometers which provide crucial information on the mechanisms at the origin of the maser phenomenon and on the physical conditions prevailing in the SiO dense clumps above the stellar photosphere. Some SiO maser pumping models including radiation, collisions and line overlaps are briefly discussed.

1 Introduction

Since their early discovery some 20 years ago, SiO masers have been observed in the circumstellar envelopes of more than 200 red stars at a late stage of stellar evolution. In this paper, we discuss, after a brief history of the discovery of the SiO maser phenomenon, the evolutionary status of the SiO stars (Section 2), the indirect (Section 3) and direct (Section 4) observational and theoretical arguments in favour of placing SiO masers near the central star, and we briefly account for some important SiO pumping mechanisms (Section 5). The observational material discussed here is mainly taken from recent high resolution studies made at mm wavelengths with either long baseline interferometers or large single dish radiotelescopes.

1.1 Historical Background

In 1973 December, Snyder and Buhl (1974) discovered an unidentified molecular emission near 86.245 GHz ($\sim$ 3.48 mm) in the direction of Orion. The new species was neither HCO^+ (the former X-ogen) nor HCN, but Snyder and Buhl found that the rest frequency of the 3.48 mm emission overlapped the v=1, J=2-1 transition of ^{28}SiO. Because this emission consisted of two narrow linewidths they also concluded that they had detected a new mm-wave maser. This first detection of the SiO molecule was shortly after confirmed by Davis et al. (1974) and Thaddeus et al. (1974) who observed the v=1, J=3-2 and 1-0 rotational transitions of the same species toward Orion. Kaifu *et al.* (1975) further discovered that the J=2-1 transition of SiO was present in several M-type Mira or semi-regular variables. Since then, several rotational transitions of silicon monoxide were observed toward many late-type stars up to J=6 in the vibrational states v=0 to 4. The v $\geq$ 1 radiation is observed in narrow emission lines and behaves as a natural maser whereas the v=0 radiation is found in both broad and narrow emission lines. The most compelling argument in favour of maser emission is the observed high brightness temperature of $10^{10} - 10^{11}$ K or more, deduced from the very long baseline interferometry (VLBI) measurements of SiO hot spots in stars and Orion. Since the kinetic gas temperature in stars is expected to be around 1000 K clearly non-thermodynamic equilibrium prevails. Non-masing v=0 SiO emission, not discussed here, is also observed in the interstellar medium. It is preferably found in hot and/or shocked regions not depleted in silicon.

1.2 Spectroscopy

SiO possesses non-zero electronic angular momentum and has $1\Sigma^+$ ground state. The main isotopic species present in space have spinless nuclei. Therefore the energy levels are described by the total quantum number J and energy constants depending on the vibrational state v,

$$E = B_v J(J+1) - D_v J^2(J+1)^2 + (\text{higher order terms})$$

Fig. 1 gives a schematic representation of some rotational levels in the first four v states. For v=1 we have enlarged the energy diagram and we show with vertical bars the maser transitions that have been observed in stars. The SiO molecule is fairly stable in the circumstellar or interstellar environment because of its relatively high energy dissociation of order 8 eV. The electric dipole moment varies slightly with the vibrational state; it is around 3.1 Debye thus implying high volume density $\sim 10^6$ cm^{-3}, at least, to collisionally excite the rotational transition lines.

2 SiO Maser Stars: Statistics and Evolutionary Path

At the time of this conference about 225 M-type stars are known to exhibit SiO maser emission. In Fig. 2, we present examples of narrow emission from stars in the v=1, J=2-1 ^{28}SiO line around 86 GHz. These stars have spectral types later than M4-M6, they are very red, have effective temperatures around 3000 K and most of them show long period time variability in the optical.The long term variability is characteristic of evolved Mira stars. On the other hand, only three star-forming regions, Orion, W51 and Sgr B2 toward the galactic center, were found to contain SiO masers. The Orion source has been extensively studied by the Berkeley group (e.g. Plambeck *et al.* 1990), and the positions and nature of

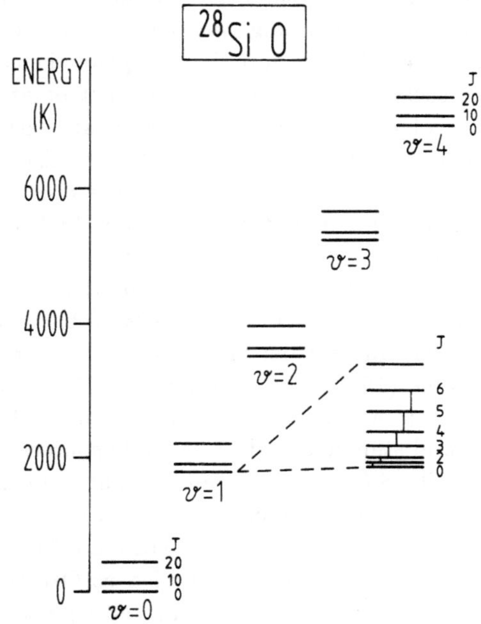

Fig. 1. The lower energy ($v \leqq 4$, $J \leqq 20$) vibration-rotation levels of ^{28}SiO. For v=1 (1450 K above the ground-state) we have expanded the rotational ladder and we show with vertical bars the transitions observed in stars. The v=0 to 1 and v=1 to 2 and the v=0 to 2 transitions lie around 8 and 4 μm respectively.

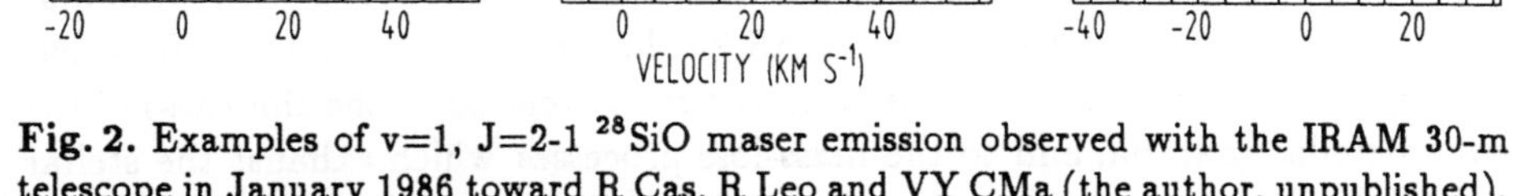

Fig. 2. Examples of v=1, J=2-1 ^{28}SiO maser emission observed with the IRAM 30-m telescope in January 1986 toward R Cas, R Leo and VY CMa (the author, unpublished). The spectral resolution was 100 kHz or 0.35 km s^{-1}. R Cas and R Leo are nearby Mira variables (~ 200-100 pc) and VY CMa is a distant (1500 pc) supergiant. The 86 GHz emission from these stars, and from late M-type stars in general, is time variable on scales of months and years. Radiointerferometry shows that various spectral features observed in these stars correspond to various SiO clumps (or hot spots) distributed around the star.

the SiO emission in W51 and Sgr B2 were measured and discussed by Morita *et al.* (1992). In W51, the long term study of the SiO emission made by Fuente (1992) suggests that the SiO is confined within a dense condensation perhaps accelerated by an embedded central young star. We do not discuss here the SiO emission in star-forming regions but we note that the general conditions required to excite SiO masers in these regions are not often met or are quite different from those prevailing in evolved stars where SiO masers are easily detected. The first compilation of SiO stars is that of Engels (1979) who listed 48 stars. More recently, Benson *et al.* (1990) completely searched the literature up to June 1989; listing a total of 205 SiO stars out of 639 stars searched for SiO emission at several observatories. (The Benson *et al.* catalog includes the SiO sources observed at 43 GHz during the period 1983-1984 by Jewell *et al.* 1991.) Other recent SiO surveys exist and one may cite in particular, the work of Allen *et al.* (1989) who searched for new southern SiO masers associated with Miras and symbiotic stars, and the work of Patel *et al.* (1992) who observed 161 Mira variables with the new 10 m antenna in Bangalore, India. These observations confirm the presence of SiO emission from several stars and add 11 new SiO sources to the Benson *et al.* list. About 225 stars with SiO emission are known at the present time when one includes the southern hemisphere detections of Le Bertre and Nyman (1990) not listed in Benson *et al.* compilation. The SiO stars are Mira or semi-regular variables, or supergiants. Few of them are classified as OH/IR stars i.e. stars without visible counterparts and with strong infrared signatures. The lack of SiO emission in OH/IR stars is partly due to the large distance of these objects. It could also be due to the fact that the SiO maser conditions are not met at the later stage of evolution reached by the OH/IR stars compared to other late-type visible stars (see below). In conclusion, the SiO maser phenomenon in M-type stars appears to be quite frequent in our Galaxy although the present SiO star source density is small; much below 1 source per square degree. The evolutionary paths of the red giants are rather well understood for stars of about one solar mass. Except for the few M supergiants with SiO emission, the SiO M-type stars have masses of order one solar mass and they lie at the top of the Asymptotic Giant Branch (AGB) which follows the ascent along the Red Giant Branch (H burning phase in a thin shell) of the H-R diagram (Fig. 3).The end of the AGB phase is characterized by thermal pulses in the He burning shell and prefaces the Planetary Nebula ejection phase. The SiO stars are red, oxygen- rich stars and lie in the regions III and IV of the IRAS 2-colour diagram of van der Veen and Habing (1988). The redness of the stars is directly related to the thickness of the circumstellar material and to the mass-loss processes which exhaust the stellar material. The mass-loss rates, in solar masses /year, are in the range 10^{-7}-10^{-6} (typical for Miras) to 10^{-4} (typical for OH/IR stars). Finally, because the SiO masing Mira variables seem to lie in a well defined region of the H-R diagram (Patel *et al.* 1992, and Fig. 3) a minimum stellar size must be attained before the SiO maser phenomenon is observed.

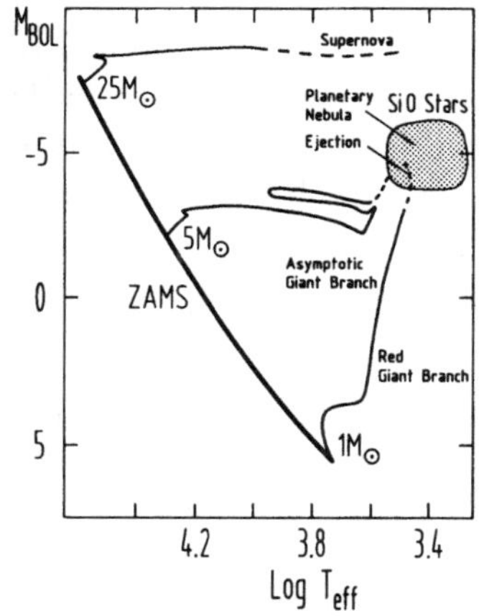

Fig. 3. Evolutionary tracks of 1, 5 and 25 solar masses stars in the luminosity-temperature diagram (adapted from Iben 1981). The one solar mass track reaches at the top of the Asymptotic Giant Branch the domain where M-type variable stars with SiO masers are observed (shaded area).

3 The Circumstellar Envelope and SiO Masers

Before discussing observations made with interferometers to locate the SiO emission within the circumstellar (CIS) envelope (Section 4) we present here some indirect arguments in favor of the SiO not being placed too far from the stellar photosphere: (1) A high degree of excitation is required to excite in many stars rotational levels up to v=3 and 4 . These levels lie around 5300 and 7000 K above the ground-state (Fig. 1); (2) The relatively low mass-loss rates derived in stars showing high v SiO masers (Alcolea *et al.* 1989) cannot be reconciled with the presence of strong masers in the circumstelllar region of uniform wind. (The high v transitions would not become optically thick.); (3) Observations show that the SiO radial velocities are statistically very close to the stellar radial velocities, and that SiO radio fluxes and stellar mass-loss rates are not correlated (e.g. Jewell *et al.* 1991). These arguments strongly suggest that the SiO masers lie near the stellar photosphere and are certainly closer to the central star than the OH masers observed toward the same stars. There are other theoretical arguments indicating that the SiO masers cannot reside far in the CIS wind. In particular, Langer and Watson (1984) show that the maser power would be increased if the SiO would reside in fragments near the star. The recent calculations made by Nedoluha and Watson (1993) on the intensity dependent circular polarization of SiO masers bring new information. They show that the circular polarization observed in the spectra of J=1-0 SiO masers (Barvainis *et al.* 1987) may be caused by weak magnetic fields of order 10 mG and does not require the standard Zeeman interpretation with fields higher than a few tens of gauss. Therefore, the lower field intensity indicated by the results of Nedoluha and Watson is not inconsistent with SiO masers in the CIS envelope. However, one cannot yet definitively exclude the standard Zeeman interpretation and high magnetic fields which would then place the SiO masers near the central star! It is also worth mentioning that the long term monitoring of the SiO radio flux (Martinez *et al.* 1988) shows large intensity variations in the SiO emission peaks. This suggests that the SiO resides in the CIS inner layers in a region not too far in the envelope and still affected by the stellar pulsation.

4 Radiointerferometry of SiO Maser Stars

Interferometer observations of the SiO masers are clearly needed to understand the complex region extending from the stellar photosphere to the CIS dust envelope. They should give direct information on the location, kinematics and spatial structure of the SiO masers. Obviously the SiO maser emission is switched-off when the silicon is removed from the gas and goes into the silicate grains of the dust envelope. In other words, the SiO maser phenomenon stops where the dust envelope begins and measuring the overall size of the SiO masing region gives an estimate of the minimum radius at which dust condenses. We have found with the IRAM mm-wave interferometer (Lucas *et al.* 1993) that quasi-thermal SiO emission in the v=0 state is observed far into the envelope with perhaps on average $\sim$ 1% of Si in SiO. These results give in turn an estimate of the dust envelope extent. In the following, we briefly describe two complementary interferometer programs undertaken in Europe toward late-type stars showing SiO maser emission. The observations were made with: (i) a European very long baseline interferometer (VLBI) consisting of 3 or 4 antennas not physically connected because of their great mutual distances and, (ii) the IRAM mm-wave connected interferometer.

4.1 VLBI Observations of v=1, J=1-0 SiO Masers

VLBI observations of cosmic masers show that the spectral channels forming the masing line spectral profile of a star correspond to clumps of gas (or hot spots) projected against the sky in the direction of the star. This is a major result of line VLBI (e.g. Reid and Moran 1981). The early VLBI observations of SiO stars were made with a relatively short baseline of 75 km in the v=1, J=1-0 transition of SiO around 43 GHz (e.g. Lane 1982). The results of Lane show that the emission breaks into hot spots distributed over a region a few stellar radii in size. The observed hot spots were partly or completely resolved with typical sizes of order 2×10^{13} to a few 10^{14} cm. However there is no a priori or theoretical reason why there should not be even more compact SiO clumps. In June 1990 we prepared a new VLBI experiment at 43 GHz with three stations in Europe and a maximum separation of about 2150 km. We detected compact SiO emission in 11 of the 19 selected Miras or supergiants (Colomer et al. 1992). The estimated sizes lie in the range 0.4-3.0 milliarcsec (mas) corresponding to less than 1AU (1AU=1.5×10^{13} cm) for the red giants and 1-4AU for the supergiants. We believe that this new result, to be compared to the typical 10 mas-size reported by Lane, is due to better sensitivity. For the strongest feature observed in the distant supergiant μ Cep (Fig. 4) the brightness temperature is $\sim 8\times10^{11}$ K, thus firmly establishing the masing nature of the SiO radiation. Although this is the highest temperature measured in an SiO late-type star it is still modest compared to the 1015 K determined in the brightest H_2O masers of the Galaxy. We performed new VLBI observations of stars in 1991 and 1992. One of our most interesting results concerns μ Cep where, using a special mapping technique called fringe rate mapping, we found that the SiO clumps are confined within a cloud (or a disk ?) 20-30 mas in size. We cannot yet define the shape of the SiO cloud but its size is comparable to the diameter of the central star as shown below. μ Cep is an M2 supergiant at a distance of 800 pc if it is a member of the Cep OB2 association. Taking M_{bol} = -9.7 (Humphreys 1983) and T_{eff} = 3000 K we derive $R^* \sim 2\times10^{14}$ cm corresponding to a diameter of about 30 mas. (This value is close to the optical interferometer measurement of Hutter *et al.* 1989 who give 20 mas.) Our VLBI observations do not provide an accurate absolute position of the SiO emission but it is likely that because of excitation requirements the 43 GHz emission is centered on the star. In fact, our IRAM observations show that the 86 GHz SiO masers coincide with the central star position within about 0.1-0.2 arc sec (see below).

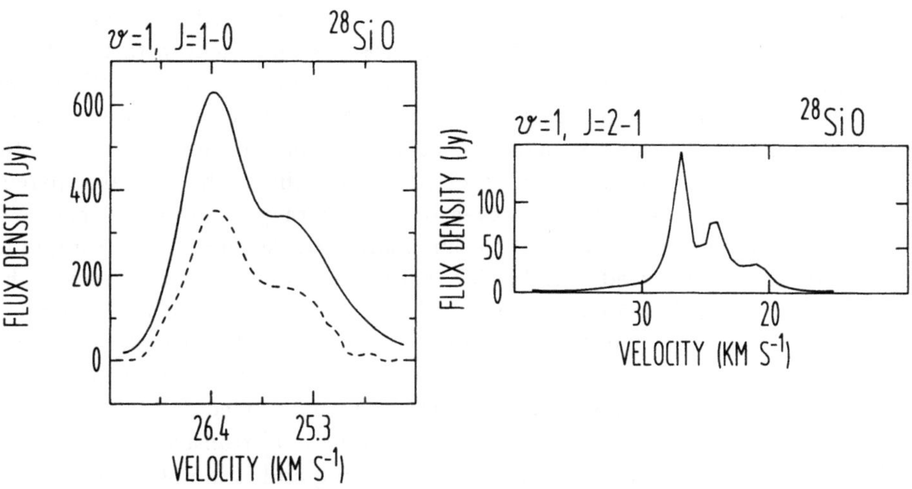

Fig. 4. SiO spectra obtained in the direction of μCep at 43 and 86 GHz (upper and lower parts of figure). In the upper part one shows the 43 GHz cross-power spectrum obtained in 1991 with the Bonn to Onsala interferometer for a projected baseline of about 800 km (crosses) and the 43 GHz total power spectrum (solid line) obtained in 1990 (after Colomer *et al.* 1992). The lower part shows the 86 GHz spectrum obtained in January 1992 for one baseline of the IRAM interferometer.

4.2 IRAM Observations of v=1, J=2-1 SiO Masers

The 43 GHz VLBI maps show that even though the individual hot spots are extremely compact, the SiO clumps are scattered over a region a few stellar radii in size. In addition to the hot spots well detected in VLBI, some other less compact components must be present since the VLBI observations do not 'see' all of the SiO flux. The most compact SiO hot spots, as well as the flux 'missed' in VLBI, should be detected with shorter sensitive interferometer baselines. Therefore, we used the IRAM interferometer on Plateau de Bure to make a new investigation of the spatial structure of 13 late-type stars and Orion. With a synthesized beam of ~ 1-2 arc sec this interferometer cannot resolve the individual hot spots. However, it has the capability to make accurate relative position measurements of spectral features across maser line profiles, and it can also give the absolute position of the SiO emission centroid. Our first observations were performed in 1991 and 1992 in the v=1, J=2-1 transition of SiO at 86 GHz. (They are followed now by an interferometer monitoring program of the spatial structure of a few 86 GHz SiO stars taken from our original sample.) A detailed account of our 1991-1992 results will be presented elsewhere and we only give here a brief summary of some major results deduced from a careful analysis of the interferometer phases. In several stars we have observed that various spectral features are spatially distinct i.e. the SiO emission shows spatial structure. This is the case for example in R Leo where our relative position measurements show that the positive and negative velocitiy features are separated by about 50 mas a result in agreement with the independent observations made with the IRAM 30-m radiotelescope during a lunar occultation (Cernicharo et al. 1993b). The structure observed for R Leo might be due to an apparent rotation or to more complex motions. With good signal to noise ratio in the interferometer data, the relative position accuracy is of order 10-20 mas (this is the typical apparent size of several Mira variables). An even better accuracy is obtained when self-calibration techniques are applied to our phase data as shown for example in Fig. 5 for the long period variable TX Cam. The SiO hot spots seem to be confined within an elongated structure which is perhaps rotating since the more and less positive velocity features are spatially separated. The overall SiO extent is ~ 30 mas in our 86 GHz spot map. This is close to the 25 mas measured by Diamond (1993) for the ring- shaped emitting region mapped at 43 GHz toward the same star with 4 antennas of the VLBA (the new US transcontinental VLBI array). In the latter work the SiO lies along a ring and it is not yet clear why the 43 and 86 GHz SiO morphologies seem to be discrepant. R Cas, an M6- M8 Mira variable, is another example showing spatial structure in the IRAM data. The SiO region extends over ~ 60 mas at 86 GHz while the diameter of the central star is around 15-20 mas. The 43 GHz VLBI maps of Lane (1982) and McIntosh *et al.* (1989) and the 86 GHz IRAM spot map for R Cas all give similar spatial extent for the SiO region although the velocity- positions of the individual hot spots differ. This is perhaps not surprising because the observation periods differ by more than 10 years, and because of the complex excitation mechanisms of the SiO molecule. Several stars observed with the IRAM interferometer do not exhibit any clear spatial structure at 86 GHz. μ Cep for example, shows no phase structure within 100 mas across the line spectrum. This is consistent with our VLBI fringe rate map made at 43 GHz (see above). Our absolute

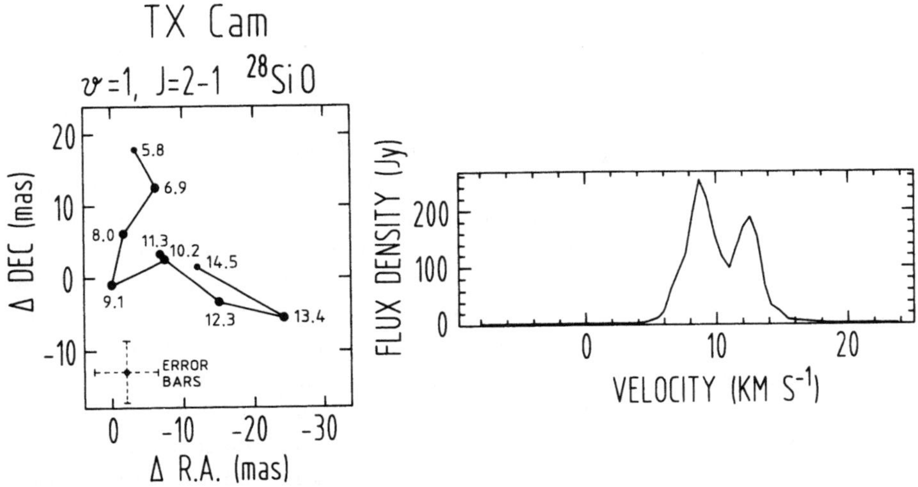

Fig. 5. SiO hot spot map for yhe 86 GHz ^{28}SiO maser emission from TXCam observed in January 1992 (top of figure). This relative position map was obtained with the most extended configuration of the IRAM interferometer on Plateau de Bure. The numbers indicate the radial velocity of the SiO emission whose spectrum, obtained with one baseline of the interferometer, is shown in the lower part of the figure.

position measurements of SiO emission from stars will be discussed elsewhere (see Baudry 1992 for a preliminary account). We simply mention here that the position accuracy obtained with the IRAM interferometer is around 0.1-0.2 arc sec; therefore, the SiO emission is closely related to the associated central star and to the stellar activity.

5 SiO Excitation and Pumping Models

The SiO map made for R Cas by McIntosh *et al.* (1989) is a particularly interesting case because it shows well identified SiO regions with distinct velocity intervals and polarization angles. These regions may be interpreted as giant turbulent cells with properties related to the magnetic field activity (Elitzur 1992). However VLBI synthesis polarization maps and connected interferometer spot maps of the SiO emission should be obtained at different epochs of the stellar light cycle to better understand the complex region extending from the photosphere to the CIS dust layers. In TX Cam at least, a model of SiO emission associated with large turbulent emission cells or with large cool dense blobs ejected from the star (Alcock and Ross 1985) does not seem viable since the emission is confined within a ring (Diamond 1993). In any case, the SiO maser emission, as suggested in Sects. 3 and 4, arises in the CIS inner layers which are coupled through shocks and complex mechanisms to the central star pulsations. SiO pumping models have been proposed since the early discovery of the SiO maser phenomenon. They solve simultaneously the statistical equilibrium equations (used to describe the SiO populations on the various energy levels) and the radiative transfer equations. A realistic solution of these coupled equations should give a range of plausible physical conditions for which some transitions are anomalously excited in the CIS environment provided that realistic collisional rates of the SiO with the dominant species H_2 are known. (Deriving accurate collision rates is always a difficult task in the astrophysical environment; see Bienik and Green 1983, for the SiO collision rates.) We make below a brief and incomplete summary of some important pumping models. Kwan and Scoville (1974) were the first to recognize the importance of the infrared radiation. When the SiO infrared transitions become optically thick then, during the vibrational deexcitation process, the photon escape probability is higher for lower rotational levels. This process is general since it tends to invert the SiO rotational levels nearly independently of the specific ways the vibrational states are excited. The collisionally dominant excitation model of Elitzur (1980) requires a high volumic density to pump the SiO molecules into high vibrational states. This in turn, implies that the SiO resides close to the stellar photosphere perhaps in large turbulent cells above the photosphere. On the other hand, if the SiO is primarily in the inner layers of the CIS envelope, radiative pumping from the star dominates over vibrational excitation (e.g. Bujarrabal and Rieu 1981). Interferometer observations tend to show indeed, that the SiO masers are a few stellar radii distant from the star but not too far in the envelope. This is also supported by the model calculations of Alcolea et al (1989) to explain the high v SiO masers. We also mentioned earlier that if the SiO masers were located too far in the wind of the CIS envelope their power would not be explained. It is interesting to note that all models have two characteristics in common: (1) They require high volumic densities with $n(H_2) \sim 10^7\text{-}10^{10}$ cm^{-3} (beyond 10^{10} cm^{-3} thermalization of the rotational states prevails); (2) Inversion of the various vibrational transitions observed in a same source cannot be obtained for the same SiO column densities. The latter property implies that the various v state masers should peak in different spatial regions a fact which could perhaps be directly observed toward the strongest SiO stars and Orion with the most sensitive interferometers. Finally, the recent observations of several vibrational and rotational states of SiO in the supergiant VY CMa show high intensity contrasts within a given vibrational ladder which cannot be explained by any of the radiation or collisional pumping models (Cernicharo *et al.* 1993b). The results of Cernicharo *et al.* suggest that line overlaps may play a role in the excitation of the SiO molecule. This mechanism was first invoked for the SiO molecule to explain the v=0, J=2-1 ^{29}SiO maser in Orion (Olofsson *et al.* 1981), and it was also invoked by Cernicharo *et al.* (1991) to explain the pumping of the v=1 ^{29}SiO masers observed in several stars. In order to appreciate the importance of line overlaps among the three main SiO isotopic species we have used the most recent spectroscopic constants to derive and compare the frequencies of the ro-vibrational transitions of ^{28}SiO, ^{29}SiO and ^{30}SiO. Limiting ourselves to v=4 and J=10 we found 11 and 16 overlaps within 5 $km\,s^{-1}$ for v-v'=2 and 1 respectively. The number of overlaps increase quickly if the overlapping domain is extended. The number of coincidences within 20 $km\,s^{-1}$ for example, increases to 40 and

65 for v-v'=2 and 1. Local turbulent effects leading to velocity dispersions of order 5 $\mathrm{km\,s^{-1}}$ are plausible in the stellar environment and would thus greatly complicate the radiative transfer equation solutions.

6 Conclusion

Because of their high brightness and compact sizes revealed by high resolution studies, SiO masers provide a powerful tool to investigate the complex physics and dynamics of gas layers extending from the pulsating photosphere of late-type stars to their distant circumstellar envelopes. However, much observational and theoretical effort is still required to understand better the mechanisms leading to stellar mass-loss and the SiO maser phenomenon. The advent of new sensitive VLBI and connected arrays, as well as regular monitoring programs of the spatial structure and radio flux of SiO stars will undoubtedly help to more closely approach the complex SiO maser reality.

Acknowledgements: This work was supported in part by the CNRS and the French-German program PROCOPE.

References

Alcock C., Ross R.R. 1985, ApJ 290, 433

Alcolea J., Bujarrabal V., Gallego D. 1989, A&A 211, 187

Allen D.A., Hall P.J., Norris R.P., Troup E.R., Wark R.M., Wright A.E. 1989, MNRAS 236, 363

Barvainis R.E., McIntosh G., Predmore R. 1987, Nature 329, 613

Baudry A. 1992, Sub-Arcsecond Radioastron. Meeting, Cambridge Univ. Press, eds. R.J. Davis, R.S. Booth

Benson P.J., Little-Marenin I.R., Woods T.C., Attridge J.M., Blais K.A., Rudolph D.B., Rubiera M.E., Keefe H.L. 1990, ApJS 74, 911

Bienick R.J. 1983, ApJ 265, L29 and ApJ 270, L101 (Erratum)

Bujarrabal V., Rieu N.Q. 1981, A&A 102, 65

Cernicharo J., Brunswig W., Paubert G., Liechti S. 1993-a, submitted to ApJ

Cernicharo J., Bujarrabal V., Lucas R. 1991, A&A 249, L27

Cernicharo J., Bujarrabal V., Santaren J.L. 1993-b, ApJ 407, L33

Colomer F., Graham D.A., Krichbaum T.P., Ronnang B.O., P.de Vicente, Barcia A., Baudry A., Booth R.S., Gomez-Gonzalez J., Alcolea J., Daigne G. 1992, A&A 254, L17

Davis J.H., Blair G.N., Van Hill H., Thaddeus P. 1974, ApJ 190, L117

Diamond P.J. 1993, private communication

Elitzur M. 1980,ApJ 240, 553

Elitzur M.1992, Ann. Rev. A&A 30, 75

Engels D. 1979, Astron. Astrophys. Suppl.36, 337

Fuente A. 1992, PhD thesis Univ. Madrid

Humphreys R.M. 1983, ApJ 269, 335

Hutter D.J., Johnston K.J., Mozurkewich D., Simon R.S., Colavita M.M., Pan X.P., Shao M., Hines B.E., Staelin D.H., Hershey J. L., Hughes J.A., Kaplan G.H.1989, ApJ 340, 1103

Iben I. 1981, Physical Processes in Red Giants,Reidel/Dordrecht,eds.I.Iben, A. Renzini
Jewell P.R., Snyder L.E., Walmsley C.M., Wilson T.L., Gensheimer P.D. 1991, A&A 242, 211
Kwan J., Scoville N.1974, ApJ 194, L97
Lane A.P. 1982, PhD thesis Univ. Mass.
Langer S.H., Watson W.D., 1984 ApJ 284, 751
Le Bertre T., Nyman L.-A. 1990, A&A 233, 477
Lucas R., Bujarrabal V., Guilloteau S., Bachiller R., Baudry A., Cernicharo J., Delannoy J., Forveille T., Guffilin M., Radford S.J. 1992, A&A 262,491
Martinez A., Bujarrabal V., Alcolea J. 1988, Astron. Astrophys. Suppl.74, 273
McIntosh G., Predmore C.R., Moran J.M., Greenhill L.J., Rogers A.E.E., Barvainis R. 1989, ApJ 337, 934
Morita K.-I., Hasegawa T., Ukita N., Okumura S.K., Ishiguro M. 1992, PASJ 44, 373
Nedoluha G.E., Watson W.D. 1993, submitted to ApJ
Olofsson H., Hjalmarson A., Rydbeck O.E.H. 1981, A&A 100, L30
Patel N.A., Joseph A., Ganesan R. 1992, J.Astro.Astr. 13, 241
Plambeck R.L., Wright M.C.H, Carlstrom J.E. 1990, ApJ 348, L65
Reid M.J., Moran J.M. 1981, Ann. Rev. A&A 19, 231
Snyder L.E., Buhl D.1974, ApJ 189, L31
Thaddeus P., Mather J., Davis J.H., Blair G.N. 1974, ApJ 192, L33
van der Veen W.E.C.J., Habing H.J. 1988, A&A 194, 125

Intergalactic Cloud and Protogalaxy? An Unusual Molecular Complex

C. Henkel

Max-Planck-Institut für Radioastronomie, Auf dem Hügel 69, D-53121 Bonn, Germany

Abstract: A molecular complex with unique properties was detected $\sim 13'$ east of M 81. Following the analysis presented by Brouillet et al. (1992) and Henkel et al. (1993), its distance and its kinematical and physical parameters are derived in Sects. 2–4. In Sects. 5 and 6, the nature of the complex and general implications for dwarf galaxy formation are discussed.

1 Introduction

With the detection of CO in the 'primeval' galaxy IRAS F 10214+4724 at z=2.286 (Brown & Vanden Bout 1991; Solomon et al. 1992a,b), searches for the elusive family of 'protogalaxies' have gained substantial momentum. By observing highly red-shifted sources, we can look back into the past and can thus hope to determine the rates of galaxy formation, the luminosity function, the number of mergers, and the probability encountering a certain Hubble type as a function of time. It is well known that stars were not only formed in the distant past but that they are born, at lower rates, in gas clouds up to the present time. Are galaxies also still forming today?

There is ample evidence for the presence of cool atomic intergalactic gas clouds. Large amounts of hot ($\sim 10^7$ K) intergalactic gas are observed in galaxy clusters. Theoretical models (e.g. Fabbiano 1989; Sarazin 1990) and observational data (e.g. Hu 1992; Johnstone et al. 1992) indicate the presence of a dense cool gas component associated with this hot tenuous medium. Molecular gas, however, was not directly detected in these sources.

Several decades ago, Zwicky (1942, 1956) suggested that dwarf galaxies can form out of the debris of highly perturbed interacting galaxies. Groups of interacting galaxies often contain prominent tidal HI arms which are thought to be torn out of the disks of gas-rich spirals (e.g. Toomre & Toomre 1972). There may be a ridge of HI emission located between NGC 7318 and NGC 7319, two members of Stephan's Quintet. This emission is associated with an elongated source of non-thermal radio continuum radiation (van der Hulst & Rots 1981; Shostak et al. 1984). An isolated HI complex with $M \sim 10^9\,M_\odot$ was detected in the M 96 group of galaxies (e.g. Schneider et al. 1986). A similar object with a few

$10^7\,M_\odot$ may be located between the giant Virgo elliptical NGC 4472 and the gas deficient dwarf irregular UGC 7636 (Henning et al. 1993). Will these HI clouds later form giant molecular clouds and stars as suggested by Gerola et al. (1983)? At the end of such an evolutionary process, we can expect to observe molecular clouds with high infrared luminosity, the signature of massive star formation and, later, dwarf galaxies with extremely blue colors. Dust enshrouded objects of this type may exist in the 'Antennae', the 'Superantennae', and NGC 7252 (e.g. Schweitzer et al. 1978; Colina et al. 1991; Mirabel et al. 1991, 1992), but the existence of an underlying older stellar population with lower luminosity cannot be ruled out. Thus, there is a striking lack of clear evidence for newly forming galaxies in the local universe. Perhaps the most convincing case is the 'Garland' detected south of NGC 3077 (Karachentsev et al. 1985). This least luminous galaxy known outside the Local Group ($M_B \sim -12^{M}.5$) contains blue stars, appears to be young, and is located in the nearest group of interacting galaxies, associated with M 81. Searching in the tidal arms of this group of galaxies for molecular gas (Becker et al. 1989; Brouillet et al. 1992), one isolated molecular complex was found. The implications of this discovery will be described in the following.

2 Observed Properties

The molecular complex detected by Brouillet et al. (1992) is centered at $\alpha_{1950} = 9^h53^m57.5^s$, $\delta_{1950} = +69°17'42''$ ($\pm 7''$), about 800'' and 160'' east of M 81 and HoIX, respectively (see Fig. 1). Its deconvolved size is $20'' \pm 4''$ (see Fig. 2). CO velocity and FWHP linewidth are $v_{\rm HEL} \sim -35\,{\rm km\,s^{-1}}$ and $\Delta V_{1/2} \sim 20\,{\rm km\,s^{-1}}$. With a bandwidth of several $100\,{\rm km\,s^{-1}}$, the lines are too narrow and too strong ($T_{\rm mb} \sim 150$ mK) to be significantly affected by baseline uncertainties. Furthermore, observations were carried out during two observing periods separated by half a year (Oct. 1991, Apr. 1992), each time on two consecutive days. All results are consistent.

Mapping the source in the CO J=1−0 and 2−1 transitions, no significant changes in $v_{\rm HEL}$ and $\Delta V_{1/2}$ are observed with the 22'' and 13'' beams of the IRAM 30-m telescope over the size of the complex. A CO J=2−1/1−0 line intensity ratio of ~0.5 indicates either optically thick emission of cool gas ($T_k \leq 15$ K) or optically thin gas with an extremely low degree of excitation. The latter is not typical for clouds of the Galactic disk.

3 The Distance to the Complex

In order to unravel the nature of the molecular source, we have to know its distance. There are five obvious possibilities:

- The complex is part of the Galactic disk.
- The complex is part of the Galactic halo.

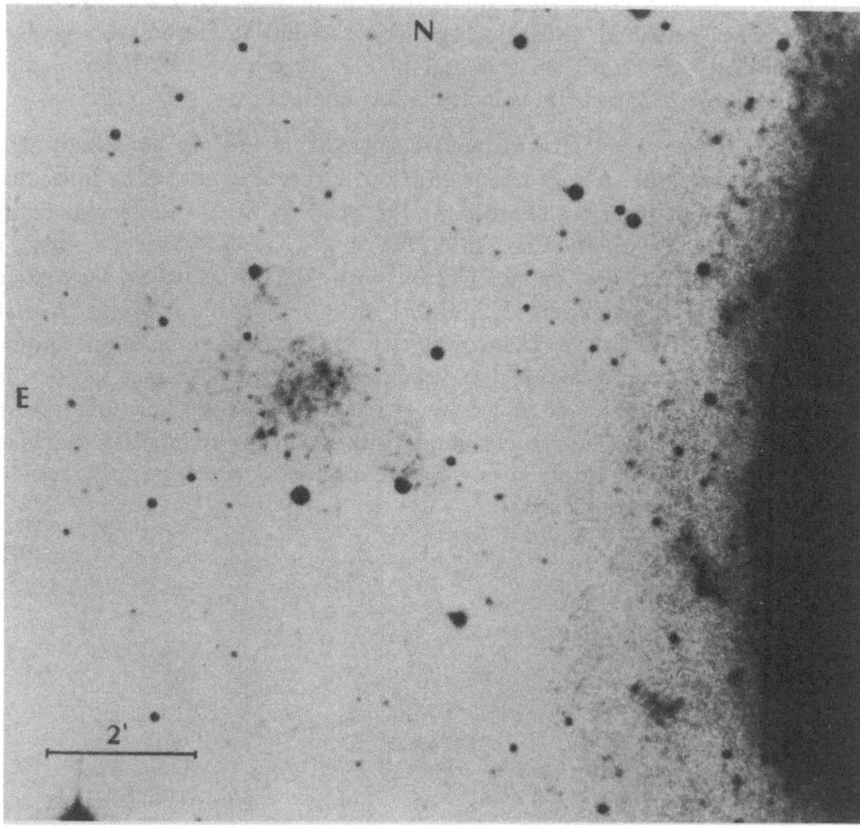

Fig. 1. A photograph (taken from Bertola & Maffei 1974) showing the eastern edge of M 81 and the dwarf irregular Ho IX. An arrow points to the position of the molecular complex. The rectangular box marks the approximate boundary of the area displayed in Fig. 2c.

- The complex is a rather nearby extragalactic source which is not associated with the M 81 group of galaxies.
- The complex is associated with the M 81 group of galaxies.
- The complex is a background object at $d > 10$ Mpc.

With a Galactic molecular disk thickness of ~600 pc at the solar circle (e.g. Combes 1991), the distance and size of the complex, located at $b^{II}=41°$ is constrained to $d \leq 460$ pc and $l \leq 0.05$ pc. The corresponding FWHP virial linewidth (e.g. Larson 1981) is then $\Delta V_{1/2} \leq 0.2\,\mathrm{km\,s^{-1}}$, at least a factor of 100 below the observed value. Furthermore, no detected radio continuum source, IRAS point source, reflection nebula, or low mass stellar object is associated with the complex. Hence, a location within the Galactic disk is not likely.

If located in the Galactic halo ($d \lesssim 100$ kpc), there would again be a discrepancy between cloud size and velocity dispersion (cf. Haud 1990). There are no known extended Galactic high velocity and intermediate velocity HI components at $-35\,\mathrm{km\,s^{-1}}$ in this region (e.g. Verschuur 1973). For a distant background object, the radial velocity is too small.

The cloud size - velocity dispersion correlation is only consistent with a distance of a few Mpc. Also, there is an HI arm, reaching out of the inner disk of M 81 with a velocity of $-35\,\mathrm{km\,s^{-1}}$ at the position of the molecular complex (for details, see Brouillet et al. 1992). With d_{M82} ~3.25 Mpc (m−M=$27^{\mathrm{m}}.6$; e.g. Jacoby et al. 1989), an association with M 81 is thus by far the most conservative, the most expected, and the most reasonable scenario. A distant background source with a radial velocity close to zero, a small quiescent Galactic disk cloud with a velocity dispersion at least 100 times the predicted value or a first detection of molecular gas in the Galactic halo may be more dramatic. Nevertheless, an association with the M 81 group of galaxies has quite a number of interesting implications. Before discussing these, we first analize the observational data in more detail.

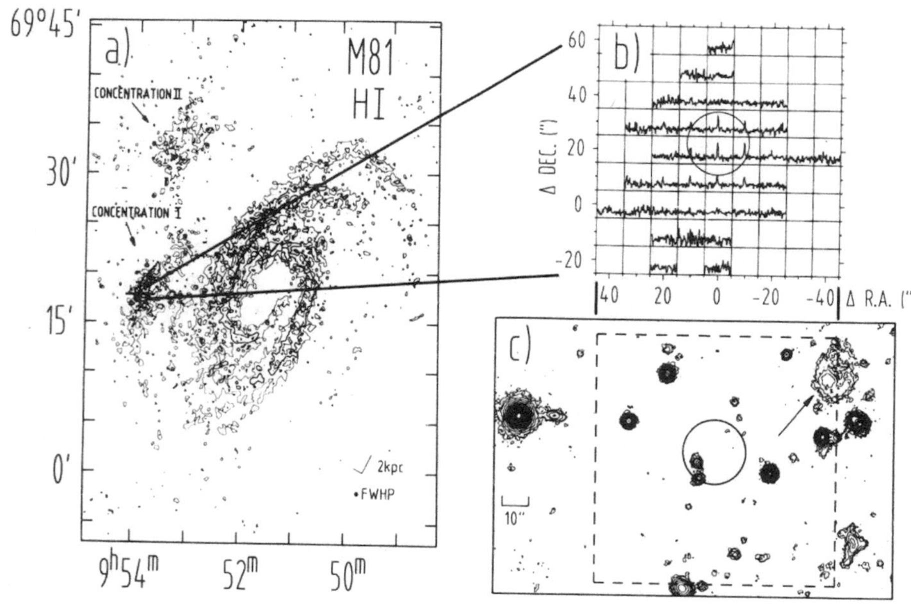

Fig. 2. a) Contour map of the HI column density distribution toward M 81 (Rots and Shane 1975); b) CO J=2−1 emission from the molecular complex (Brouillet et al. 1992); c) An R-band contour plot (Henkel et al. 1993). The entire figure was taken from Mauersberger & Henkel (1993).

4 Physical Parameters of the Complex

With d=3.25 Mpc and an H_2 column density to integrated CO J=1−0 intensity ratio of $N(H_2)/I_{CO} = 2.3\,10^{20}\,cm^{-2}/(K\,km\,s^{-1})$ (Strong et al. 1988), the cloud size is 300 pc and the mass becomes $10^6\,M_\odot$ (see Table 1). The virial mass of $10^7\,M_\odot$ is an order of magnitude larger, leading to an average density of a few particles per cm^3 and to a surface density of $150\,M_\odot\,pc^{-2}$. The mass discrepancy may be explained by a $N(H_2)/I_{CO}$ ratio which is higher than the Galactic disk value. Such excessive values are also obtained in the outer Galaxy and the Magellanic clouds (e.g. Digel et al. 1990; Rubio et al. 1993).

The IRAS 60 and 100μm fluxes of our complex can be constrained in the following way (see Brouillet et al. 1992): For the fluxes integrated over the IRAS field of view, $S_{60\mu m}$= 0.1 Jy and $S_{100\mu m}$= 0.4 mJy. Because these fluxes are integrated over a much larger region than that covered by our source, they are upper limits. Assuming that the molecular complex is not located in a local minimum of the far infrared (FIR) distribution, we can also restrict the FIR emission to the angle subtended by the molecular complex and obtain $S_{60\mu m}\geq 0.9$ mJy and $S_{100\mu m}\geq 2.7$ mJy. From these limits we derive with $M=3\,10^6\,M_\odot$ the following physical parameters: $3\,10^4\,L_\odot \leq L_{IR} \leq 4\,10^6\,L_\odot$, $0.01\,L_\odot/M_\odot \leq L_{IR}/M(H_2) \leq 1\,L_{IR}/L_\odot$, and star formation rates (see Eq. 3 of Thronson & Telesco 1986) of $2\,10^{-5}\,M_\odot\,yr^{-1} \leq$ SFR$\leq 2.5\,10^{-3}\,M_\odot\,yr^{-1}$. With $L_{IR}/M(H_2)\sim 8$ in the solar neighborhood, it is clear that the cloud is in an extremely quiescent state. With a minimum lifetime of 10^6 to 10^7 years (e.g. Larson 1994), however, there is still sufficient time for the formation of several late O or early B stars which should be detectable at optical wavelengths. The essential question is thus: Is there a low surface brightness galaxy associated with the complex?

CCD images covering the 20″ region show no extended optical counterpart down to a surface brightness of 27^m/arcsec^2 in B for $B-R$=$1\overset{m}{.}5$ (see Fig. 2 and Henkel et al. 1993). Similarly, deeper optical images and a K-band image (U. Hopp, S. Beckwith, priv. comm.) do not show a low surface brightness galaxy. Since the analysis of the latter data is not yet complete, we instead refer to the results presented by Henkel et al. and conclude that stellar sources as luminous as $-5\overset{M}{.}0$ ($22\overset{m}{.}6$) do not exist in the relatively unobscured parts of the object. With evolutionary tracks of Iben & Renzini (1983) and Maeder (1987) this implies an absence of main sequence objects with $M \geq 10\,M_\odot$ as well as of lower mass stars reaching the tip of the asymptotic giant branch. Furthermore, there is no background of unresolved less luminous ($> -5\overset{M}{.}0$) stars even though this is a characteristic feature of the nearby dwarf irregular HoIX (see Fig. 1).

With the known size (~300 pc) and mass ($10^6 - 10^7\,M_\odot$) of the complex, we find an average extinction of $A_v \sim 1^m - 10^m$ across its core. For the Galactic disk, $A_v \sim 10^m$ in many clouds (e.g. Larson 1981). About 15% of the Galactic O stars are embedded in molecular gas and are thus not accessible at optical wavelengths (Churchwell 1990). Since the average extinction of the complex is not larger than in local molecular clouds, we can cautiously assume that at least one out of ten young massive stars should be optically visible (η_v=0.1). Applying a Salpeter initial mass function (IMF)

$$\Phi(m) \propto m^{-2.35}$$

with boundary conditions

$$\int_{M\mathrm{l}}^{M_\mathrm{u}} \Phi(m)\, m\, \mathrm{d}m = 1, \quad M_\mathrm{l} = 0.1\, M_\odot, M_\mathrm{u} = 100\, M_\odot$$

(M_l: lower mass cutoff, M_u: upper mass cutoff), a mass fraction

$$\eta_\mathrm{m} = \frac{\int_{10M_\odot}^{M\mathrm{u}} \Phi(m)\, m\, \mathrm{d}m}{\int_{M_\mathrm{l}}^{M_\mathrm{u}} \Phi(m)\, m\, \mathrm{d}m} \sim 0.1$$

of the newly formed stellar material goes into high mass stars ($M > 10\,M_\odot$) of average mass

$$M_\mathrm{m} = \frac{\int_{10M_\odot}^{M\mathrm{u}} \Phi(m)\, m\, \mathrm{d}m}{\int_{10M_\odot}^{M_\mathrm{u}} \Phi(m)\, \mathrm{d}m} \sim 15\, M_\odot.$$

These have a hydrogen and helium burning lifetime of τ_m $\sim 1.5\,10^7$ yrs (e.g Maeder 1987). The star forming rate (SFR) is then

$$\mathrm{SFR} < \eta_\mathrm{v}^{-1}\, \eta_\mathrm{m}^{-1}\, M_\mathrm{m}\, \tau_\mathrm{m}^{-1} \sim 10^{-4}\, M_\odot\, yr^{-1}.$$

Combining this limit with that obtained from the FIR data, we find a fairly narrow range of values, $2\,10^{-5}\,M_\odot\,\mathrm{yr}^{-1} \leq \mathrm{SFR} \leq 10^{-4}\,M_\odot\,\mathrm{yr}^{-1}$. In view of the rather cautious estimates of upper stellar luminosity limit and percentage of visible high mass stars and accounting for the possible influence of cirrus clouds on the measured FIR fluxes, the actual value may be closer to the lower than to the upper limit. The total resulting gas consumption times are then $10^{10} - 5\,10^{11}$ yrs which should be compared with $\sim 5\,10^8$ yrs for the Galactic disk and $\sim 3\,10^7$ yrs for a typical starburst.

The estimates of the SFR and gas consumption times were so far based on a Salpeter IMF. Using instead a Miller-Scalo IMF does not significantly alter the results. More drastic deviations from a standard IMF are, however, possible. In the Galaxy, there exist extended molecular clouds devoid of any recently formed massive stars. A good example is the Taurus Molecular Cloud ($d \sim 150$ pc) which is with $10^4\,M_\odot$ (e.g. Cernicharo 1991) much less massive than the complex near M 81. Only one giant molecular cloud in the Galactic disk, that observed by Maddalena & Thaddeus (1985), is known to have a mass almost as large as that of the complex and to be 'quiescent', i.e. not forming massive stars. It is likely that we have found a similarly quiescent source near M 81.

Table 1. Cloud parameters

α_{1950}	$09^h\,53^m\,57.5^s$
δ_{1950}	$69°\,17'\,42''$
v_{HEL}	$-35\,km\,s^{-1}$
$\Delta V_{1/2}$	$20\,km\,s^{-1}$
Distance	3.25 Mpc
Cloud size	300 pc
Surface density	$15 - 150\,M_\odot\,pc^{-2}$
L_{IR}	$3\,10^4 - 4\,10^6\,L_\odot$
$L_{IR}/M(H_2)$	$0.01 - 1\,L_\odot/M_\odot$
M	$10^6 - 10^7\,L_\odot$
Salpeter SFR	$2\,10^{-5} - 10^{-4}\,M_\odot\,yr^{-1}$
Gas consumption time	$10^{10} - 5\,10^{11}$ yrs

5 The Nature of the Complex

If the molecular complex east of M 81 is associated with gas torn out of M 81, it cannot be extremely metal poor. Then the column density thresholds for CO self-shielding and self-gravitation are similar (see Elmegreen 1985), implying that, although no high mass stars are apparent, low mass stars are being formed.

It may be debatable whether the molecular complex is part of the outermost regions of M 81 or an intergalactic object. If located in the plane of the M 81 disk, its radial distance is ~20 kpc. While the Galactic HI disk is more extended (e.g. de Geus et al. 1993), the complex is located well outside the HI disk of M 81 (e.g. Fig. 1) and it is also situated beyond HoIX (Fig. 1), commonly classified as a dwarf irregular galaxy. Note that the displacement from M 81 is much more convincing than the separation of the 'Garland galaxy' from NGC 3077 (cf. Karachentsev et al. 1985). The complex is also not associated with HoIX. Kinematically, it belongs instead to an HI arm with a velocity differing by $\sim 100\,km\,s^{-1}$ from that of the HI gas along the line-of-sight to HoIX. Dynamically, there may have been a close encounter of M 82 and NGC 3077 at the eastern side of M 81, $\sim 10^8$ yrs ago (cf. Rieke et al. 1980; Brouillet et al. 1991). The time elapsed since the encounter is larger than the expected lifetime of a Galactic molecular cloud (e.g. Larson 1994). The complex may thus have formed after the encounter in a region only weakly linked to M 81. Only being connected by an HI bridge with this galaxy, there are thus good reasons to call the molecular complex ***intergalactic***. Likely forming low mass stars, it may be the first detected ***protogalaxy***. The terms 'intergalactic' and 'protogalaxy' (as 'interstellar' and 'protostar') refer to the location and evolutionary state, ***not*** to the metallicity as it is sometimes wrongly inferred. Like 'protostars', 'protogalaxies' are ***not*** confined to objects of high redshift.

6 Outlook

The formation of molecular clouds in tidal HI arms requires high column densities to allow self-shielding against UV-radiation as well as self-gravitation to become effective. According to Toomre & Toomre (1972), direct passages of dwarfs with small inclinations relative to the disk of a much larger gas-rich spiral can provide massive tidal arms. This may well describe the case of M 81, where tidal arm HI column densities exceed $10^{21}\,\mathrm{cm}^{-2}$, the molecular column density threshold in the local interstellar medium. It remains to be seen whether tidal counterarms, which may be most massive in interactions between two similarly massive gas-rich spirals (e.g. Toomre & Toomre 1972) also allow the formation of molecular clouds.

Modelling dwarf galaxy formation in interacting disk galaxies, accounting for central bulges, exponential disks, spheroidal dark matter halos (including $\sim 10^5$ particles), Barnes & Hernquist (1992) found that bound structures develop even before the disk galaxies merge. These structures become more conspicuous while the merger evolves and the tails spread out. Cloud masses and surface densities are $\leq 4\,10^8\,\mathrm{M}_\odot$ and $\geq 10^2\,\mathrm{M}_\odot\,\mathrm{pc}^{-2}$, which is consistent with the parameters of the molecular complex (see Table 1). The newly formed dwarfs can capture large amounts of moderately enriched gas but retain little dark matter from their progenitor galaxies. Lower mass objects near the colliding partners are destroyed due to tidal forces. Thus, the least massive objects should be confined to the outer tidal tails.

While our data appear to fit well into this scenario, one such object is not yet very impressive. Assuming a star formation efficiency of 10% (e.g. Evans 1985), the stellar mass of the dwarf formed out of the complex will only be $10^5 - 10^6\,\mathrm{M}_\odot$. With $M = 3\,10^5\,\mathrm{M}_\odot$, a Salpeter IMF, $L_\mathrm{B}/\mathrm{L}_{\mathrm{B}\odot} = (M/\mathrm{M}_\odot)^{3.5}$, $\mathrm{M}_{\mathrm{B}\odot}=5\overset{\mathrm{M}}{.}47$, and $M=10\,\mathrm{M}_\odot$ as the most massive main sequence stars, we obtain $L_\mathrm{B}\sim 3\,10^6\,\mathrm{L}_{\mathrm{B}\odot}$ and $M_\mathrm{B} = -10\overset{\mathrm{M}}{.}7$ (cf. Henkel et al. 1993). While this result depends sensitively on the luminosity of the most massive stars, the luminosity is similar to that of the 'Garland' near NGC 3077 (see Sect. 1) which may also be young.

With one molecular complex and one blue dwarf within a timescale of perhaps 10^8 yrs (the time since the possible encounter of M 82 and NGC 3077), the number of such objects will only be significant (a few 100 over a Hubble time) if the present state is typical for the entire lifetime of M 81. This, however, can be excluded since then all the gas would have been lost a long time ago. The recent formation of dwarfs out of tidal debris, significantly more massive than the molecular complex, is also not likely. Small dwarfs tend to have low metallicities (cf. van den Bergh 1993). A more attractive alternative is the possibility that the molecular complex and the 'Garland' are only two of the most conspicuous members of a numerous family of similar objects (see also Barnes & Hernquist 1992). To quantitatively investigate this scenario, detailed CO and optical surveys are highly desirable. Note that to date the area covered by the tidal arms has not been fully mapped in HI. Accounting for the expected small angular sizes and intensities of the hypothesized molecular clouds, a large mm-wave telescope with a beam size $\ll 1'$ has to be employed. In view of the required sensitivity and extent of the region (> 1°), surveys of the M 81 group or similar groups at somewhat larger distances require too much time to be feasible. Nevertheless, the single detected complex near M 81 is one of the most significant observational hints for ongoing galaxy formation in the local universe.

References

Barnes, J.E., Hernquist, L., 1992, Nat. 360, 715
Becker, R., Schilke, P., Henkel, C. 1989, A&A 211, L19
Bertola, F., Maffei, P., 1974, A&A 32, 117
Brouillet, N., Baudry, A., Combes, F., Kaufman, M., Bash, F. 1991, A&A 242, 35
Brouillet, N., Henkel, C., Baudry, A. 1992, A&A 262, L5
Brown, R.L., Vanden Bout, P.A. 1991, AJ 102, 1956
Cernicharo, J. 1991, The Physics of Star Formation and Early Stellar Evolution, eds. J. Lada, N.D. Kylafis, Kluwer Academic Publishers, Dordrecht, p287
Churchwell, E. 1990, A&AR 2, 79
Colina, F., Lipari, S., Macchetto, F. 1991, ApJ 379, 113
Combes, F. 1991, ARA&A 29, 195
De Geus, E.J., Vogel, S.N., Digel, S.W., Gruendl, R.A. 1993, ApJ 413, L97
Digel, S., Bally, J., Thaddeus, P. 1990, ApJ 357, L29
Elmegreen,B.G. 1985, Protostars & Planets II, eds. D.C. Black, M.S. Matthews, University of Arizona Press, Tucson, p33
Evans, N.J. 1985, Protostars & Planets II, eds. D.C. Black, M.S. Matthews, University of Arizona Press, Tucson, p175
Fabbiano, G. 1989, ARA&A 27, 87
Geogiev, Ts.B., Tikhonov, N.A., Karachentsev, I.D., Bilkina, B.I. 1991, A&AS 89, 529
Gerola, H., Carnevali, P., Salpeter, E.E. 1983, ApJ 268, L75
Haud, U. 1990, A&A 230, 145
Henkel, C., Stickel, M., Salzer, J.J., Hopp, U., Brouillet, N., Baudry, A. 1993, A&A 273, L23
Henning, P.A., Sancisi, R., McNamara, B.R. 1992, A&A 268, 536
Hu, E.M. 1992, ApJ 391, 608
Iben, I., Renzini, A. 1983, ARA&A 21, 271
Jacoby, G.H., Ciardullo, R., Ford, H.C., Booth, J. 1989, ApJ 344, 704
Johnstone, R.M., Fabian, A.C., Edge, A.C., Thomas, P.A. 1992, MNRAS 255, 431
Karachentsev, I.D., Karachentseva, V.E., Börngen, F. 1985, MNRAS 217, 731
Larson, R.B. 1981, MNRAS 194, 809
Larson, R.B. 1994, this volume
Maddalena, R.J., Thaddeus, P. 1985, ApJ 294, 231
Maeder, A. 1987, A&A 173, 247
Mauersberger, R., Henkel, C. 1993, Rev. Mod. Astron. 6, 69
Mirabel, I.F., Lutz, D., Maza, J. 1991, A&A 243, 367
Mirabel, I.F., Dottori, H., Lutz, D. 1992, A&A 256, L19
Rieke, G.H., Lebofski, M.J., Thompson, R.I., Low, F.J., Tokumaga, A.T. 1980, ApJ 238, 24
Rots, A.H., Shane, W.W. 1975, A&A 45, 25
Rubio, M., Lequeux, J., Boulanger, F. 1993, A&A 271, 9
Sarazin, C.L. 1990, in 'The Interstellar Medium in Galaxies', Second Wyoming Conference, eds. H.A. Thronson, J.M. Shull, Kluwer, Dordrecht, p201

Schneider, S.E., Salpeter, E.E., Terzian, Y. 1986, AJ 91, 13
Schweitzer, F. 1978, IAU Symp. 77, Structure and Properties of Nearby Galaxies, eds. E.M. Berkhuijsen, R. Wielebinski, Reidel, p279
Shostak, G.S., Sullivan, W.T., Allen, R.J. 1984, A&A 139, 15
Solomon, P., Downes, D., Radford, S.J.E. 1992a, ApJ 398, L29
Solomon, P., Radford, J.S.E., Downes, D. 1992b, Nat 356, 318
Strong, A.W. et al. 1988, A&A 207, 1
Thronson, H.A., Telesco, C.M. 1986, ApJ 311, 98
Toomre, A., Toomre, J. 1972, ApJ 178, 623
van den Bergh, S. 1993, Nature 364, 490
van der Hulst, J.M., Rots, A.H. 1981, AJ 86, 1775
Verschuur, G.L. 1973, A&A 22, 139
Zwicky, F. 1942, Phys. Rev. 61, 489
Zwicky, F. 1956, Erg. d. exakt. Naturwiss., Band XXIX, p344

Dense Gas in Galactic Nuclei

R. Mauersberger and C. Henkel

Max-Planck-Institut für Radioastronomie, Auf dem Hügel 69, D-53121 Bonn, Germany

Abstract: Recent results on the molecular gas component in the central regions of galaxies are reviewed. Following the introduction, Sect. 2 discusses the most important tracers of molecular gas mass and describes the detailed spatial distribution of the dense molecular gas component. Sect. 3 summarizes our knowledge on molecular chemistry in external galaxies.

1 Introduction

What causes stars to form in large amounts in the inner few 100 pc of certain galaxies? Is the Milky Way a dormant starburst galaxy? Any answer to these questions requires some knowledge about the fuel for star formation: the dense molecular clouds. Most of the information about their mass, motion, and makeup must be extracted from observations of trace molecules. Observations of the relatively strong $\lambda = 3\,\mathrm{mm}$ line emission of Galactic and extragalactic carbon monoxide (CO) with moderate angular resolution ($\gtrsim 1'$) make up the most extensive database. Recent years brought the detection of a number of useful molecules as well as a dramatic increase in angular resolution. In this review, we highlight some of the more important consequences of these developments.

2 Tracers and distribution of molecular mass

2.1 Mass tracers

At first glance it seems paradoxical that a transition that is optically thick in most molecular clouds and, therefore, should be observed only from their surfaces, can give information about their total masses. It has been the subject of a major controversy (see the impressive number of references in Maloney 1990) that the intensity of CO lines, $I_{\mathrm{CO}} = \int T_{\mathrm{MB}}(\mathrm{CO})\mathrm{d}v$ (where T_{MB} is the beam averaged brightness temperature of the CO line) is proportional to the column density of molecular hydrogen in Galactic and extragalactic molecular clouds: $N(\mathrm{H_2}) = XI_{\mathrm{CO}}$ or, equivalently, $M(\mathrm{H_2}) = \alpha A_{\mathrm{b}} I_{\mathrm{CO}}$, where A_{b} is the beam area.

Independent empirical methods yield values of $\alpha = 2 - 10\,M_\odot/(K\,km\,s^{-1}pc^2)$ ($X = 1 - 5\,10^{20}cm^{-2}/K\,km\,s^{-1}$).

The solution of the CO paradox involves a chain of arguments with three links. A heuristic approach is sketched in Fig. 1. The same arguments can also be used for any other transition known or suspected to be optically thick.

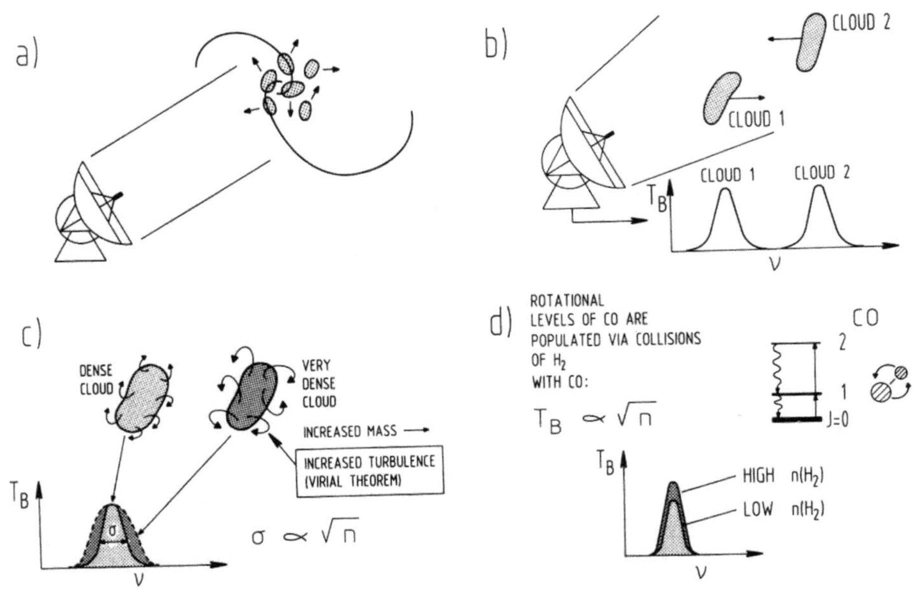

Fig. 1. A heuristic approch to the $N(H_2) \propto I_{CO}$ relation. a) Since extragalactic clouds are smaller than the beam, the measured intensity depends on the number of clouds in the beam (if they do not overlap), that is "CO counts clouds". b) The clouds may even overlap spatially if they have different Doppler velocities. c) In virial equilibrium, dense clouds have more turbulence and, hence, larger linewidths: $\sigma \propto n(H_2)^{0.5}$. d) For the bulk of the gas, CO will be subthermally excited. The source averaged brightness temperature is roughly proportional to the root of the H_2 density: $T_B(CO) \propto n(H_2)^{0.5}$.

This is probably the case for HCN, which is, after CO, one of the strongest emitters of thermally excited interstellar lines. While the critical density of CO is sufficiently low to trace *all* molecular gas components, HCN requires densities of more than $10^4\,cm^{-3}$ to be excited, and probes dense cores within molecular clouds that may be directly related to star formation. $C^{18}O$ mm-wave transitions are nearly always optically thin and the integrated intensity of its easily observable $J = 2 - 1$ line is independent of temperature and density over a wide range of parameters (see e.g. the discussion in Wild et al. 1992). The gas phase abundance of $C^{18}O$ relative to H_2 is approximately known in the Milky Way but has not been determined for external galaxies.

Although dust particles make up only about 1% of the mass of the interstellar medium, they play an important role in the energy balance of the ISM. At mm and sub-mm wavelengths, thermal emission from interstellar dust is almost certainly optically thin. Therefore the flux density of the dust emission is proportional to the dust mass and, hence, to the H_2 mass, if the dust-to-gas mass ratio is known.

Table 1. Mass conversion formulae from different tracers.[a]

Tracer	$(M/M_\odot)$	Ref.
^{12}CO, HCN	$3.9\,10^{-5}(\sqrt{n}/T_B)\theta''^2 D^2_{kpc}\int T_{MB}\,dv$	
	(for ^{12}CO: $\sqrt{n}/T_B = 2(\pm0.5)\,cm^{-3/2}K^{-1}$)	b
	(for HCN: $\sqrt{n}/T_B = 35(\pm17)\,cm^{-3/2}K^{-1}$)	c
$C^{18}O$	$1.4\,10^{-3}\theta''^2 D^2_{kpc}\int T_{MB}\,dv$	d
Dust	$410 S_{Jy}\lambda^4_{mm} D^2_{kpc}(Z/Z_\odot)^{-1} b^{-1} T_d^{-1}(\exp(x)-1)/x$	e
	$(x = 14.4/\lambda_{mm}T_d)$	

a) Explanation of symbols used: $M/M_\odot$: gas mass in solar masses; n: H_2 density in cm^{-3}; T_B: brightness temperature of the molecular emission (K); θ source size convolved with the beam (arcseconds): $\theta = \sqrt{\theta^2_{beam} + \theta^2_{source}}$; $\int T_{MB}\,dv$ molecular main-beam brightness temperature integrated over velocity ($K\,km\,s^{-1}$); λ_{mm} wavelength in mm; S: flux density integrated over the source in Jy; $Z/Z_\odot$: metallicity relative to solar value; T_d: dust temperature in K; b is a parameter, which is ~ 1.9 for moderately dense molecular gas.
b) Radford et al. (1991)
c) Mauersberger & Henkel (1993)
d) Wilson & Mauersberger (1990), Mauersberger et al. 1992
e) Mezger (1990)

Of all mass tracers mentioned above, ^{12}CO line emission is the most useful since it is not very sensitive to changes in metallicities and since it can easily be observed in hundreds of sources, while $C^{18}O$ line emission is too weak to be measured in more than a few strong sources. Note, that the dust emission, which can be detected in many galaxies, provides no velocity information.

2.2 The distribution of dense circumnuclear gas

Molecular gas has been detected toward more than 650 galaxies (Verter 1990), in most cases with a resolution of $\sim 1'$. High-resolution $(10 - 25'')$ data of CO and higher density tracers show that many spiral galaxies exhibit line intensity maxima toward the nuclear regions which are found to contain gaseous circumnuclear rings, bars, and spirals. With mm-wave interferometers, it is possible to map the centers of galaxies with angular resolutions of order 2″. At a distance of ∼ 3 Mpc (e.g. M 82 and NGC 253), this corresponds to a linear scale of 30 pc, so we would just resolve the largest Giant Molecular Clouds of our Galaxy. In Fig. 2, we show the gas distribution in the centers of four prominent galaxies, including our own, drawn on approximately the same linear scale. It is evident that the size of the molecular gas distribution in each of the four objects is similar, namely a few hundred parsec.

Fig. 2. (See next page.) The molecular gas distributions of four prominent galaxies are drawn on the same linear scale. a) Integrated CO emission from the the center of IC 342 (adopted distance: D=1.8 Mpc; McCall 1990) shows a bar and a compact spiral composed of individual molecular clouds. The data were collected with the Nobeyama interferometer by Ishizuki et al. (1990). For comparison, we show dense gas and free-free emission in the center of IC 342 mapped by Downes et al. (1992) with 2.7″ resolution using the IRAM interferometer (thin contours: velocity integrated HCN (J=1−0) emission; thick contours: 3.4 mm continuum emission). b) Integrated CO emission from the center of NGC 253 (D=3.4 Mpc; Sandage & Tammann 1975), measured with the Owens Valley interferometer, shows a bar-like morphology (Canzian et al. 1988). c) Toward the SW lobe of M 82 (D=3.3 Mpc; Tammann & Sandage 1968), Brouillet & Schilke (1993) used the IRAM interferometer to map the emission of the (J=1−0) line of the high density tracer HCN, plotted as contours. The triangle denotes the kinematic center of M 82. The "radiophotography' is the 6 cm continuum mapped by Kronberg et al. (1985) using the VLA. The point sources are supernova remnants that seem to group around the molecular clouds. d) Finally, we adopt the sketch of the center of the Milky Way Galaxy (Güsten 1989). The rectangle shows where the distribution of ^{13}CO was mapped using the Bell Labs 7-m antenna (Heiligmann 1987). The H_2CO line emission was mapped with the Effelsberg 100 m telescope (Zylka et al. 1992). Thin contours show the 6 cm continuum (Altenhoff et al. 1978 with the 100 m telescope).

M 82. The CO distribution in the interacting irregular M 82 (e.g. Lo et al. 1987, Nakai et al. 1987, Carlstrom 1988, Loiseau et al. 1988, 1990) can be interpreted as a rotating ring of 200 pc radius. Toward the SW part this ring, Brouillet & Schilke (1993) have mapped the emission of the $J = 1 - 0$ line of the high density tracer HCN. The "ring" is now resolved into a number of individual molecular clouds (Fig. 2). The supernovae, seen in the VLA map by Kronberg et al. (1985), are grouped around these molecular clouds. A comparison of the distributions of dense molecular gas and supernovae illustrates the possible importance of photodissociation regions (PDRs) at the surfaces of the molecular clouds in M 82.

IC 342. Only 10° from the plane of the Galaxy, IC 342 is highly obscured by Galactic dust. Recently, it was suggested (Mc Call 1989, Madore & Freedman 1992) that the distance toward this face-on spiral is 1.8 Mpc and not 4 Mpc as previously believed. In that case, IC 342 is a galaxy at the edge of the Local

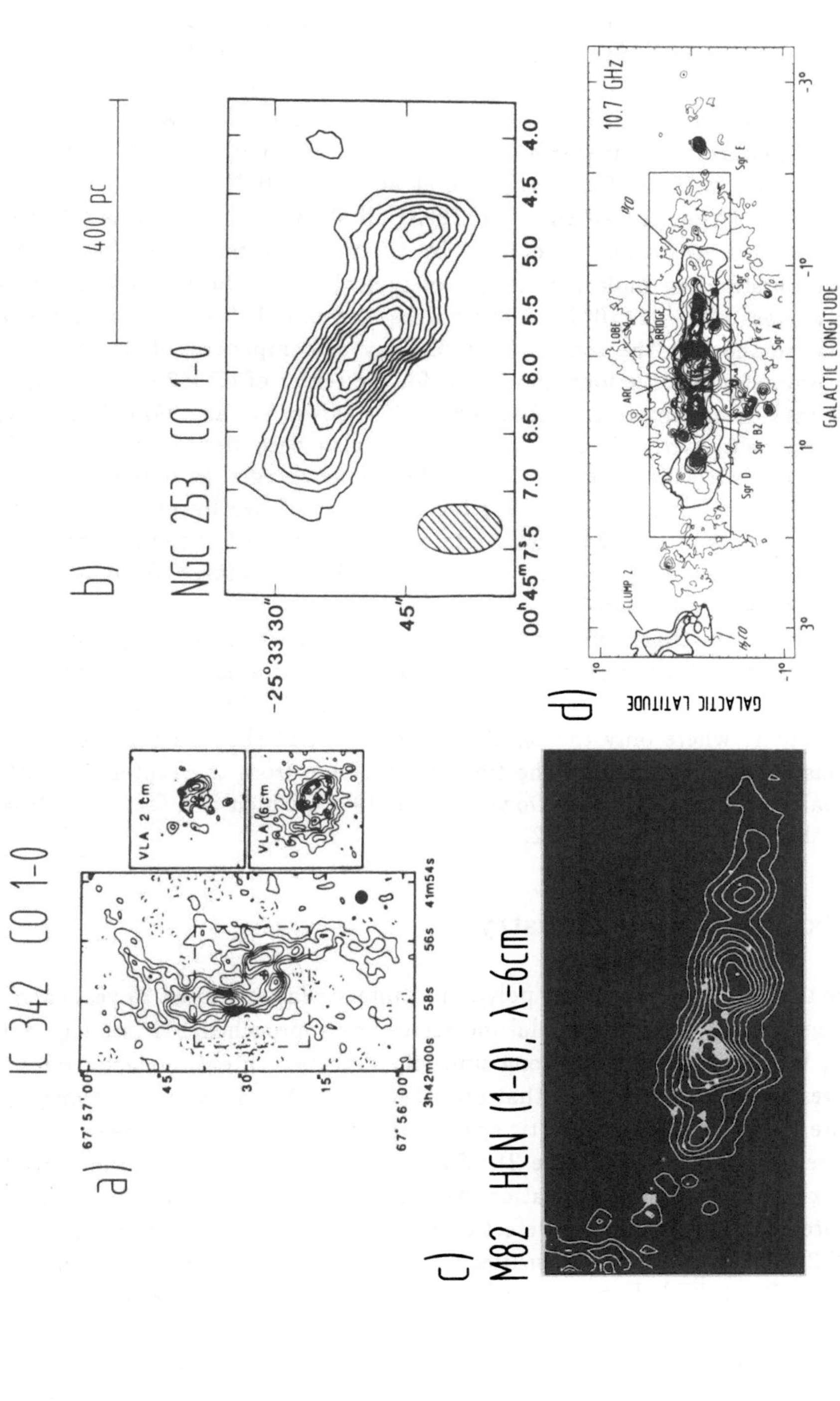
a) IC 342 CO 1-0
VLA 2 cm
VLA 6cm
b) NGC 253 CO 1-0
400 pc
c) M82 HCN (1-0), λ=6cm
d)
10.7 GHz
ARC
BRIDGE
LOBE
CLUMP 2
Sgr A
Sgr B2
Sgr C
Sgr D
Sgr E
GALACTIC LONGITUDE
GALACTIC LATITUDE

Group, and its infrared and molecular emission are more 'normal' than thought: In the central 400 pc, its FIR luminosity and H_2 mass of $8\,10^8\,L_\odot$ (HBM) and $3\,10^7\,M_\odot$ (Eckart et al. 1990b) are similar to the corresponding values for the Milky Way ($10^9\,L_\odot$ and $\sim 1\,10^8\,M_\odot$; Cox & Mezger 1989, Güsten 1989). Also the sizes of the central gas structures appear to be similar (compare Fig. 2).

CO is found in a $15'' \times 70''$ (130×600 pc) molecular bar (e.g. Lo et al. 1984). Interferometric maps of CO (Ishizuki et al. 1990), ^{13}CO (Ishizuki et al. 1991, Turner & Hurt 1992), NH_3 (Ho et al. 1990), HCN (Downes et al. 1992), and HCO^+ (Nguyen-Q-Rieu et al. 1993) all show that the gaseous bar is S-shaped. Its inner region is confined by a ring of 100 pc radius. This consists of five outstanding molecular clouds of typically 20 – 30 pc in diameter, each with a mass of $3 - 12\,10^5\,M_\odot$. These dense (several $10^4\,cm^{-3}$) clouds are very similar to those in the center of our Galaxy. A comparison of single dish and interferometer observations shows that for the center of IC 342 there are at least three molecular components observed in CO (Downes et al. 1992). The presence of components with different densities has also been derived from multilevel studies of CO (Güsten et al. 1993), CS (Mauersberger & Henkel 1989), and HC_3N (Mauersberger et al. 1990) for a number of nearby galaxies. Only the densest clumps in IC 342, which contain about one third of the gas mass in the inner 500 pc, are also observed in HCN. For these clumps, CO/HCN intensity ratios are between 5 and 9.

The 3.4 mm continuum emission from IC 342, which is generated by free-free emission (Downes et al. 1992), is mainly associated with one of the five molecular clouds (Fig. 2). In this regard, the center of IC 342 resembles the Milky Way center, where only the Sgr B2 molecular cloud shows a considerable rate of massive star formation. The free-free emission from the center of IC 342 is equivalent to 300 O5 stars (Downes et al. 1992), which is a factor of 30 more than the emission from Sgr B2.

3 Extragalactic chemistry

Since the detection of the first polyatomic interstellar molecules 25 years ago, the number of known extraterrestial molecules has approached 100 (see e.g. Herbst 1988, Williams 1993). For a long time, the detection of extragalactic molecules was restricted to few species that are either very abundant or are strong maser emitters. Today 23 extragalactic species have been identified, some of them also in rare isotopic species (Table 2). Most of them are detected in the mm-range and exhibit quasithermal radiation. Not lack of sensitivity but confusion between the broad lines prevents the detection of more molecular species in galaxies like NGC 253 (integrated line intensities are only an order of magnitude fainter than towards Orion-KL). Here we are in a similar situation as toward the Eldorado of molecular astronomy, the Orion Hot Core.

If one wants to detect or study extragalactic molecules, M 82 and NGC 253 are as outstanding as the hot cores in Orion or Sgr B2 for Galactic molecular

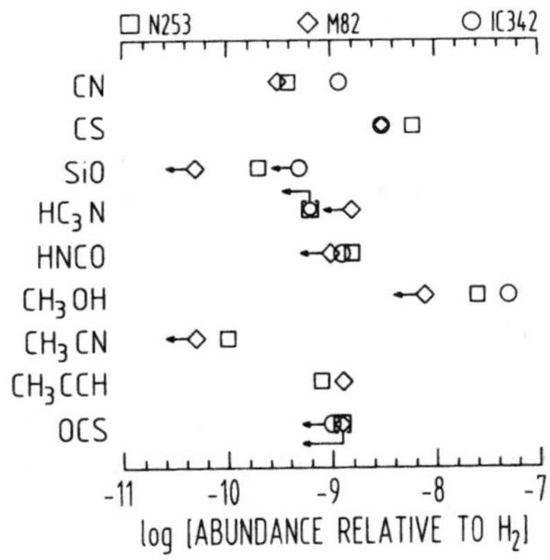

Fig. 3. Relative abundances of a number of molecules in the centers of M 82 and NGC 253. Adopted H_2 column densities are $10^{23.6}$ and $10^{23.8}$ cm^{-2} respectively.

Table 2. Molecules identified outside the Galaxy

Molecule	Detected galaxies	Ref.	Molecule[a]	Detected galaxies	Ref.[a]
OH^b	~ 75	1	H_2CO^b	~ 12	2
CO	~ 1000	3,4	H_2O^b	~ 15	5
HCN	~ 20	6	H_2	~ 60	7
NH_3	1	8	HCO^+	~ 20	9
CH^b	8-9	10	CS	~ 20	11
C_3H_2	4	12	CH^+	1	13
CH_3OH^b	6	14	CN	5	15
C_2H	4	15	HNC	9	15
HC_3N	$2-4$	15,16	HNCO	5	17,18
SO	2	19	N_2H^+	6	20
SiO^c	2	20	CH_3CCH	2	21
CH_3CN	1	21	OCS	1	22

a) (1) Weliachew (1971); (2) Gardner & Whiteoak (1974); (3) Rickard et al. (1975); (4) Solomon & de Zafra (1975); (5) Churchwell et al. (1975); (6) Rickard et al. (1977); (7) Thompson et al. (1978); (8) Martin & Ho (1979); (9) Whiteoak et al. (1980); (11) Henkel & Bally (1985); (12) Seaquist & Bell (1986); (13) Magain & Gillet (1987); (14) Henkel et al. (1987); (15) Henkel et al. (1988); (16) Mauersberger et al. (1990); (17) Nguyen-Q-Rieu et al. (1989); (18) Nguyen-Q-Rieu et al. (1991); (19) Johansson (1991); (20) Mauersberger & Henkel (1991); (21) Mauersberger et al. (1991); (22) Mauersberger et al. (1994), in press

b) molecule with maser emission

c) also detected from SN 1987a at $\lambda = 8\mu m$ (Roche et al. 1991) spectroscopists. It has been known for a long time that the Orion and Sgr B2 star forming regions have different chemistries, i.e. chemical compositions, and both have some spatial fine structure, with different molecules observed at different locations (see e.g. Ungerechts et al. 1992). This was not a great surprise, in view of their very different locations in the Milky Way and in view of the presence of energetic outflows, density fluctuations and temperature gradients caused by nearby stars. One would naively expect such variations to cancel out when one compares the chemical mixture averaged over larger volumes. Unfortunately, there is not much information about the large-scale distribution of rare molecular species in our Milky Way. For galaxies like M 82 and NGC 253, on the other hand, even the smallest single dish beams collect the radiation averaged over a scale of more than 100 pc, which is larger than most Giant Molecular Clouds in the Milky Way.

Surprisingly, the variations in chemical composition do *not* cancel out, even if one observes two galaxies that have as much in common (e.g. IR luminosity, molecular mass) as M 82 and NGC 253. Fig. 3 compares the relative abundances of a number of molecules in these two galaxies. Of those molecules, only CN and CH_3CCH have comparable relative abundances in both galaxies. All other molecules considered are at least an order of magnitude rarer in M 82 than in NGC 253. By analyzing several transitions of each molecule, we have determined that this is not an artifact due to line excitation. Possible explanations for the different abundances have been discussed for some specific molecules.

SiO is a molecule that in our Galaxy has been detected in stellar envelopes (usually via maser emission from vibrationally excited states) and in Orion-like compact and hot molecular cores (usually quasi-thermal emission). Despite deep integrations, SiO emission has not been detected from Galactic clouds colder than about 30 K (Ziurys et al. 1989). This cannot be entirely explained by depletion onto grains. According to neutral gas phase models (Langer & Glassgold 1990), the synthesis of SiO requires that the first excited fine structure line of the Si atom is populated and predicts that the SiO abundance scales with $\exp(-111\,\mathrm{K}/T_{\mathrm{kin}})$. Using SiO as a "chemical thermometer" we can thus explain its underabundance in M 82 relative to NGC 253 by a higher temperature of the dense gas in the core of NGC 253.

Methanol (CH_3OH) has been easily detected in half a dozen galaxies (Hüttemeister 1993) but despite deep integration, it was not seen in M 82. If elevated methanol abundances can be explained by evaporation from dust grains in a warm medium (Brown et al. 1988), this would be another hint that the dense molecular gas in M 82 is colder than in NGC 253.

Unfortunately, only upper limits for *ammonia (NH_3)* inversion lines exist for M 82 and NGC 253. This may partly be explained by the difficulties of observing broad lines in the cm-range. Also, the fragile NH_3 molecules may be destroyed in the violent ISM of galaxies like NGC 253 and M 82.

Time dependent models of chemical evolution of molecular clouds (Herbst & Leung 1989) predict that complex molecules should be produced effectively within about 10^5 years. In steady state, which is reached after 10^6 years, the chemical composition favors less complex species. It is possible that the molecular clouds in NGC 253 never reach this steady state because they are destroyed and regenerated frequently, e.g. due to cloud-cloud collisions in its very compact nuclear gas concentration.

The centers of starburst galaxies stand out by their immense radiation density and the number of massive stars and supernovae. The UV radiation field in M 82 is 250 times stronger than in the solar neighborhood (Maloney & Black 1988). The molecular clouds in those regions remain opaque to this radiation since they must have high densities to withstand tidal disruption. They should, however, possess photon dominated regions (or photodissociation regions, PDRs), whose volume should depend on the surface area of the nuclear molecular clouds, i.e. on the degree of clumping. Evidence for these PDRs comes from observations of ionized carbon, C II. Stacey et al. (1991) report that up to 40% of the circumnuclear gas may be photodissociated. Toward the center of NGC 6946, Madden et al. (1993) find a C II line emitted from a very dense (a few $10^5\,\mathrm{cm}^{-3}$) gas component.

There are more processes that may determine molecular abundances (e.g. the initial ion density; Herbst & Leung 1989), but we are far from having a single and consistent scenario to explain the different chemistries of NGC 253 and M 82.

4 Conclusions

While the integrated line intensity of CO is becoming accepted by most astronomers as a good tracer of the total molecular masses of galaxies, HCN and other high density tracers can be used to trace the mass of dense ($n(\mathrm{H}_2) > 10^4\,\mathrm{cm}^{-3}$) molecular cores, the potential sites of star formation.

Variations in the chemical mixture of the gas are not only found on a small scale but are unexpectedly large on a 100 pc scale.

Interferometric observations of the central regions of galaxies like M 82 or IC 342 show that the relationship between the available gas mass and the star forming activity is more complex than it seems from the "correlations" between the gas mass and luminosity. They also point out our ignorance about the distribution and physical and chemical parameters in the inner few 100 pc of the Milky Way.

References

Altenhoff, W.J., Downes, D., Pauls, T., Schraml, J., 1978, A&AS 35,23

Brouillet, N., Schilke, P., 1993, A&A, in press

Brown, P.D., Charnley, S.B., Millar, T.J., 1988, MNRAS 231, 409

Canzian, B., Mundy, L.G., Scoville, N.Z., 1988, ApJ 333, 157

Carlstrom, J.E., 1988 in: *Galactic and extragalactic star formation*, eds. R.E. Pudritz, M. Fich, Kluwer, p. 571

Churchwell, E., Witzel, A., Huchtmeier, W., Pauliny-Toth, I.I.K., Roland, J., Sieber, W., 1977, A&A 54, 969

Cox, P., Mezger, P.G., 1989 A&AR 1, 49
Downes, D., Radford, S.J.E., Guilloteau, S., Guélin, M., Greve, A., Morris, D., 1992, A&A 262, 424
Eckart, A., Downes, D., Genzel, R., Harris, A.I., Jaffe, D.T., Wild, W., 1990b, ApJ 348, 434
Gardner, F.F., Whiteoak, J.B., 1974, Nat. 247, 526
Güsten, R., 1989, in *The center of the Galaxy*, ed: M.Morris, Kluwer Dordrecht, p89
Güsten, R., Serabyn, E., Kasemann, C., Schinckel, C., Schneider, G., Schulz, A., Young, K., 1993, ApJ 402, 537
Heiligman, G.M., 1987, ApJ 314, 747
Henkel, C., Bally, J., 1985, A&A 150, L27
Henkel, C., Jacq, T., Mauersberger, R., Menten, K.M., Steppe, H., 1987, A&A 188, L1
Henkel, C., Mauersberger, R., Schilke, P., 1988, A&A 201, L23
Herbst, E., 1988, Rev. Modern Astron. 1, 114
Herbst, E., Leung, C.M., 1989, ApJS 69, 271
Ho, P.T.P., Martin, R.N., Turner, J.L., Jackson, J.M., 1990, ApJ 355, L19
Hüttemeister, S., 1993, PhD Thesis, Bonn University
Ishizuki, S., Kawabe, R., Ishiguro, M., Okumura, S.K., Morita, K.I., Chikada, Y., Kasuga, T., 1990, Nat 344, 224
Ishizuki, S., Kawabe, R., Ishiguro, M., Okumura, S., Morita, K.I., Chikada, Y., Kasuga, T., Wright, M.C.H., 1991, in *Dynamics of galaxies and their molecular cloud distributions*, IAU-Symp. 146, eds. F. Combes, F. Casoli, Kluwer, p.272
Johansson, L.E.B., 1991, in *Dynamics of galaxies and their molecular cloud distributions*, IAU-Symp. 146, eds. F. Combes, F. Casoli, Kluwer, p1
Kronberg, P.P., Biermann, P., Schwab, F.R., 1985 ApJ 291, 693
Langer, W.D., Glassgold, A.I, 1990, ApJ 352, 123
Lo, K.Y., Ball, R., Masson, C.R., Phillips, T.G., Scott, S.L., Woody, D.P., 1987 ApJ 317, L63
Lo, K.Y., Berge, G.L., Claussen, M.J., Heiligman, G.M., Leighton, R.B., Masson, C.R., Moffet, A.T., Phillips, T.G., Sargent, A.I., Scott, S.L., Wannier, P.G., Woody, D.P., 1984, ApJ 282, L59
Loiseau, N., Reuter, H.-P., Wielebinski, R., Klein, U., 1988, A&A 200, L1
Loiseau, N., Nakai, N., Sofue, Y., Wielebinski, R., Reuter, H.-P., Klein, U., 1990, A&A 228, 331
Madden, S. C., Geis, N., Genzel, R., Herrmann, F., Jackson, J.M., Poglitsch, A., Stacey, G.J., Townes, C.H., 1993, ApJ in press
Madore, B.F., Freedman, W.L., 1992, PASP 104, 362
Magain, P., Gillet, D., 1987, A&A 184, L5
Maloney, P., 1990, in: *The interstellar medium in galaxies*, eds: H.A. Thronson, J.M. Shull, Kluwer, p493
Maloney, P., Black, J.H., 1988, ApJ 325, 389
Martin, R.N., Ho, P.T.P., 1979, A&A 74, L7
Mauersberger, R., Henkel, C., 1989, A&A 223, 79
Mauersberger, R., Henkel, C., 1991, A&A 245, 457
Mauersberger, R., Henkel, C., 1993, Rev. Mod. Astron. 6, 69
Mauersberger, R., Henkel, C., Sage, L.J., 1990, A&A 236, 63
Mauersberger, R., Henkel, C., Sage, L.J., Wiklind, T., 1991, A&A 247, 307
Mauersberger, R., Wilson, T.L., Mezger, P.G., Gaume, R., Johnston, K.J., 1992, A&A 256, 640
McCall, M.L., 1989, AJ 97, 1341

Mezger, P.G., 1990, in: *Physics and Composition of Interstellar Matter*, Inst. of Astronomy, Nicolaus Copernicus University, Torun, eds.: J. Krelowski and J. Papaj
Nakai, N., Hayashi, M., Handa, T., Sofue, Y., Hasegawa, T., 1987 PASJ 39, 685
Nguyen-Q-Rieu, Henkel, C., Jackson, J.M., Mauersberger, R., 1991, A&A 241, L33
Nguyen-Q-Rieu, Nakai, N., Jackson, J.M., 1989, A&A 220, 57
Nguyen-Q-Rieu, Viallefond, F., Combes, F., Jackson, J.M., Lequeux, J., Radford, S.J.E., Truong-Bach, 1993, in: *Astronomy with mm and sub-mm wave interferometry*, IAU Coll. 140, ed: M. Ishiguro
Radford, S.J.E., Solomon, P.M., Downes, D., 1991, ApJ 368, L15
Rickard, L.J., Palmer, P., Morris, M., Zuckerman, B., Turner, B.E., 1975, ApJ 199, L75
Rickard, L.J., Palmer, P., Turner, B.E., Morris, M., Zuckerman, B., 1977, A&A 214, 390
Roche, P.F., Aitken, D.K., Smith, C.H., 1991, MNRAS 252, 39p
Sandage. A., Tammann, G.A., 1975, ApJ 196, 313
Seaquist, E.R., Bell, M.B., 1986, ApJ 303, L67
Solomon, P.M., de Zafra, R., 1975, ApJ 199, L79
Stacey, G.J., Geis, N., Genzel, R., Lugten, J.B., Poglitsch, A., Sternberg, A., Townes, C.H., 1991, ApJ 373, 423
Tammann, G.A., Sandage, A., 1968, ApJ 151, 825
Thompson, R.I., Lebovski, M.J., Rieke, G.H., 1978, ApJ 222, L49
Turner, J.L., Hurt, R.L., 1992, ApJ 384, 72
Ungerechts, H., Bergin, E.A., Carpenter, J., Goldsmith, P.F., Irvine, W.M., Lovell, A., Mcgonagle, D., Schloerb, F.P., Snell, R.L., 1992, IAU-Sympos. 150: *Astrochemistry of Cosmic Phenomena*, ed: P.D. Singh, p271
Verter, F., 1990, PASP 102, 1281
Weliachew, L., 1971, ApJ 167, L47
Whiteoak, J.B., Gardner, F.F., Höglund, B., 1980, MNRAS 190, 17p
Wild, W., Harris, A.I., Eckart, A., Genzel, R., Graf, U.U., Jackson, J.M., Russell, A.P.G., Stutzki, J., 1992, A&A, 265, 447
Williams, D.A., 1993, Rev. Modern Astron., 6, 49
Wilson, T.L., Mauersberger, R., 1990, A&A 239, 305
Ziurys, L.M., Friberg, P., Irvine, W.M., 1989, ApJ 343, 201
Zylka, R., Güsten, R., Henkel, C., Batrla, W., 1992, A&AS 96, 525

CO at High Redshifts

Paul A. Vanden Bout, Robert L. Brown

National Radio Astronomy Observatory***, 520 Edgemont Road, Charlottesville, VA 22903-2475

Abstract: The prospects for molecular astrophysics at high redshifts are presented. The discussion is concentrated on the sources IRAS 10214+4724.

1 Introduction

Before the discovery of complex molecules, our knowledge of the galactic interstellar medium (ISM) was heavily based on absorption spectra obtained along lines of sight to nearby early-type stars. Together with radio observations of HI and studies of optical extinction these data led to a model of the ISM that we now know was remarkably incomplete. The discovery twenty-five years ago of complex interstellar molecules signaled a revolution in our understanding of diffuse matter and star formation in the Galaxy. This history may be repeating itself in the context of extragalactic space. Our knowledge of potentially large reservoirs of gas in the "late early-universe", at redshifts $z \sim$ 1-5, is almost totally dependent on the optical absorption spectra obtained towards QSOs. The detection of high-z CO emission, first occasioned by the identification of the IRAS source F10214+4724, signals the possibility of directly observing concentrations of gas at these early epochs in the universe when, presumably, galaxies were formed. The detection of CO emission toward damped Lyman-α absorption systems, the deepest absorption features seen in QSO spectra, strengthens these prospects. Damped Lyman-α systems may well be to protogalaxy molecular gas clouds what diffuse interstellar clouds are to giant molecular cloud complexes in the Galaxy.

2 IRAS FSC 10214+4724

The Faint Sky Survey was intended to provide the last word on the presence of faint point sources as seen by the Infrared Astronomical Satellite (IRAS), and the Faint Source Catalog is a potentially rich mine for the identification of high-redshift infrared sources. Actually doing so, however, is apparently as

*** Operated by Associated Universities, Inc., under Cooperative Agreement with the National Science Foundation.

daunting a task as finding the proverbial needle in a haystack. After taking spectra of some 1500 candidate IRAS source identifications, a team led by M. Rowan-Robinson identified IRAS FSC 10214+4724 with an optical object having emission lines redshifted by $z = 2.3$ (ref. a). The identification was strengthened by the detection of faint radio emission at the optical position using the Very Large Array (VLA). The IRAS flux together with the distance implied by the redshift leads to the conclusion that this object is one of the most luminous objects in the universe: $L \simeq 10^{14}\ L_\odot$ in the rest frame of the object at 6-36 μm with $H_o = 100\ \mathrm{km\,s^{-1}Mpc^{-1}}$, $q_o = 1/2$, and $\Lambda = 0$. Fitting models of starburst galaxies to the observed 12-100 μm fluxes, Rowan-Robinson, et al. found that dust masses of order 10^9 to 10^{10} $M_\odot$were required. Alternatively, a model with an obscured nonthermal source of energy required 4x10 $M_\odot$of dust. These two models provided the title for their discovery paper – "A High-Redshift IRAS Galaxy with Huge Luminosity – Hidden Quasar or Protogalaxy?" Recognizing that CO observations could in principle reveal which of these models was correct, we used the NRAO 12 Meter Telescope to search for CO emission at the redshift and in the direction of the Rowan-Robinson, et al. "Object F." The established relation between CO and far-infrared flux for starburst galaxies indicated that the CO flux for 10214+4724 was detectable. The CO(3-2) emission line was detected after 16.7 hours of integration (ref. b). The weather was relatively poor, access to the telescope having been gained during what would normally have been the start of summer shutdown; it had been kept operational to observe the solar eclipse of July 11, 1991. A comparable amount of observing time on 3C 380 with the identical receiver setup yielded the expected null result. Of particular importance to these and future observations of high-z spectral lines are the flat baselines at zero intensity made possible by the chopping secondary mirror on the 12 Meter Telescope. No baseline, not even a DC offset, has been removed from the data. The observed line has a peak intensity of 3 to 4 mK on the T_R^* scale and a full width at zero intensity of 800 $\mathrm{km\,s^{-1}}$. The integrated line intensity is 0.74 K $\mathrm{km\,s^{-1}}$ or 26 Jy $\mathrm{km\,s^{-1}}$ using a point source gain of 35 Jy/K. This flux in the CO line implies a large amount of molecular gas is resident in the source assuming conditions are similar to those of galactic molecular clouds, of order 10^{12} $M_\odot$ depending on the assumptions made. If this gas is confined to the 4" extent of the optical image, a starburst is a virtual certainty. The detection of this emission line in F 10214 and its identification as CO(J=3-2) have been confirmed,. Brown and Vanden Bout (ref. c) observed both the J = 3-2 and J = 4-3 lines using the IRAM 30 Meter Telescope. Solomon, Downes, and Radford (ref. d) also used the 30 Meter Telescope to observe J = 3-2 and J = 6-5. Vanden Bout and Brown (ref. c) have used the NRAO 140 Foot Telescope to observe the J = 1-0 line and the 12 Meter Telescope to observe the J = 2-1 and J = 4-3 lines. Solomon (ref. f) has observed the J = 3-2 line on the 12 Meter Telescope. And Tsuboi and Nakai (ref. g) report observations of the J = 1-0 and J = 3-2 lines with the Nobeyama 45 Meter Telescope. In addition, Brown and Vanden Bout (ref. h) have reported detections of the CI J = 1-0 and J = 2-1 lines. Encrenaz et al. have published a possible detection of the H_2O ($J = 2_{11} - 2_{02}$)

lines (ref. i). The number of different lines all observed at the same redshift, that of the optical object, establish the reality of the highly redshifted CO emission beyond any doubt. By comparison, the mass of the molecular gas associated with this emission has been somewhat controversial. A miscalculation in Brown and Vanden Bout's original paper was quickly corrected by Solomon, Radford, and Downes (ref. j), but inconsistencies between the integrated flux in the CO ($J =$ 3-2) line as seen with the 12 Meter and 30 meter Telescopes has been a more persistent problem. The 12 Meter spectrum of the original line detection has two to six times the flux of the 30 Meter spectra. The prospect this raises of extended CO emission is more important than the exact amount of molecular gas in the source, that amount being large in any event. Solomon (ref. f) has repeated the detection observation of $J =$ 3-2 made by Brown and Vanden Bout on the 12 Meter Telescope and finds a line flux intermediate between that of Brown and Vanden Bout and what would be expected on the 12 Meter based on his own 30 Meter observations. The flux in his 12 Meter line is consistent with Brown and Vanden Bout's 30 Meter observations. Accepting this 12 Meter data as giving the most correct value of the line flux together with Brown and Vanden Bout's 30 meter observation, means the discovery spectrum is too strong by a factor of 2.5 and all other spectra, including those obtained at Nobeyama, are too weak by the same factor. Differences of factors of two to three should be expected. The observations are difficult and the data is noisy. The spectrometer bandwidths are generally inadequate, and it is difficult to set the baseline. The implications for future instrumentation are also clear: effective chopping secondaries and wide bandwidth spectrometers are a must. Within a factor of three the best value for the mass of molecular gas associated with IRAS 10214 is $M(H_2) = 3x10^{11}$ $M_\odot$. This estimate assumes the gas is similar to Galactic molecular gas, most importantly, that the carbon abundance is solar. Single dish observations say little about the location or distribution of the CO emission. Fortunately, IRAS 10214 is observable with existing millimeter wavelength interferometers. The first reported observations (ref. k and ref. l) were made with the Nobeyama Millimeter Array. Maps with a resolution 89x60 detect the CO emission in an unresolved source located at the position of the optical object. Using the IRAM Interferometer, Radford et al. barely resolved the CO source with a $23'' \times 21''$ beam. The interferometers record different total line fluxes; the Nobeyama flux is consistent with Solomon's 12 Meter and Brown and Vanden Bout's 30 meter line flux. The IRAM flux is weaker by a factor of about two. Brown and Vanden Bout (ref. n) have used the Owens Valley Interferometer to observe ($J =$ 3-2) CO emission in IRAS 10214. The source is again seen at the optical position with a total flux somewhat larger than is in the IRAM image. The source is just resolved and the image is suggestive of a tidally interacting system. The emerging picture of IRAS 10214 is that of an extreme example of a merger starburst system, as was first suggested by Solomon, Radford, and Downes (ref. j), and as can now be seen in near infrared images from the Keck Telescope taken by Matthews et al. (ref. o). Other observations support this model: imaging at many wavelengths by Lawrence, et al. (ref. u), the initial Hα imaging in the near-IR by Soifer, et al. (ref. v), and submillimeter observations by Downes et al. (ref. w). What are the prospects for finding more sources like IRAS 10214? The search technique that led to the identification of IRAS 10214 is obviously very difficult and apparently required luck as well. An improvement on this basic technique might make it easier. The tight correlation between radio continuum and far-infrared emission from starburst galaxies suggests taking VLA snapshots of IRAS candidate fields. Optical spectra would be taken for only those optical candidates coincident with radio sources. This reduces the number of optical spectra required by a large factor. The VLA Sky Survey being made by Condon et al. will make this technique available to anyone willing to pursue the optical spectra. Meanwhile, Low, et al. (ref. p) have applied this to their most likely dozen IRAS Faint Object Catalog sources and discovered a source at a redshift of $z = 0.93$ that appears very similar to IRAS 10214. Observations of CO lines are yet to be made.

3 Damped Lyman Alpha Systems

An alternate search technique for high redshift molecular gas is one familiar to the pioneers of molecular line spectroscopy in the Galaxy - simply guess where the prospects might be favorable. At redshifts of z 2-3, the most obvious reservoirs of gas are the damped Lyman-α systems. There are absorption features seen in the spectra of quasars with HI column densities sufficient to produce damping wings on the Ly-α line. Damped Ly-α systems have been studied extensively by Wolfe and his collaborators (refs. r, s, t), and he argues that these systems are signposts of mass concentrations sufficiently large to form a cluster of galaxies. While it is not obvious that the gas sampled along the line of sight to the quasar contains molecular material, it is possible that molecular gas is present off the line of sight at the redshift of the Ly-α feature. Brown and Vanden Bout have searched for CO emissions in directions toward fifteen so-called confirmed damped Ly-α systems and report (rf. u) a detection at $z = 2.14$ to PKS 0528-250. A number of other lines of sight show spectra worthy of more telescope time. Since then a number of searches have been conducted by several groups and there are rumors of other detections. If these detections are confirmed, astronomers will be presented with an opportunity of particular significance. The process of galaxy formation is one of low temperature gas and thermal processes. This is the domain of millimeter, submillimeter, and far-IR astronomy. The capability to make images of thermal line emission provides both positional and dynamical information, along with the physical parameters of temperature, density, and abundance typical of spectroscopic analysis. This capability is barely achieved by present instruments for the first source discovered. A large millimeter wavelength array, such as the Millimeter Array proposed by NRAO, is needed to reveal the galaxy forming history of the Universe.

References

a. Rowan-Robinson, M., Broadhurst, T., Lawrence, A., McMahon, R. G., Lonsdale, C. J., Oliver, S. J., Taylor, A. N., Hacking, P. B., Conrow, T., Saunders, W., Ellis, R. S., Efstathion, G. P., and Condon, J. J. 1991, Nature, 351, 719.
b. Brown, R. L., and Vanden Bout, P. A. 1992, AJ, 102, 1956.
c. Brown, R. L., and Vanden Bout, P. A. 1992, ApJ, 397, L19.
d. Solomon, P. M., Downes, D., and Radford, S.J.E. 1992, ApJ, 398, L29.
e. Brown, R. L., and Vanden Bout, P. A. in preparation.
f. Solomon, P. M., private communication.
g. Tsuboi, M., and Nakai, N. 1992, Publ. Astron. Soc. Japan, 44, L241.
h. Brown, R. L., and Vanden Bout, P. A. 1992, ApJ, 397, L11.
i. Encrenaz, P. J., Combes, F., Casoli, F., Gerin, M., Pagani, L., Horellou, C., and Gue, C. 1993, A&A 273, L19.
j. Solomon, P. M., Radford, S.J.E., and Downes, D. 1992, Nature, 356, 318.
k. Sakamoto, K., Ishizuki, S., Kawabe, R., and Ishiguro, M. 1992, ApJ 397, L27.
l. Kawabe, R., Sakamoto, K., Ishizuki, S., and Ishiguro, M. 1992, ApJ, 397, L23.
m. Radford, S.J.E., Brown, R. L., and Vanden Bout, P. A. 1993, A&A , 271, L21.
n. Brown, R. L., and Vanden Bout, P. A., in preparation. Matthews, K., Soifer, B. T., Nelson, J., Boesgard, H., Graham, J. R., Harrison, W., Irace, W., Jernigan, G., Larkin, J. E., Lewis, H., Lin, S., Neugebauer, G., Siroto, M., Smith, G., and Ziomkowski, C., preprint.
o. Cutri, R. M., Huchra, J., and Low, F., private communication.
p. Wolfe, A. M., Turnshek, D. A., Smith, H. E., and Chen, R. D., 1986, ApJS, 61, 249.
q. Wolfe, A. M. 1988, in QSO Absorption Lines Probing the Universe, ed. C. Blades, D. Turnshek, and C. Norman, NY, Cambridge Univ. Press, 291.
r. Larzetta, K. M., Wolfe, A. M., Turnshek, D. A., Lu, L., McMahon, R. G., and Hazard, C. 1991, ApJS, 77, 1.
s. Lonsdale, C. J., Hacking, P., Conrow, T., Efstathiou, G. P., and Saunders, W. S. 1993, NMRAS, 260, 28.
t. Soifer, B. T., Neugebauer, G., Matthews, K., and Lawrence, C., ApJ, 399, L55.
u. Downes, D., Radford, S.J.E., Greve, A., Thum, C., Solomon, P. M., and Wink, J. E. 1992, ApJ, 398, L25.

Lecture Notes in Physics

For information about Vols. 1–403
please contact your bookseller or Springer-Verlag

Vol. 404: R. Schmidt, H. O. Lutz, R. Dreizler (Eds.), Nuclear Physics Concepts in the Study of Atomic Cluster Physics. Proceedings, 1991. XVIII, 363 pages. 1992.

Vol. 405: W. Hollik, R. Rückl, J. Wess (Eds.), Phenomenological Aspects of Supersymmetry. VII, 329 pages. 1992.

Vol. 406: R. Kayser, T. Schramm, L. Nieser (Eds.), Gravitational Lenses. Proceedings, 1991. XXII, 399 pages. 1992.

Vol. 407: P. L. Smith, W. L. Wiese (Eds.), Atomic and Molecular Data for Space Astronomy. VII, 158 pages. 1992.

Vol. 408: V. J. Martínez, M. Portilla, D. Sàez (Eds.), New Insights into the Universe. Proceedings, 1991. XI, 298 pages. 1992.

Vol. 409: H. Gausterer, C. B. Lang (Eds.), Computational Methods in Field Theory. Proceedings, 1992. XII, 274 pages. 1992.

Vol. 410: J. Ehlers, G. Schäfer (Eds.), Relativistic Gravity Research. Proceedings, VIII, 409 pages. 1992.

Vol. 411: W. Dieter Heiss (Ed.), Chaos and Quantum Chaos. Proceedings, XIV, 330 pages. 1992.

Vol. 412: A. W. Clegg, G. E. Nedoluha (Eds.), Astrophysical Masers. Proceedings, 1992. XX, 480 pages. 1993.

Vol. 413: Aa. Sandqvist, T. P. Ray (Eds.); Central Activity in Galaxies. From Observational Data to Astrophysical Diagnostics. XIII, 235 pages. 1993.

Vol. 414: M. Napolitano, F. Sabetta (Eds.), Thirteenth International Conference on Numerical Methods in Fluid Dynamics. Proceedings, 1992. XIV, 541 pages. 1993.

Vol. 415: L. Garrido (Ed.), Complex Fluids. Proceedings, 1992. XIII, 413 pages. 1993.

Vol. 416: B. Baschek, G. Klare, J. Lequeux (Eds.), New Aspects of Magellanic Cloud Research. Proceedings, 1992. XIII, 494 pages. 1993.

Vol. 417: K. Goeke P. Kroll, H.-R. Petry (Eds.), Quark Cluster Dynamics. Proceedings, 1992. XI, 297 pages. 1993.

Vol. 418: J. van Paradijs, H. M. Maitzen (Eds.), Galactic High-Energv Astrophysics. XIII, 293 pages. 1993.

Vol. 419: K. H. Ploog, L. Tapfer (Eds.), Physics and Technology of Semiconductor Quantum Devices. Proceedings, 1992. VIII, 212 pages. 1993.

Vol. 420: F. Ehlotzky (Ed.), Fundamentals of Quantum Optics III. Proceedings, 1993. XII, 346 pages. 1993.

Vol. 421: H.-J. Röser, K. Meisenheimer (Eds.), Jets in Extragalactic Radio Sources. XX, 301 pages. 1993.

Vol. 422: L. Päivärinta, E. Somersalo (Eds.), Inverse Problems in Mathematical Physics. Proceedings, 1992. XVIII, 256 pages. 1993.

Vol. 423: F. J. Chinea, L. M. González-Romero (Eds.), Rotating Objects and Relativistic Physics. Proceedings, 1992. XII, 304 pages. 1993.

Vol. 424: G. F. Helminck (Ed.), Geometric and Quantum Aspects of Integrable Systems. Proceedings, 1992. IX, 224 pages. 1993.

Vol. 425: M. Dienes, M. Month, B. Strasser, S. Turner (Eds.), Frontiers of Particle Beams: Factories with $e^+ e^-$ Rings. Proceedings, 1992. IX, 414 pages. 1994.

Vol. 426: L. Mathelitsch, W. Plessas (Eds.), Substructures of Matter as Revealed with Electroweak Probes. Proceedings, 1993. XIV, 441 pages. 1994

Vol. 427: H. V. von Geramb (Ed.), Quantum Inversion Theory and Applications. Proceedings, 1993. VIII, 481 pages. 1994.

Vol. 428: U. G. Jørgensen (Ed.), Molecules in the Stellar Environment. Proceedings, 1993. VIII, 440 pages. 1994.

Vol. 429: J. L. Sanz, E. Martínez-González, L. Cayón (Eds.), Present and Future of the Cosmic Microwave Background. Proceedings, 1993. VIII, 233 pages. 1994.

Vol. 430: V. G. Gurzadyan, D. Pfenniger (Eds.), Ergodic Concepts in Stellar Dynamics. Proceedings, 1993. XVI, 302 pages. 1994.

Vol. 431: T. P. Ray, S. Beckwith (Eds.), Star Formation and Techniques in Infrared and mm-Wave Astronomy. Proceedings, 1992. XIV, 314 pages. 1994.

Vol. 432: G. Belvedere, M. Rodonò, G. M. Simnett (Eds.), Advances in Solar Physics. Proceedings, 1993. XVII, 335 pages. 1994.

Vol. 433: G. Contopoulos, N. Spyrou, L. Vlahos (Eds.), Galactic Dynamics and N-Body Simulations. Proceedings, 1993. XIV, 417 pages. 1994.

Vol. 434: J. Ehlers, H. Friedrich (Eds.), Canonical Gravity: From Classical to Quantum. Proceedings, 1993. X, 267 pages. 1994.

Vol. 435: E. Maruyama, H. Watanabe (Eds.), Physics and Industry. Proceedings, 1993. VII, 108 pages. 1994.

Vol. 436: A. Alekseev, A. Hietamäki, K. Huitu, A. Morozov, A. Niemi (Eds.), Integrable Models and Strings. Proceedings, 1993. VII, 280 pages. 1994.

Vol. 437: K. K. Bardhan, B. K. Chakrabarti, A. Hansen (Eds.), Non-Linearity and Breakdown in Soft Condensed Matter. Proceedings, 1993. XI, 340 pages. 1994.

Vol. 438: A. Pękalski (Ed.), Diffusion Processes: Experiment, Theory, Simulations. Proceedings, 1994. VIII, 312 pages. 1994.

Vol. 439: T. L. Wilson, K. J. Johnston (Eds.), The Structure and Content of Molecular Clouds. 25 Years of Molecular Radioastronomy. Proceedings, 1993. XIII, 308 pages. 1994.

Vol. 440: H. Latal, W. Schweiger (Eds.), Matter Under Extreme Conditions. Proceedings, 1994. IX, 243 pages. 1994.

New Series m: Monographs

Vol. m 1: H. Hora, Plasmas at High Temperature and Density. VIII, 442 pages. 1991.

Vol. m 2: P. Busch, P. J. Lahti, P. Mittelstaedt, The Quantum Theory of Measurement. XIII, 165 pages. 1991.

Vol. m 3: A. Heck, J. M. Perdang (Eds.), Applying Fractals in Astronomy. IX, 210 pages. 1991.

Vol. m 4: R. K. Zeytounian, Mécanique des fluides fondamentale. XV, 615 pages, 1991.

Vol. m 5: R. K. Zeytounian, Meteorological Fluid Dynamics. XI, 346 pages. 1991.

Vol. m 6: N. M. J. Woodhouse, Special Relativity. VIII, 86 pages. 1992.

Vol. m 7: G. Morandi, The Role of Topology in Classical and Quantum Physics. XIII, 239 pages. 1992.

Vol. m 8: D. Funaro, Polynomial Approximation of Differential Equations. X, 305 pages. 1992.

Vol. m 9: M. Namiki, Stochastic Quantization. X, 217 pages. 1992.

Vol. m 10: J. Hoppe, Lectures on Integrable Systems. VII, 111 pages. 1992.

Vol. m 11: A. D. Yaghjian, Relativistic Dynamics of a Charged Sphere. XII, 115 pages. 1992.

Vol. m 12: G. Esposito, Quantum Gravity, Quantum Cosmology and Lorentzian Geometries. Second Corrected and Enlarged Edition. XVIII, 349 pages. 1994.

Vol. m 13: M. Klein, A. Knauf, Classical Planar Scattering by Coulombic Potentials. V, 142 pages. 1992.

Vol. m 14: A. Lerda, Anyons. XI, 138 pages. 1992.

Vol. m 15: N. Peters, B. Rogg (Eds.), Reduced Kinetic Mechanisms for Applications in Combustion Systems. X, 360 pages. 1993.

Vol. m 16: P. Christe, M. Henkel, Introduction to Conformal Invariance and Its Applications to Critical Phenomena. XV, 260 pages. 1993.

Vol. m 17: M. Schoen, Computer Simulation of Condensed Phases in Complex Geometries. X, 136 pages. 1993.

Vol. m 18: H. Carmichael, An Open Systems Approach to Quantum Optics. X, 179 pages. 1993.

Vol. m 19: S. D. Bogan, M. K. Hinders, Interface Effects in Elastic Wave Scattering. XII, 182 pages. 1994.

Vol. m 20: E. Abdalla, M. C. B. Abdalla, D. Dalmazi, A. Zadra, 2D-Gravity in Non-Critical Strings. IX, 319 pages. 1994.

Vol. m 21: G. P. Berman, E. N. Bulgakov, D. D. Holm, Crossover-Time in Quantum Boson and Spin Systems. XI, 268 pages. 1994.

Vol. m 22: M.-O. Hongler, Chaotic and Stochastic Behaviour in Automatic Production Lines. V, 85 pages. 1994.

Vol. m 23: V. S. Viswanath, G. Müller, The Recursion Method. X, 259 pages. 1994.

Vol. m 24: A. Ern, V. Giovangigli, Multicomponent Transport Algorithms. XIV, 427 pages. 1994.

FSC
www.fsc.org
MIX
Papier aus verantwortungsvollen Quellen
Paper from responsible sources
FSC® C105338

If you have any concerns about our products,
you can contact us on
ProductSafety@springernature.com

In case Publisher is established outside the EU,
the EU authorized representative is:
Springer Nature Customer Service Center GmbH
Europaplatz 3, 69115 Heidelberg, Germany

Printed by Libri Plureos GmbH
in Hamburg, Germany